Philosophy of Psychiatr

This is the first introductory textbook of its kind devoted to philosophy of psychiatry, offering a thorough and accessible investigation of the conceptual and philosophical problems at the heart of psychiatric practice and research. While it applies some of the long-standing concerns of philosophy to the mental health professions, it also investigates philosophical problems and issues that have arisen more recently from careful examination of psychiatric phenomena. Divided into two parts, *Philosophy of Psychiatric Practice and Research* and *Philosophy and Psychopathology*, the book's 12 chapters cover topics like the ontological status of mental illness, philosophical issues in diagnosis, the role of culture in psychiatry and the relationship between mental illness and personal identity, as well as explore foundational problems in studying well-known psychopathologies like schizophrenia, depression and addiction. All chapters include initial overviews and concluding summaries and a list of suggested readings.

Key features

- Two-part structure – divided between (1) philosophy of psychiatric practice and research, and (2) philosophy and psychopathology – presents a clear, yet distinctive picture of the field
- Offers a unified style and vision throughout, with easy-to-follow segues from chapter to chapter
- Pedagogical features include chapter overviews and summaries, discussion questions and sections for further reading

Sam Wilkinson is Senior Lecturer in Philosophy in the Department of Sociology, Philosophy and Anthropology at the University of Exeter, whose research focuses on hallucinations, delusions, psychological trauma, brain injury and the nature of health and well-being. He has published articles in leading journals, including *Mind and Language, Consciousness and Cognition, Review of Philosophy and Psychology* and *Philosophy, Psychiatry and Psychology*.

Routledge Contemporary Introductions to Philosophy
Series editor: Paul K. Moser, Loyola University of Chicago

This innovative, well-structured series is for students who have already done an introductory course in philosophy. Each book introduces a core general subject in contemporary philosophy and offers students an accessible but substantial transition from introductory to higher-level college work in that subject. The series is accessible to non-specialists and each book clearly motivates and expounds the problems and positions introduced. An orientating chapter briefly introduces its topic and reminds readers of any crucial material they need to have retained from a typical introductory course. Considerable attention is given to explaining the central philosophical problems of a subject and the main competing solutions and arguments for those solutions. The primary aim is to educate students in the main problems, positions and arguments of contemporary philosophy rather than to convince students of a single position.

Recently published volumes:

Philosophy of Perception
2nd Edition
William Fish

Business Ethics
Jeffrey Moriarty

Philosophy of Social Science
2nd Edition
Mark Risjord

Philosophy of Psychiatry
A Contemporary Introduction
Sam Wilkinson

Philosophy of Emotion
Christine Tappolet

For a full list of published *Routledge Contemporary Introductions to Philosophy*, please visit www.routledge.com/Routledge-Contemporary-Introductions-to-Philosophy/book-series/SE0111

Philosophy of Psychiatry

A Contemporary Introduction

Sam Wilkinson

Routledge
Taylor & Francis Group
NEW YORK AND LONDON

Cover image: © Shutterstock.com

First published 2023
by Routledge
605 Third Avenue, New York, NY 10158

and by Routledge
4 Park Square, Milton Park, Abingdon, Oxon, OX14 4RN

Routledge is an imprint of the Taylor & Francis Group, an informa business

ISBN: 978-1-138-21266-4 (hbk)
ISBN: 978-1-138-21268-8 (pbk)
ISBN: 978-1-315-45012-4 (ebk)

DOI: 10.4324/9781315450124

Typeset in Times New Roman
by Apex CoVantage, LLC

Contents

Preface

There are many interesting and important domains where the tools and training of philosophy can be put to good use. Psychiatry, conceived broadly as the attempt to understand and treat mental illness, is undoubtedly one of those domains.

I have written this book mainly for advanced undergraduates and graduates studying philosophy who would like to branch out into philosophy of psychiatry from more "core areas" of philosophy, such as philosophy of mind, epistemology and ethics, as well as from (what is sometimes construed as a different tradition) phenomenology. It is also hoped that it will function well as a textbook for anyone teaching an advanced undergraduate or graduate course in philosophy of psychiatry. Indeed, the idea for the book grew out of my own efforts in designing and teaching an advanced undergraduate course in philosophy of psychiatry at Durham University and then at the University of Exeter.

Although some philosophy of psychiatry acquires its meaning and importance once set against existing issues in other parts of philosophy, pre-existing knowledge of these issues is not a necessary requirement for reading and understanding this book. It is in fact a hope of mine that clinicians, whether they be psychiatrists, psychologists, psychotherapists or mental health nurses, will read, enjoy and learn from this book and come to appreciate first-hand the important contribution that philosophy can make to clinical thinking and to theoretical questions of import to clinical practice. Some philosophy of psychiatry has been more insulated from the daily concerns of the clinic than perhaps it ought to have been. An unfortunate result of this is that some philosophers may cite as new and radical ideas that have long been implicit in clinical practice. The converse is also true: clinicians may sometimes publish articles purporting to put forward new and controversial claims that, in the philosophical literature, have been mainstream for decades. I cannot claim to have fully rectified this division in this book, but I will try to mitigate it as best I can.

Overall, the aim of this book is to introduce readers, in a gently opinionated way, to the current state of philosophy of psychiatry and to point them in directions where they can explore the topic further and perhaps even make their own contributions. It is not an exhaustive compendium, however, and a great deal of important work has been omitted. This does not reflect in any way a low opinion of that work on my behalf: it is simply that, in the service of achieving a readable and intuitive argumentative flow through the chapters, I have not found an obvious place to incorporate it.

Another inevitable feature of this book is that it "comes from" the sort of philosophical tradition that I come from, which is broadly speaking Anglo-American and "Analytic," rather than (to highlight the most salient contrast) "Continental" (although the "Analytic/Continental" distinction is far from clear-cut). Methodologically speaking, although I would never (knowingly or willingly) abandon the analytic aims of clarity and rigor, I am more open-minded than many analytic philosophers to the sorts of data and subject-matters that have long been thought to be the purview of continental philosophy. I have in mind two things, the first of which is especially worthy of mention.

The first is "phenomenology," which is (roughly) the skilled and careful reflection on and theorising about the quality and structure of experience. I love phenomenology, and I find it immensely important to philosophy of psychiatry. Mental illness matters in large part because of what it's like to have it. What's more, our theoretical accounts must also do justice to the phenomenology of what they are seeking to be accounts of. A neuroscientific account of, say, thought insertion, that doesn't square with what it's like to experience an inserted thought, needs to try again. As a result, I believe that some of the most exciting work in philosophy of psychiatry is in what is known as "phenomenological psychopathology." A testament to this is Stanghellini et al.'s (2019) masterful handbook. While I do appeal to phenomenology in various places, where it makes sense to do so, it is, inevitably, given what I am trying to achieve, not front-and-centre.

The second thing with typically continental associations is what we might broadly call "critical theory." I would distance myself from much of what falls under this moniker, especially at the methodological level (clarity and rigor are often sacrificed for profundity, which is fine, but not what comes naturally to me). However, there is a kernel to this work that is hugely important to philosophy of psychiatry as I present it here, and it is the following. When we examine and seek to understand particular institutions and practices (e.g. psychiatric ones), we need to be alive to their contingency and fragility, to the fact that if things had turned out slightly differently, they too would look rather different. We need to think about the forces at play in this. Where are the vested interests, the power imbalances, the dumb procedural features that "lock-in" sub-optimal practices and so on? And, once this has been ascertained, how can we make things better?

As the reader will see in Chapter 1, the Introduction, "philosophy of psychiatry" can mean two broadly different things. This is not in any way problematic, since many theorists who would think of themselves as specialists in philosophy of psychiatry either produce research that falls into both camps or, if they only work within one of these, tend overwhelmingly to recognise the value and importance of the other. In short, these are not conflicting factions, just two different but important ways of bringing philosophy to bear on psychiatry. These different ways, which I elucidate in the Introduction, correspond to the two parts of the book, entitled, *Philosophy of Psychiatric Practice and Research* and *Philosophy and Psychopathology,* respectively.

In Chapter 2, I address what is perhaps the most fundamental question in philosophy of psychiatry, which is, "What is mental illness?" As ever, things are more complicated than they seem, but the two opposing sides are roughly those who are realists about mental illness and those who are not. It is an interesting feature of this debate that it mirrors a (perhaps more mature) debate in meta-ethics between realists and anti-realists about morality, and some of the insights and advancements in that debate apply here.

Chapter 3 moves from the question of what mental illness is generally to the question of psychiatric diagnosis. Should psychiatry proceed as general medicine does, via diagnosis? If so, what is diagnosis? What does it mean to diagnose someone with one mental illness as opposed to another? There are some who will claim that a diagnosis will be a good one if, and only if, the diagnosis picks out a "natural kind." Other theorists are less demanding: a diagnosis is a good one, and a diagnostic category is worth using, if it has practical utility, for example, insofar as it allows us to group together patients who would benefit from the same kinds and course of treatment.

One important aspect of mental illness, and of illness generally, is that it is something that the person who has it undergoes, confronts or is a victim of. This way of thinking is in line with the notion that someone's illness is not a part of them, as a moral agent, and therefore has the ability to morally (and legally) exculpate. For example if a man kills his father, this will not be viewed as a straightforward patricide if the man did it because he believed his father was an evil robot. This is the topic of Chapter 4, which amounts to answering the question: Where does one draw the line between somebody being a victim of their condition and somebody being simply a bad person? In other words: When is a person to be viewed as a victim in need of care or a perpetrator in need of punishment?

Building on a recurring theme from previous chapters, Chapter 5 asks: What is the role that culture plays, both in the generation of psychiatric phenomena and in our evaluations of those phenomena as pathological or otherwise? Nowhere is this issue more clearly visible than in the question of when, or indeed whether, religious beliefs and religious experiences should be viewed as pathological, and a substantial part of this chapter examines this.

After the first five chapters, which mainly concern matters of clinical practice, namely, the diagnosis and treatment of mental illness, we move towards philosophy of scientific clinical research. Arguably the central question here is: How does scientific explanation work in psychiatry? In this chapter, I draw on general philosophy of science to clarify different ways of thinking about explanation in general, with a view to applying this to scientific psychiatry. For a long time, there has been an emphasis on *levels* of explanation. I draw attention to a different constraint concerning *kinds* of explanation. I also explore related issues about reductionism and the notion of explanatory power. Explanatory power is an important notion since it is an explanatory virtue that operates (in principle) separately from the virtue of accuracy. We need our theories, and the explanations they generate, to be not just accurate, but explanatorily powerful.

Then we move from psychiatric practice and research to the topic of Part 2, namely, the psychiatric phenomena themselves. In particular, we look at specific psychiatric conditions and explore a two-way interaction between them and certain core areas of philosophy. The interaction is "two-way" since, in one direction, our philosophical thinking might help us to better understand the conditions in question, while, in the other direction, these conditions might force us to revise and rethink long-held concepts that are central to philosophical theorising.

The first chapter of Part 2 (Chapter 7) focuses on what is perhaps the quintessential psychiatric disorder of the past hundred years, namely, schizophrenia. But what is it? How are we to understand its symptoms? And does it tell us something interesting about the nature of the mind in general? In this chapter, I show how the concept of schizophrenia is to be distinguished from psychosis. I then show how there is a reciprocal influence between our understanding of psychosis and a seismic shift in computational neuroscience in our understanding of the mind and brain. The brain, on this view, is a prediction machine, and psychosis arises when the mechanisms of prediction go awry (and this can be caused by different things, hence the different causes of psychosis, over and above the so-called disease "schizophrenia").

One symptom, which is typically (and not always helpfully) associated with both psychosis and schizophrenia, is the phenomenon of hearing voices or auditory verbal hallucinations (AVHs). This is the topic of Chapter 8. AVHs occur in a wide variety of diagnostic contexts, as well as in people who are not in need of psychiatric care. Why do people hear voices? This chapter presents and criticises existing theories and presents fruitful future directions. In particular, an overlooked but important question is: Why is it that AVHs are so often experienced as coming from specific agents?

Another archetypal symptom of mental illness is delusion, and this is the topic of Chapter 9. This chapter concerns what delusions are (when does something have delusional status?), and, rather differently, how we are to understand certain well-known delusions, such as the Capgras, Fregoli and Cotard delusions. It is often assumed that delusional claims express beliefs. However, theorists have questioned this assumption, and it is important to see why they have done so, whether they are right to do so and what the practical and theoretical consequences of this are.

Although it lacks the cultural prominence of psychosis, perhaps due to it seeming more on a continuum with normality, depression has a claim to being the most common and destructive form of mental illness. But what is depression? How does it differ from severe sadness? Is it just the lack of an appropriate or proportionate cause, or is there something in the phenomenon itself that makes it intrinsically different? In this chapter, first-person accounts take centre stage, and the nature of depression is characterised. A preferred view is suggested, according to which our experience of the world is shot through with possibilities, and degrees of possibility, in way that is more primitive than the attitudes of belief and desire. This fits well with the predictive theory of cognition used to make sense of psychosis.

Just as delusion interests philosophers as an instance of pathological belief, so addiction interests them as an instance of pathological desire. Analogously, it raises interesting issues for rationality and the nature of desire, and this is the topic of the final chapter of Part 2. It is a foundational assumption, not just of how we make sense of each other, but also of many of the social sciences, that human beings will take into account their beliefs and act in the service of their desires and goals. What addiction presents us with is an instance of someone apparently freely acting in direct opposition to his or her professed desires. How should we make sense of this?

The concluding chapter of the book presents some hopes and predictions for the future of philosophy of psychiatry. First, I envisage that conditions thus far under-examined by philosophers will begin to be more fully examined. For example, PTSD and its impact on our understanding of the nature of trauma and of memory could be a fruitful area of future research. Another important group of conditions in urgent need of further philosophical research (although there is some work on this already) are eating disorders. Both, I believe, will benefit from a cutting-edge framework that I drew upon at several junctures in the book, namely, the predictive processing framework. I also mention the category of personality disorders, which I think raise profound and only partly explored philosophical issues.

A timely issue is the question of whether the Internet is generally a good or bad thing for our mental health. One might think that unprecedented interpersonal connectivity is a promising way of combatting loneliness and social isolation. It might also be used for support groups, offering advice and helping those who have rare mental health conditions not feel so alone in the world. But of course, there are pernicious influences too. The over-glamourised presentation of unrealistic life goals, exposure of toxic subcultures in discussion forums, and so on, all present threats to psychological well-being, especially among children and teenagers.

I then make wishful predictions of about general methodologies and frameworks. The first of these, which is more a comment on what I already feel is promising and cutting-edge research, is the rise of computational psychiatry. The term "computational psychiatry" simply points to the use of powerful computational tools in the advancement of scientific psychiatry. It therefore encompasses predictive processing, and more generally computational approaches to *cognition*, but it also includes computational approaches to big-picture psychiatric epidemiology. This broader, environmentally and socially embedded approach to mental health and well-being leads me to point to an exciting family of approaches that we might call "externalistic psychiatry."

Another strand of research is also socially embedded, but in an importantly different way, exploring exciting synergies between the philosophy and sociology of psychiatry. It exemplifies a kind of "sociological turn" in philosophy of psychiatry that was prefigured by some of the Ordinary Language Philosophy-inspired theorists of the 1980s. In short, instead of asking, "What is mental illness?" (or indeed, "What is delusion?", "What is ADHD?") as if this were a metaphysical question (a matter of characterising a natural kind, for example), we ask ourselves, "What are we *doing* when we diagnose someone with a mental health condition?" One consequence of this would be a tight rapprochement between theorising about mental illness and mental health policy. Our attributions of psychiatric conditions, to the extent that we'd want to make them at all, will be determined by the sort of society that we want to live in and the kinds of people that we deem in need of care. This does not obviate important interactions between factual scientific advancements and normative judgments but rather makes them more nuanced.

I'd like, finally, to say something about the style of this book. Some textbooks neutrally and impartially present the lie of the land: how core debates have been and are being played out. Although here the key debates will be presented, this is not how I have written this book. I treat the core issues in an accessible way but from a somewhat opinionated perspective. I therefore don't simply present ideas, but also praise and criticise them. An upshot of this is that I present lines of questioning that seem natural to me, but that might not seem so to everybody.

The reason for this stylistic choice is threefold. First, I want to make the book engaging and driven endogenously by argument. Claims are always motivated by context, and a claim can be perfectly plausible (or at least not implausible) but still baffling if its motivation, the argumentative context, is not known. What does a given claim stand in opposition to, for example? Second, I want to encourage in my reader a critical rather than merely encyclopaedic engagement with the issues at hand. It is one thing to know everything that has ever been said within a certain debate; it is quite another to have one's own opinion about it. In my experience, prioritising the former encourages students to simply defend what is defensible within the game that philosophy can sometimes become. Prioritising the latter encourages students to find their own way to positions that, though they have been stated before, or may be unfashionable, are a better reflection of what they really think. Third, I do this out of a kind of necessity. Philosophy of psychiatry, unlike "core" areas of contemporary Anglo-American philosophy like metaphysics, epistemology, philosophy of language and mind, is a polymorphous, multidisciplinary hive of theorising, with people from different disciplinary backgrounds and different traditions within each discipline, all contributing in their own way. In short, while I have tried to be a pluralist, I have also had to be selective, and some might say biased, because there simply is no clear canon that everyone can agree on; at least not *yet*.

Acknowledgements

As I have mentioned, this book grew out of repeated attempts to design, and annually improve, courses in philosophy of psychiatry, for the best part of a decade. I'd therefore like to thank, first and foremost, all of my students in philosophy of psychiatry at both Durham and Exeter. Thanks for keeping me on my toes, and for being such a valuable sounding board for what works as an intuitive and accessible route into these issues and debates. I would also like to thank Matthew Ratcliffe for being a source of support and inspiration. He has always been very generous with his time, feedback and conversation. From that same time in Durham, I would also like to thank all of the members of the *Hearing the Voice* team for initiating me into deeply interdisciplinary research. Special thanks, among this crew, to Charles Fernyhough, Ben Alderson-Day, Angela Woods and Marco Bernini, with whom I spent many happy hours developing ideas. I would like to thank my colleagues at Exeter, especially Tom Roberts, Joel Krueger, Shane Glackin, Adam Toon and Giovanna Colombetti, for discussing many of the ideas and ways of presenting them that are contained in this textbook. Finally, I would like to extend heartfelt thanks to Andy Beck and Marc Stratton at Routledge for being so patient with me, and to the two anonymous referees who gave me such helpful feedback.

1 Introduction

What is philosophy of psychiatry and why does it matter?

Overview

This book is an introduction to philosophy of psychiatry. But what is philosophy of psychiatry? And why does it matter? In order to answer the first of these questions, we need to examine what it means for something to count as philosophy of something. We also need to examine what the target phenomenon – in this case psychiatry – is. It may not be obvious at first, but there are relatively broad and narrow understandings of what psychiatry is. Since the narrow understanding presupposes answers to questions that we will be in the business of asking, we will adopt a broad construal. Then we ask: Why does this matter at all? Why should anyone study philosophy of psychiatry, or indeed conduct research in it? At first glance, it might seem like a very niche area, but it is in fact (in my humble and perhaps biased opinion) one of the most important and far-reaching areas in the whole of contemporary philosophy. I aim in this chapter to give you a sense of precisely why.

1.1 Philosophy of x

A useful first step in answering the question "What is philosophy of psychiatry?" is to reflect on what the general schema "philosophy of x" usually means.

DOI: 10.4324/9781315450124-1

Where x is a human practice, whether this be science in general ("philosophy of science") or a specific science ("philosophy of biology") or something else ("philosophy of law"), this tends to mean, in the broadest sense, a philosophical engagement with that practice.

In line with this, two rather different kinds of thing fit under the broad category of "philosophy of psychiatry." These correspond to different ways in which philosophy can engage with psychiatry, and, at least to start with, this can be helpfully put in terms of the direction of influence between the two. On the one hand, "philosophy of psychiatry" can involve philosophical scrutiny of psychiatry. Here, the direction of influence is from philosophy to psychiatry: it is hoped here that philosophy will improve psychiatry. On the other hand, much of what gets called philosophy of psychiatry involves precisely the opposite direction of influence, namely, from psychiatric phenomena to philosophy. What can various mental disorders tell us about philosophical theories? In light of these rather different enterprises, this book will be divided into two parts. The first part will deal with the philosophy-to-psychiatry direction of influence, and the second part will deal with the psychiatry-to-philosophy direction of influence.

Some might insist that "philosophy of psychiatry" in the strict sense ought only really to refer to what we will examine in the first part, namely, philosophical scrutiny of psychiatry. This would be in keeping, certainly, with some other philosophies of x. Philosophy of science, for example, is clearly philosophical reflection on science. However, other philosophies of x (e.g. philosophy of psychology, philosophy of economics) have both directions of influence, and a large proportion of the exciting research that is produced under the auspices of "philosophy of psychiatry" involves the sorts of things that we will examine in Part 2. Thus, while I see some merit to the purely terminological point about the label "philosophy of psychiatry" in the strict sense, I am unapologetically including the second part of this book since (i) it constitutes a large proportion of the interesting work that gets classified under philosophy of psychiatry, and (ii) there is a perfectly defensible, broader use of "philosophy of psychiatry" that simply means philosophical engagement, of whatever kind, with psychiatry and psychiatric phenomena. This clearly includes, for example, philosophers of mind realising that clinical case studies of delusional patients threaten to undermine the theories of belief that they constructed from the armchair (see Chapter 8).

Aside from this clear division between the two parts of the book, there is a further bifurcation within Part 1. Psychiatry is a human practice, but it bifurcates in terms of what kind of human practice it is seen to be (you get a similar bifurcation in philosophy of medicine). "Psychiatry," broadly construed, can refer to a kind of therapeutic, patient-oriented practice. Its aim, broadly speaking, is to make people better. But "psychiatry," especially in the context of "philosophy of psychiatry," can also refer to psychiatric *research* – to the attempts that scientists and theorists make towards understanding psychiatric phenomena. In line with this bifurcation, I have entitled Part 1 *Philosophy of Psychiatric Practice and Research*. In partial acceptance that the contents of Part 2 don't fall under "philosophy of psychiatry" in the strictest possible sense, I have entitled it *Philosophy and Psychopathology*.

1.2 Psychiatry broadly construed

But what sorts of things count as psychiatry? Psychiatry, in the narrow sense that describes "what psychiatrists in the Western World do," is a branch of medicine, at least in the following way. Psychiatrists are medically trained doctors who, instead of specialising in, say, urology or haematology, specialise in psychiatry. However, this is not, or at least not obviously, what we mean by "psychiatry" here, at least not exclusively. There are at least three reasons for this.

First, philosophy is arguably (at least in part) about scrutinising, about critical thinking, and it would be distinctly *un*critical (and hence un-philosophical) to stick unquestioningly to the *status quo*. To be sure, we might well *end up* supporting it, but we shouldn't *start* there.

Second, it is not a very profound or philosophical enterprise to take as one's subject matter something that seems so contingent. In a nearby possible world (namely, had human history turned out slightly differently), psychiatry might have looked rather different. Whether that would be a good or bad thing is an open question, but that is precisely what we are interested in here. What does the *landscape of possible psychiatries* look like? What should our psychiatry aspire to? We don't even need to engage in imaginative counterfactual thinking to see the different forms psychiatry can take: just look at the differences in the ways that mental disorders are conceptualised and treated, not only in other societies and cultures, but also in different factions within our own. So not only is there contingency, there is also a huge amount of variety. Even between two card-carrying psychiatrists you will find *some* disagreement, if you look hard enough, about the nature and boundaries of mental illness, let alone between a biologically minded psychiatrist and, say, an existential psychotherapist. The (not-too-distant) past also provides an informative contrast. As will become clear throughout this book, our attitudes towards mental illness and towards appropriate response to it are not only varied, but constantly evolving.

The third and most important reason why the concept of psychiatry that we are dealing with here is broader than simply "what psychiatrists happen to do" is that this would presuppose answers to several questions that we will be in the business of asking, and which remain highly controversial and contested. Most prominently, this includes whether psychiatry *ought* to be thought of in medical terms. In the next two chapters, in particular, we will look at what mental illness is (Chapter 2), and at what issues and problems arise from the very practice of psychiatric diagnosis (Chapter 3). Both topics, especially the latter, have a direct impact on whether, and in what sense, psychiatry should be thought of as a branch of medicine.

As a result, we are concerned, at least in the first instance, with psychiatry *broadly construed* so as to include any *clinical* and *theoretical* engagement with mental illness. Whether the (actual or hypothetical) persons who engage in this are called "psychiatrists" is largely irrelevant (except to the extent that a certain role and status is attached to that professional label).

When we talk about clinical engagement with mental illness, "the clinical" is to be carefully distinguished from "the medical." The clinical involves the helping of people in a certain context, and who have a certain status, namely, are "patients" (there are non-clinical reasons for helping someone, and non-clinical ways of doing so, e.g. charity work or social care). The medical (or at least, the clinically medical rather than the theoretically/scientifically medical, e.g. medical research) is the attempt to do this via the practice of medicine, and, of course, much more needs to be said about what medicine is, and we will touch upon this in Chapter 3.

Suffice it to say that there are (or at least certainly seem to be) things that fall under the non-medically clinical, and also under the non-clinically medical. The various forms of psychotherapy might fall under the former category. Medical research falls under the latter category. Also, the medical profession routinely deals with cases like childbirth that aren't quintessentially clinical in that nobody is in need of cure, nobody is *ill*, nothing (in the good case) has gone wrong or needs to be put right, but where the offerings of modern medicine (e.g. pain relief, hygiene, sterilisation) are extremely welcome. (Within medicine, there also is a category of work that falls somewhere between the clinical and the scientific, and that is diagnostic lab work. If you are analysing a sample (e.g. bloods) you are not in a clinical, patient-facing role, but it does pertain to a given patient and you are not seeking to extract generalisable findings, so you are not doing medical science in that sense.)

A consequence of all of these considerations is that "psychiatry," broadly construed in the *clinical* branch, includes all work done by mental health practitioners, including clinical psychologists and psychotherapists, and in the *theoretical/research* branch includes any research that deals with mental illness (from neurobiology all the way up to epidemiology and sociological mental health research). Of course, when we focus on the clinical branch, we are not just in the business of reflecting on all of the possible ways of engaging with mental illness, but of ascertaining the *best* way of doing so. Thus, it may turn out that clinical engagement with mental illness is best achieved through (the tools and concepts of) medicine, but this is not something that should be presupposed (as Chapter 3 will partly demonstrate). Notice that mental health research divides into medical research and other forms (e.g. social science research). The various subdivisions of psychiatry broadly construed can be represented as given in Figure 1.1.

Different positions in philosophy of psychiatry prioritise different boxes in this diagram. Some positions may even suggest that some of these boxes shouldn't really exist. For example as we'll see in Chapter 3, opponents of "the medical model" (sometimes under the banner of "anti-psychiatry") will be critical of the medical branches of mental health practice and research. Proponents of the medical model, in contrast, will insist on their importance and may be critical

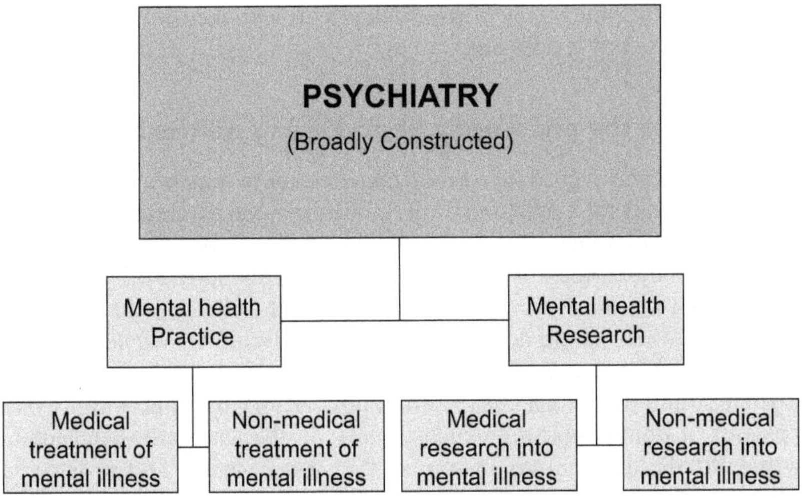

Figure 1.1 The subdivisions of "psychiatry broadly construed"

of their non-medical counterparts. This is all fine, but it is important to acknowledge that these are substantive philosophical positions that need to be carefully argued for.

Before moving on, it is worth mentioning another consequence of interpreting "psychiatry" broadly. Quite simply, if psychiatry is just a branch of medicine, this implies that *philosophy of* psychiatry is merely a branch of *philosophy of* medicine. And yet to think this would be to grossly misinterpret the relationship between the two. Most obviously, this would suggest that *more* work counts as philosophy of medicine than does philosophy of psychiatry, since the former would subsume the latter. This is not reflected in the current landscape of philosophy. If anything, philosophy of psychiatry, with all of the conceptual issues that arise from dealing with the mind (and its relationship to brain, body, physical and social environment, etc.), is the larger discipline.

Someone might object that this is simply a matter of pragmatics. Philosophy of psychiatry is literally a sub-branch of philosophy of medicine, and it just isn't relevant or informative to explicitly say so. Since philosophy of psychiatry takes up such a large proportion of philosophy of medicine, calling it philosophy of medicine is simply misleading. That would imply that we are talking precisely about the philosophy of medicine that is not philosophy of psychiatry, since otherwise we would have stipulated. Undoubtedly there is some truth to this observation, and a significant amount of philosophy of psychiatry counts as philosophy of medicine, but for these pragmatic reasons wouldn't usually be called such.

Nevertheless, putting these pragmatic effects aside, I would suggest that the real relationship is one of overlapping categories rather than one being a subset of the other. Some philosophy of psychiatry deals with medical issues, but a significant proportion of it does not.

1.3 Why does the philosophy of psychiatry matter?

I have presented and argued for a broad construal of psychiatry, at least when we are thinking about the "psychiatry" in "philosophy of psychiatry."[1] To clarify: I am certainly not suggesting that people who aren't considered psychiatrists (e.g. psychoanalysts, mental health nurses) suddenly start calling themselves such. Rather I am merely suggesting that we need to be open in the scope of our enquiries when we are doing philosophy of psychiatry. We need to do this on pain of begging the question against some key players in a number of central debates (and handing a cheap and hollow victory to others). But what are the payoffs of doing philosophy of psychiatry in the sense just sketched? I will now present the different contributions that can be made and provide some illustrations.

1.3.1 It improves understanding and treatment of mental illness

In Part 2 of this book, *Philosophy and Psychopathology*, we will see how careful philosophical reflection helps us to understand specific mental health conditions. There are three very different ways in which philosophy can make important contributions in this way.

The first is a contribution of conceptual work, fixing confusion, clarifying vagueness or ambiguity, analysing concepts, etc. For example in Chapter 8, we will see that delusions are defined in terms of belief. But if we examine the concept of belief, it looks like some of these delusions might not actually be beliefs (Berrios 1991; Currie and Jureidini 2001). This has potentially serious consequences for how we are to treat them. If delusions aren't beliefs, we shouldn't interact with them as if they are. Another example is with the hearing of voices. If we unthinkingly assume that those who say that they are hearing voices are simply undergoing a sensory auditory experience of a voice in the absence of the usual environmental stimulus (namely, a speech sound), then we are likely to look for explanations of auditory hallucinations, and presumably we would expect activation in auditory cortex. However, a closer look at the experiences of hearing voices shows us that many instances aren't about straightforward auditory experiences. Some, for example, are actually soundless voices. This sounds paradoxical, but a bit of conceptual analysis allows us to understand what the experience might be like and what might need explaining (Wilkinson 2019).

The second, which can be related to this, is the contribution that phenomenology can make. For example if we look at depression, we might think that it's just an extreme form of sadness. However, more careful phenomenological analysis reveals that depression is further removed from everyday experience than we might first have thought. In particular, it may involve a fundamental disturbance to our background sense of being in the world, what Ratcliffe (2014) calls our existential feelings. We'll see more on this in Chapter 10.

The third is to do with division of labour and the freedom to be interdisciplinary, to read across disciplines and to see the "Big Picture," to think laterally and creatively. For example one might see that some advancements that are being made in one area can be fruitfully applied to another. For example general theories of brain function (Friston 2005) can be applied to delusions and psychosis (Fletcher and Frith 2009), to voice hearing (Wilkinson 2014), depression (Chekroud 2015), PTSD (Wilkinson, Dodgson and Meares 2017) or addiction (Miller, Kiverstein and Rietveld 2020). These new ways of thinking about certain symptoms and conditions could lead to innovative and effective treatments.

1.3.2 It improves the way our institutions function

The consequences of our thinking in philosophy of psychiatry don't only touch upon the moral, but also on the legal and institutional. It has been common in Western institutions to distinguish mental disorders that have physical causes from those that have "merely psychological" causes. The two will be treated differently in the eyes of the law, and, more mundanely, employers and insurance companies. Putting mental illness on a par with somatic illness, either by showing mental illness to have physical underpinnings or by showing illness in general to not be centrally about underlying physical underpinnings, impacts on this significantly.

A parallel issue exists in matters of funding for our research institutions. Somatic health gets far more funding than mental health. According to the UK-based mental health charity MQ, 25 times more funding goes into cancer research than into mental health research. Why is that? Is it because cancer itself is seen as more "real" that mental illness? Is it because cancer research is seen as more "scientific," more legitimate or more likely to yield concrete and practical results? Or is it because people fear cancer more? If so, is this because it is thought to exact a more negative impact on humankind? Or is it because it is seen as a far more prolific cause of death?

Any of these reasons would be grounded on questionable assumptions, and the last of these is simply inaccurate. The WHO estimates that there are 8 million deaths every year that are attributable to mental illness, compared to 9.6 million deaths from cancer. The rates are comparable, and certainly not different by a factor of 25! Perhaps more accurately, cancer is seen as a *direct* cause of death in a way that mental illness (e.g. leading to suicide) is not. But why should this make a difference? These are all questions that warrant careful, critical reflection. On the basis of this reflection, it may well turn out that our funding distribution is not how it should be.

Here is a related issue. Currently in the UK, healthcare and social care are conceptualised differently and funded separately. Health care is administered by the National Health Service (NHS) and is entirely free at the point of use and funded through general taxation. Social care is largely private and has to be paid for by the user (if they are deemed able to afford it). Is this difference between illness (whether mental or physical) and something more vague, but which entails as much (and often more) need of care, which we might call a "problem with living," really justifiable and non-arbitrary? Infirmity is no more the sufferer's fault than illness. And as we shall see in the next chapter, several theorists don't think that illness has a special nature or status: it just *is* a certain problem with living. This would make the healthcare/social care divide not only hard to morally justify, but theoretically arbitrary too. But even if this isn't the case, and there is a non-arbitrary theoretical distinction to be drawn between illness and infirmity, why is treatment in the face of the former seen as being more of a basic right than being looked after in the face of the latter? These are all issues of immense importance, and the debates in this book (especially Chapters 2 and 4) speak to them directly.

1.3.3 It improves our societal responses to mental illness

The way we talk and think about people (others and ourselves) has a profound impact on their lives, and this is especially so in the context of mental illness. There are different ways of thinking about mental illness in general and about specific conditions in particular, and both have major implications for how we respond, as a society, to those who are deemed to be in need of psychiatric help.

The attribution of mental illness is not a fixed thing. In the past, it has been accompanied by certain other negative evaluations that we now deem inappropriate (think of the moral failings associated with WW1 Shell Shock). When we attribute mental illness in general, we are negotiating a landscape of norms, often walking a line between medicalisation and moralisation. It can be a difficult balancing act. Not blaming the depressed individual for absenteeism is surely a good thing, but going too far in the other direction and depriving them of all autonomy, or objectifying them and making them feel like they are nothing but a chemical imbalance is hardly optimal either. Viewing addiction as a "brain disease" (Leshner 1997) might be better than just thinking of it as a moral failing, or even weakness of will, but it can lead to an abdication of responsibility ("It's not me, it's my brain") or prognostic pessimism ("I'll never get better, this is just the way I am") (see Chapter 11).

Various mental health diagnoses are accompanied by social stigma (indeed one can argue that all are to at least some degree). Stigma is not simply the result of trivial choices about how to react to mental illness. It is based on factual ignorance and conceptual confusion. Some of it is based on folk versions of the medical model ("The Broken Brain" view). Some of it is based on a natural human aversion to what is different and hard to comprehend. In general, we want our ways of talking and thinking about mental illness to reflect and promote attitudes that are informed, appropriate, humane and sympathetic. Philosophy of psychiatry, with its reflection on both the nature of mental illness in general, and specific mental health conditions in particular, the discourse surrounding them, and questions pertaining to autonomy, responsibility and so on, has the power to help us achieve this goal.

1.3.4 It gets to the heart of what makes us human

In a closely related manner, philosophy of psychiatry gets to the heart of what makes us human. Arguably the most profound question in philosophy of psychiatry (I would even say in all of philosophy) is the following: When is somebody ill, and when are they a bad person? Put more generally: When is someone a victim to be pitied and helped, and when are they a perpetrator to be blamed and punished?

Many mental illnesses (including cognitive decline in the context of dementia) cause people to behave in ways that flout social norms, often in ways that, taken in isolation, would normally be found to be reprehensible. When do we deem someone unwell, a victim of a condition, and does this necessarily, and to what extent, lower or remove blameworthiness? And where do we draw the boundaries of the person? Is a difficult childhood that leads to delinquency different only in degree from the dangerous behaviour of someone experiencing delusions? Or is the former simply what it is to be a bad person? After all, people don't become bad by magic; *something* must cause it. But what about cases like foetal alcohol spectrum disorder (FASD) (Carel 2016), whose symptoms closely resemble (indeed one might argue *are*) character flaws of impulsivity, aggression, etc.? It's not their fault that their mother drank while they were in utero. Arguably the same applies to difficult, abusive, traumatic childhoods.

However, the counter-argument goes, if we start forgiving people because of childhood circumstance, or for FASD, then why not forgive people on account of their genes, too? No one asks to be born and to have the parents they end up having. Speaking more broadly, there are, surely, myriad factors beyond the individual's control. But, of course, this is what our moral evaluations, at least to the extent that they target individuals as well as actions, are all about. We do talk about bad people as well as bad actions. Is this a mistake? After all, we can imagine someone protesting (as I'm sure some people have) "It's not my fault that I'm a bad person!" These are some of the thorniest issues, not just in philosophy of psychiatry, but in all of philosophy, and we will touch on them throughout, but most explicitly in Chapter 4.

These may also get to the heart of how we think of ourselves and our own flaws and undesirable behaviours. Although these may well be of "sub-pathological" severity (or within the bounds of "normality"), it makes an important difference to our self-evaluations and our subsequent course of action, whether we think of, for example, our tendency for unacceptable bouts of explosive anger as something that is a part of who we are but that we can improve or control, or some kind of affliction that we suffer from. An orthogonal issue is then whether this is something that we can treat or whether this would involve fundamental changes in who we are. All of these questions, and ways of thinking, loop back into who we take ourselves to be, whether we feel guilty about certain things, feel we can improve in certain ways, etc. In short, it really gets to the heart of what makes us human.

Summary

- Philosophy of psychiatry involves two quite separate enterprises: (i) philosophical scrutiny of psychiatric practice and research and (ii) philosophical engagement with specific mental health conditions.
- The term "psychiatry" must be, at least initially, taken broadly, on pain of assuming answers to several core questions in the field.
- Philosophy of psychiatry has a number of important real-world applications and also gets to the heart of what makes us human.

Questions

- What do you think of when you think of psychiatry? Do you think of prescription drugs or therapists' couches?
- Try to imagine radically different societies: what does mental illness (and psychiatry) look like in them? Ancient Greece, the Aztecs, Feudal Japan, Elizabethan England: how might they think about the phenomena that we think of as mental illnesses?
- Under what circumstances might you seek professional help with your mental health? How would you like to be engaged with in such circumstances?

Note

1. Some might object that I am actually talking about "Philosophy of Mental Illness." However, this doesn't capture the level of engagement with a human practice. Mental illness is a phenomenon with which certain human practices engage: it is not a practice in itself.

Further reading

Bentall, R.P. 2003. *Madness Explained*, London: Penguin.

Cooper, R. 2007. *Psychiatry and Philosophy of Science*, Stocksfield: Acumen Publishing.

Graham, G. 2010. *The Disordered Mind: An Introduction to Philosophy of Mind and Mental Illness*, London: Routledge.

Murphy, D. 2006. *Psychiatry in the Scientific Image*, Cambridge, MA: MIT Press.

PART 1

Philosophy of Psychiatric Practice and Research

2 What is mental illness?

Overview

The question "What is mental illness?" is one of the central questions in philosophy of psychiatry. It delineates the subject matter of the entire field, and one's answer to it feeds into many other important debates. This question, however, is ambiguous. In particular, what the question means depends on where you place the emphasis in asking it. Compare the following two questions.

Question 1: "What is mental *illness* (as opposed to mental *health*)?"
Question 2: "What is *mental* illness (as opposed to *physical* illness)?"

The first question requires us to examine concepts of illness and health generally. The second question effectively asks: "Where do we draw the line between the mental and the physical, when it comes to illness?"

We address these two questions in turn, addressing, first, the notion of illness as opposed to other related notions, most notably disease, disorder and disability, and then running through some philosophical accounts of illness. The second question is usually taken to amount to the issue of whether properly *mental* illness makes sense or, to use the terminology of one particularly prominent theorist, "is a myth." On one reading, this is a question about whether it exists, perhaps, for example, because the only intelligible kind of illness is physical. On a second reading, we

DOI: 10.4324/9781315450124-3

can bracket questions about myth entailing non-existence and focus on whether mental illness has some of the positive narrative features of myth. In other words, our final question becomes: what does mental illness discourse achieve?

2.1 Illness, disease and neighbouring notions

What is it for someone to be ill *in general*? At first glance, it might seem as though the question "What is mental illness?" is only difficult because the mental is conceptually a treacherous territory. Our understanding of somatic illness in contrast is rather straightforward. One might think the real challenge is about how to extend this well-understood notion to the mental domain. However, this assumption that we understand somatic illness relatively well is mistaken: there are in fact very thorny issues concerning the notion of illness generally, and therefore including somatic illness.

Before we start looking at specific accounts of illness, let's make a preliminary distinction between illness and related concepts. In the conceptual vicinity of "illness" are concepts expressed by words like ailment, sickness, disorder, and disease. It is worth noting that different natural languages (English, French, Japanese, etc.) will have different words with different nuances and may have a relative lexical richness or poverty that enables more or less fineness of grain.[1]

2.1.1 Illness, disease, disorder and sickness

One concept related to illness that is especially relevant to existing literature in this area is the concept of disease. Several theorists (e.g. Wakefield 1992; Cooper 2002; Murphy 2006) take illness and disease to be synonymous and interchangeable. There is nothing obviously wrong with this. It is, after all, not clear what different definitions of "disease" or "illness" are answerable to, beyond mere stipulation. If you make the synonymy clear and explicit, then, by all means, use them interchangeably. However, there are two things to note that may count in favour of keeping disease and illness separate.

First, it might be that there are two interesting and important phenomena that should be carefully distinguished and that the stipulation of synonymy fails to make visible. Second, one might argue that differences between illness and disease are already reflected in our linguistic intuitions (as English speakers). In other words, one reason for resisting the idea that "illness" and "disease" are synonymous is that the contexts in which they are felicitously used are different.

Following Cassell's (1976) observation, we say that people are ill, we don't say that people are diseased (to call a person diseased is infelicitous, it would be ungrammatical in Wittgenstein's (1953) sense or a category mistake in Ryle's

(1949) sense). In direct contrast, we talk about a diseased body part, a diseased lung, for example. But it is not the lung that is ill. The person to whom the lung belongs might well be ill as a result of having a diseased lung.

Another relevant distinction might be between personal and subpersonal vocabularies (Dennett 1969). Dennett is concerned with pointing out that pain and belief are "personal-level" concepts. It is people who are in pain, not nervous systems (or any other parts of people) that are. Similarly, it is people who believe things (and perceive things etc.), not brains or any other parts of people. Granted, brains and nervous systems might need to be in a particular state in order for people to be in states of pain or believing or perceiving. Something similar might be said about illness and disease. Illness is personal; disease is subpersonal.

What we *do*, however, say is that a person *has* a disease, usually in virtue of a part of them being diseased or, perhaps in the most evocative case, in virtue of being infected by a pathogen. Taking these linguistic intuitions into account, we can still retain a rather strong link between illness and disease. You might, for example, think that, although they are different kinds of concept, they depend on each other. Perhaps if you have a disease, you are ill, and vice versa. The first way around does not seem that plausible. It seems that people can have a disease (e.g. diabetes), but they can manage it well and therefore not be ill. The other way around is perhaps more plausible. You might claim that someone is only truly ill if they have a disease, just like you might think that someone is only in pain if they are in a particular neural state. Of course, some theorists will want to deny this link. Pain, one can plausibly argue, is a particular experience, and that's what makes it pain. You can't over-ride someone's claim to be in pain by looking at his or her nervous system. Someone could say the same thing about illness (especially if, like Carel 2007, you take a phenomenological approach to illness).

Just like with pain, there is an important difference between what *constitutes* a given instance of pain (and it might just so happen that all such instances are similarly constituted, but that's an empirical, not a conceptual issue) and what makes us grant something the *status* of pain, and the same can be said about *illness*. This difference between "the status question" and the "constitution question" has been most clearly articulated by Roberts, Krueger and Glackin (2019) and helps to cut through some unhelpful ambiguities when we ask "What is x?" What we are interested in when we ask "What is illness?" is the status question. We are asking: "what grants a given state or condition the *status* of being illness?" rather than asking: "what constitutes this particular state or condition?"

Another distinction between illness and disease (put forward by Chris Boorse (1977), a figure who looms large in debates on the nature of disease) is that, while illnesses are *particulars*, diseases are *universals*. This somewhat technical metaphysical vocabulary roughly means that each illness is an individual event in space and time. A consequence of this is that two individuals cannot have the same illness (this is reflected in the familiar statement, "He died after a brief illness"). Two illnesses can be similar, but they cannot literally be the same.

Diseases, however, are the sorts of things that you can have (or catch, in the case of infectious disease) again and again. Two people can have the very same disease and will be diagnosed accordingly. This is Boorse's view, and whether it fits with our linguistic intuitions, or even has to, is up for dispute. After all, this would seem to suggest that, strictly speaking, you cannot diagnose someone with an illness, since this wrongly suggests that illnesses are universals. Diagnoses should be about diagnosing diseases. But it doesn't strike me, for one, as obviously confused to talk of diagnosing someone with an illness. "Illness," like "game," can be used as a universal or as a particular. Tennis is a game that you can play again and again, but a given game of tennis can only be played once.

There is a third, related concept in the semantic vicinity of illness and disease, and that is the concept of "disorder." Whereas illness and disease apply as readily, and arguably more so, to the somatic (bodily) realm than the mental realm, "disorder" is a word that you are perhaps more likely to encounter in the context of mental problems than somatic ones. Take, for example, the bible of Western psychiatric diagnosis (that we will examine in more detail in the next chapter), the DSM. That acronym stands for "The Diagnostic and Statistical Manual of Mental *Disorders*." The WHO's ICD, on the other hand, which covers both somatic and psychiatric conditions is "The International Classification of *Disease*." So what is a disorder, and how does it differ from disease and illness?

Again, I'm afraid, disorder can be used in different ways by different people, and it's often not clear what, if anything, the usage is answerable to. One thing that such usage may be answerable to (or at least might guide one in deciding to use a word in one way as opposed to another) is usefulness. The usefulness of a word, however, cannot be ascertained by considering the word in isolation, since it is determined by the way in which other words cover relevant conceptual territory. Thus, many theorists use "disorder" and "disease" interchangeably (e.g. Cooper (2007, p. 29) stipulates this explicitly). However, if you look closely, that is because they are using disease in a very broad sense. My preference, based on the expressive usefulness of the words at my disposal, is to use disease for an underlying subpersonal pathogenic condition (an infectious disease caused by a virus or bacterium being a quintessential example, although any example where a pathogenic mechanism is detectable and can lead, but is not conceptually tied, to certain symptoms would fit too), and a disorder for a condition that is not necessarily "underlying," but which is subpersonal, and which interferes with daily functioning. Thus Post-Traumatic Stress Disorder (PTSD) is a disorder in the sense that something is wrong with your nervous system and that interferes with your daily functioning. But it is not a disease like Alzheimer's or cancer. An asymptomatic disorder (e.g. asymptomatic PTSD) makes little or no sense. Diseases, on the other hand, are frequently detected before symptoms appear.

Our scientific, diagnostic and clinical practices thereby feed into how we think and talk about various unwanted conditions, and it seems to me that something like this reflects a useful distinction between disease and disorder. It also, for reasons that will become clearer later, accounts for why we more commonly talk about mental disorders rather than diseases. Mental disorders are fundamentally *about* "symptoms": there is no asymptomatic or pre-symptomatic detection of schizophrenia and depression (which is not to say that risk factors cannot be ascertained – that's the whole idea behind early intervention services).

So much for the differentiation between disease and disorder. Disorder seems closer (on my stipulated usage) in semantic space to illness. Indeed, one might say that they are near-synonyms. Even so, they differ in emphasis. First, it is worth reiterating that people are ill, but it is not people who are disordered. Second, and in a related manner, illness is more of a daily, practical, even "folk" concept, whereas a "disorder" has (some would say questionable) connotations of specialist and scientific authority. What is more, the claim of near-synonymy needs to be approached with caution. Some things are commonly called disorders that I would be very wary of calling cases of illness. For example autism, now commonly referred to as ASD (Autism Spectrum Disorder), is called a disorder, but I wouldn't think of it as an illness. One move that allows us to retain the synonymy is to say that autism is wrongly called a disorder. That may be true, but it is worth unpacking a bit further why we might be tempted to call ASD a disorder (or indeed some other disorders, like FASD (Foetal Alcohol Spectrum Disorder)).

Now is not the time to get into advanced philosophy of language, but if you were inclined (as I am) to think that the meaning of a word is about what the word does practically (pragmatism) (or, in a related manner, that it is about all of the inferences – practical and theoretical – you can make with it (inferentialism)), then you might think that "illnesses" and "disorders" differ in a few respects. For example you might think that "illnesses" are the sorts of things that you can in principle try to cure. Disorders, on the other hand, can be congenital, incurable and can perhaps even be thought of as "part of who the individual is." Calling something a disorder is more general and seems to simply flag that something is "wrong." This means that many (though perhaps not all) illnesses (namely, instances of people being ill) may be the result of disorders. However, a high proportion of disorders are not illnesses and do not underpin or give rise to them. It certainly sounds very odd to say of somebody with, for example, a personality disorder, that that disorder makes them *ill*.

These observations are also in keeping with what makes the labelling of a condition as a disorder or as an illness problematic in different ways. Take the label ASD as applied to those with autism. The "disorder" element is problematic because it deems it to be "bad" or to indicate a negative valence, a below normal functioning. Deeming autism to be an illness is even worse since it implies not just negativity, but also suffering and curability, or at least the desirability and conceivability of cure. However, as we are about to see, neurodiversity activists say that it is misguided, even confused, to talk of "curing someone of autism" or "getting your son back from autism." Your son is not suffering (except to the extent that society is failing to accommodate him), and his autism is part of who he is. In contrast, this way of thinking does, however, make very good sense for psychosis, depression, generalised anxiety and so on.

Finally, another term within this semantic vicinity is "sickness" (Twaddle 1994; Hofmann 2002). Twaddle (1994) distinguishes sickness from both disease and illness: "Sickness is a social identity. It is the poor health . . . of an individual, defined by others with reference to the social activity of that individual" (Twaddle 1994, p. 11). As Hofmann (2002) writes, approvingly summarising Twaddle, "the paradigm case is when a *disease* leads to *illness*, which then results in *sickness*" (p. 653). Note that this is only a paradigm case, rather than a necessary state of affairs. This is not to claim that sickness needs disease or even illness. The point is rather, in a way that very much echoes sociological work on "the sick role" (Parsons 1951), that there are different lenses through which we can view poor health. These are roughly: objective/biological (disease), subjective/ personally undesirable (illness), and societal/socially undesirable (sickness).

2.1.2 Illness versus disability

Another concept that is informatively compared to illness is disability. Unlike the contrast between illness and disease, where the two concepts can be seen as categorically different, and yet overlapping (people can, but need not, be ill in virtue of having a disease), illness and disability are concepts that operate at the same categorical level (it is *people* who are ill and also *people* who are disabled). However, they operate in interestingly different ways.

But first, there are important and illustrative similarities. At first glance, both are, all other things being equal, *bad* states to be in, at least in the sense that being ill is worse than not being ill, and being disabled is worse than not being disabled. Indeed, it is the inherent negativity of the term "disabled" that has pushed many (both theorists and activists) to advocate a move away from "bad difference" towards "mere difference." Another similarity is that both concepts raise difficult threshold questions: there are not only canonical examples of illness and disability but also tricky borderline cases. A third similarity, related to the previous two, is that both areas encourage a theoretical debate about whether these are objective biological categories or socially constructed ones (more on this later). In short, both illness and disability are normative categories (whether biological or social) that admit of difficult borderline cases. But how do they differ? Since we will give illness a more thorough treatment later, I'll focus on disability here, gesturing to political differences with illness.

The sorts of conditions that are commonly classified as disabilities are hugely varied. For example there are physical disabilities involving missing or deformed limbs (both congenital and acquired), neurological or neuropsychological disabilities, such as ataxia, prosopagnosia, aphasia, blindness (cortical and retinal), and, perhaps most interestingly for our purposes, conditions that are classified as mental illnesses can also be classified as disabilities, for example individuals with psychiatric diagnoses (e.g. depression and schizophrenia) may well count as disabled. In addition to these, there are developmental conditions like Autism or Down's syndrome that are not happily classed as illnesses (although "Autism Spectrum Disorder" (ASD) does feature as a diagnostic category in the DSM). What could possibly tie all of these things together?

The issue of how to define disability has given rise to its own field in philosophy and disability studies, and I cannot begin to do it justice here. Nevertheless, a few things are worth reflecting on. First, the distinction between *constitution* and *status* is just as important when thinking about disability, as it is for when thinking about disease and illness. In other words, we might ask what it is about a given individual that gives them the condition that we call a disability. This amounts to answering the constitution question. But we might also ask: what gives any condition the *status* of a disability? This allows in principle that conditions that are very differently constituted elicit disability status. So one way of addressing the concept of disability is to reflect not so much on the conditions themselves, but the sorts of criteria that might form the basis of a judgement of disability status.

A distinction that is related to the one I here refer to as constitution versus status is that between "impairments" and 'limitations." Although the distinction itself and the two concepts independently have come under scrutiny and criticism, the idea is roughly as follows. The *impairment* is the characteristic (mental or physical) that is seen as dysfunctional and often as contributing to, and associated with, the *limitation*. In turn, the limitation is seen as the way in which the individual is prevented from flourishing in life.

As with illness, philosophers thinking about the nature of disability can adopt a "medical model." This is similar to the medical model in psychiatry in that it takes the appropriate conceptualisation of, and by extension response to, disability to be medical. The usual opposition to the medical model of disability is often thought to be the *social model*. The social model comes in various strengths and flavours, but, roughly speaking, it downplays the role, or even the necessity, of an underlying objective medical condition. Social factors are central to whether someone counts as disabled, and our response to them should be social (e.g. social and ergonomic accommodation, facilitation, support, preservation of dignity).

Our focus here is clearly not on disability itself, but it is certainly useful to reflect on relationship between illness and disability or, at least, on the sorts of circumstances that lead to the attribution of illness and disability. One important observation is that some people are deemed disabled in virtue of some illnesses. Other illnesses, however, do not grant disability status (e.g. the common cold). Furthermore, many disabilities aren't thought of as illnesses. One obvious, if over-simplistic, first response to this is to say that disability is an overtly socially and institutionally embedded concept. Indeed, disability, it could be argued, is more full-bloodedly a *status* in that it carries with it overt social, legal, political consequences. Viewed in this way, it is hardly surprising that one might be deemed in need of this status based on a number of conditions. Some, but not all, of them might be illnesses, and some illnesses do not warrant the individual receiving this status, especially if they are relatively short-lived. I could be socially and personally limited in relevantly similar ways by (i) the loss of a leg to a road-side bomb, (ii) severe arthritis, (iii) chronic fatigue, (iv) a severe anxiety disorder, and so on.

Two challenges for this view arise immediately. The first is that people are personally and socially limited in ways that typically do not encourage us to attribute disability status: poverty is one, unfortunate appearance, or tone of voice, might be another. The general point is this: if the conditions that elicit disability status are so multifarious, where do we draw the line? For all its faults (and there are many), the medical model can at least say: it's the conditions that are medical in nature (or a more pragmatic medical model might say: it's those that benefit from medical interventions). Conversely, however, the medical model is seen as objectifying, disempowering, and presenting as overly pitiable those who fall under the category "disabled."

The second challenge, which is especially relevant here, is that there are several accounts of illness that take it, too, to be a social status in a sense very similar to the way in which disability is (see, again, Parsons' concept of "the sick role" (Parsons 1951)). On such a view, attributing illness and disability are both bound up with certain societal roles and responses, which have some similarities, but are importantly different. The challenge, if these views are correct, then becomes providing a good account of these statuses, of how they differ, of how they interact, etc. Having said this, there is no obvious reason to think that this challenge cannot be met.

2.1.3 Illness, difference and diversity

Illness and disease need to be distinguished from diversity in general, and neurodiversity in particular. The word "neurodiversity" was first introduced by the sociologist Judy Singer in reference to autism, but it has since been applied more broadly. Its role, as a concept, is to refer to robust differences in the brain and in cognition, which are not pathological. The closely associated "Neurodiversity Movement" is precisely concerned with promoting the view that autism is a matter of diversity, not of illness nor indeed disability. The movement itself can't really be adequately understood out of its historical context in response to "mother blaming" accounts. Though extremely interesting, I must not digress too far.

Nevertheless, the very introduction of the concept of neurodiversity raises profound philosophical issues. What is the relationship between diversity, on the one hand, and illness or disability on the other? If you look across the entire global human population, there is a wealth of difference. In particular, there are many different kinds and degrees of difference. How do we choose (and who gets to choose), which of those (whether rightly or wrongly) are deemed to be instances of illness or disability?

Again, one starting point involves reflecting on the *attribution* of diversity, as starkly and emphatically opposed to the attribution of either illness or disability. Regardless of whether there is an objective line to be drawn between the diverse and the ill, what does thinking of someone in a certain way *do*? More specifically, what does thinking of someone as neurodiverse as opposed to mentally ill, or disabled, achieve?

One obvious thing that is achieved is a move away from thinking about the relevant individuals in negative terms and, to the extent that this has consequences for actions, in terms of wanting to fix or cure them. This is particularly important for understanding the distinction between neurodiversity and mental illness. This also (as we will see in Chapter 4) feeds into how we think of personality and personhood. Compare one parent, whose son, in late adolescence, develops psychosis or depression. The parent will, understandably, want to get her son back. An underlying thought behind the idea of neurodiversity is that this same idea of "getting my son back" is neither desirable, nor even *coherent*. My son, the parent in support of neurodiversity will say, just is this way. Getting rid of his autism would not be relieving him of an ailment: it would be changing him as a person. A number of factors fuel these intuitions. One is: how debilitating is the condition? Others include: When is the age of onset? (The later the onset, the more likely one is to form a conception of the individual prior to the condition/state in question.) How long does this condition last? What are the hopes of treatment? How is it conceptualised by the medical profession?

The line between diversity and disability is less sharp. That is in part because disability status, like diversity status, doesn't involve "cure" (at least not in the same way). It is about assistance, management, accommodation and facilitation. The real locus of difference between disability and diversity is, first, the perception of negativity. Disability just is (if even only etymologically) a negative term that seems to encourage an intervention involving a "bringing up" of the disabled individual to "our" level. Diversity is emphatically value-neutral: there is no implication of negativity, and the appropriate intervention is deemed to be an inclusive negotiation among equals. Of course, the proponent of neurodiversity is not going to deny that a great many people need that status of disability and the level of help that comes with this. The plea is rather to keep that within its proper bounds. (Here, the distinction between "high-functioning" and "low-functioning" varieties of autism is key, with the latter far more likely to attract disability status and the former more likely to attract diversity status.)

2.2 Philosophical accounts of illness

Enough, for now, about the relationship between illness and related concepts: What do philosophical accounts of illness look like? How we should divide different accounts of illness is deceptively complicated (see, e.g., Simon 2007; Kingma 2017 for nice engagement with this complexity). In this section, we start by mapping the terrain, which is to say, clarifying different positions that can be taken and their philosophical relevance. Then, starting with an extreme end of a spectrum, namely, with "objectivism," we rehearse concerns and objections until we have presented and evaluated all of the major positions available.

2.2.1 Mapping the terrain

We will begin with a very broad and oversimplified divide that we will subsequently subdivide and enrich further. At first glance, we can distinguish theorists who think of illness in objective terms from those who think of it in terms of values. Robert Kendell expresses this distinction in the following passage.

> The most fundamental issue, and also the most contentious one, is whether disease and illness are normative concepts based on value judgments, or

> whether they are value-free scientific terms; in other words, whether they are biomedical terms or sociopolitical ones.
>
> (Kendell 1986, p. 25)

I agree that this is certainly one fundamental issue in this debate, and it is a good place to start. I will call the two broad camps on either side of this initial divide, *objectivist* and *values-based accounts*.

Sometimes, this distinction between objectivists and values-based theorists is put in terms of "naturalism" and "normativism" (e.g. Schwartz 2007). However, I find this a somewhat unhelpful terminological choice for the following reasons.

"Normativism" is a misleading label in this context, because it implies that its opponents ("naturalists") don't think of illness in terms of norms. But this is not the case. Every theorist in this debate thinks that illness is a normative concept, at least in the sense that illness involves a deviation from a norm. The debate is about how we are to think of these norms. Are they objective, value-neutral statistical or biological norms, or are they based on perspective-dependent values?

"Naturalism" is also a bit misleading since it suggests that those in opposition (those I've called values-based accounts) are somehow not naturalists. Now, "naturalism" means different things to different theorists in different contexts, but, roughly, it is to adhere to the world-view that the natural sciences adhere to. Of course what we then count as the "natural sciences" is up for debate, but there is nothing, in principle, unnatural about values. We can tell a story from the Big Bang to the present day that sees the emergence of values from social organisms without appeal to any "funny stuff."

To sum up, then, in light of these considerations, I think a good starting point is the following opposition.

Objectivist accounts: The norms of illness can be cashed out objectively. In other words, someone being ill is independent of what any individual or collective thinks (i.e. human perspective).

Values-based accounts: The norms of illness fundamentally reflect the values of an individual or collective. In other words, someone being ill is heavily dependent on (individual or societal) human perspective.

This is a good place to start. However, one might ask two things at this point. First: What are values? Second, does this dichotomy between objectivist and values-based accounts exhaust the possibilities? Let's take these questions in turn.

The question "What are values?" is one of the biggest questions in all of philosophy, so our answer to this here will unfortunately have to be relatively superficial. Values, in the relevant sense that we are talking about here, are, very broadly speaking, the things (whatever they may be) that make certain individuals or collectives evaluate certain things positively or negatively (or neutrally). This is somewhat circular, since I've smuggled values into the very term "evaluate," but think of it as a sort of thumbs up or thumbs down. In a related manner, there is an issue of primacy. Are "values" the things that make you evaluate things a particular way, or are they something that you "read off" the evaluative practices of groups and individuals? I haven't the space to answer this adequately here. For what it's worth, I think that if we are happy to think of "values" as abstract entities that model the dispositions of individuals and societies, then the answer is: "kind of both." The values I attribute to someone function as an updateable hypothesis that both predicts and responds to that person's evaluative practices.

Without wanting to get too bogged down in the metaphysics of normativity, values are often opposed to facts. Indeed, one broad area of philosophical debate can be referred to as the "fact/value distinction" (see the edited collection by Marchetti and Marchetti 2018). Of course there are many different views about the fact/value distinction, and there are even theorists who reject the distinction altogether (Putnam 2002). There are many other areas of technical theorising and dispute (e.g. What is the relationship between values and obligations? What is the relationship between words like "good" and "ought"?). Nevertheless, there are two features of values that are of particular relevance for our purposes here. They are as follows:

1. Values (in the relevant sense) *vary* across cultures, subcultures and individuals.
2. Values are intrinsically *action driving*. (A positive evaluation promotes, encourages or attracts you to that activity or thing, and a negative evaluation does the opposite).

Because of 1, a consequence of the values-based account is a degree of *relativism*. Now, as we shall see, this can either be seen as an observable fact that needs accommodating (in which case it's good news for the account) or it's an unpalatable threat (in which case it's bad news for the account). Viewing relativism as a threat lies at the heart of the opposing view; the heart of the objectivist position, of whatever form it may take in detail, is the idea of *independence of perspective*. For the objectivist, in whatever possible world, even one where there are no suitably intelligent perspectives, there just is a fact of the matter as to whether an organism is ill or has a disease. Organisms (plant and animal) get diseases, have their fitness negatively impacted upon and die.

We are now in a good position to ask that second question: do values-based and objectivist accounts exhaust the possibilities? One can answer this by asking: does the negation of objectivism (i.e. perspective-independence) necessarily entail a values-based account? Logically speaking, this is not the case. Although a focus on values yields perspectival variations (because values vary), the reverse is not necessarily the case. In other words, there may well be perspectival variations (i.e. forms of relativism) that are not so clearly values-based. For example there might be variations, say, in institutional structures that may result in cross-cultural variations in what counts as ill but which are not clearly explained by differences in values. They might instead be best explained by brute historical contingencies.

In a related manner, the idea behind the values-based account that norms "reflect our values" is ambiguous, pointing to a divergence within the values-based camp. One way of seeing this divergence is to reflect on what it is that we are doing when we say that someone is ill. What does discourse about illness *do*?

With regard to this question, the objectivist is a descriptivist, namely claims that saying that someone is ill is describing the world and, more specifically, attributing an objective (perspective-independent) property to that someone. Another way of putting this is to say that it is "fact-stating" discourse in the sense that it is simply in the business of trying to represent the facts, to state what is true. Now, some values-based theorists will agree with the objectivist that saying that someone is ill is describing the world, that it is (at least purportedly) fact-stating, but will disagree with cashing this out in terms of objective, perspective-independent properties. Here, we need to introduce the notion of *social construction* (see Hacking 1999). The "social constructivist," or simply "constructivist" for short, thinks that illness discourse is fundamentally in the business of describing reality (a.k.a. "representing the facts"), but that the reality described is (facts represented are) socially constructed. As a result, it will potentially vary from culture to culture and so is clearly at odds with objectivism. Notice that this is not, at least not yet, to deny that it is real. This depends on what you mean by "real." In any case, it is certainly not to belittle it. Currencies, monetary values and sovereign borders are quintessential examples of social constructs, but it is hard to think of anything more consequential than something being worth a certain amount, or a country being invaded. The same can be said, the constructivist will insist, for mental illness in general, and specific mental illnesses in particular. Being described in terms of a socially constructed category is anything but inconsequential.

The constructivist is usually thought of as a descriptivist about illness discourse, but, since this is a socially constructed category, it depends on society. If society were relevantly different, then the way in which we thought about illness (if we thought about it at all) would be different. Is this value-based? Well, it remains values-*based* to the extent that a society's constructs often *reflect* the values of that society, but the relationship between values and illness (and illness discourse) is indirect.

However, there is a more radical departure from objectivism, which also goes beyond constructivism. It is the view that illness discourse is more *intimately, directly*, tied to values. It isn't merely that a society's values feed into the construction of what is fundamentally a descriptive category. Rather, it is that illness discourse is not descriptive at all. It merely *looks* descriptive. It is actually evaluative. Let's call this *evaluativism*. To call someone ill is not to describe them as having a certain property (whether objective or constructed); rather, it is to evaluate them in a certain way. Prominent evaluativists include Hare (1986), Reznek (1987) and Fulford (1989).

Evaluativism comes in various different forms. Roughly, once you have claimed that illness discourse does not plainly describe reality (constructed or otherwise), but does something else, the burden is on you to explain what that something else is. We will see some of these accounts in more details as we put them to the test.

For now, let's take stock of the conceptual terrain. The objectivists are, at least about the deep philosophical issues, relatively united, except in the finer details of what they take the objective property of illness to be. They agree that illness discourse is in the business of describing reality (i.e. is fact-stating) and agree that that reality described is objective. There is an objective fact of the matter, independent of cultures, values and perspectives, about whether an organism is ill or has a disease. The anti-objectivists (namely all those who aren't objectivists, which we might also call relativists) subdivide depending on whether they take illness discourse to be descriptive (like the objectivist) but constructed (so society-dependent, and hence relative, and potentially a reflection of that society's values), or whether they take illness discourse to be non-descriptive, namely, more intrinsically evaluative and in the business of doing something other than coldly describing reality. We see this represented in the flow chart given in Figure 2.1.

From here, things get even more complicated, with different varieties of objectivism, constructivism and evaluativism, as well as positions that seem to straddle these categories. Having said that, I hope that I have presented the landscape of positions sufficiently for now. What I have not done yet is present any clear reason or argument for adopting one position over another. A useful strategy at this point is to start from one far side of the spectrum and to chip away at that. So, for example, we can usefully start by asking: "how viable is objectivism?"

In answering this, we start by presenting pure objectivism. Then we look at milder forms of objectivism, which we might call *hybrid accounts* (e.g. Wakefield 1992) that take illness to involve elements of both the objectivist and values-based accounts. On such a view, one might hold that these elements are individually *necessary* and jointly *sufficient* for illness. In other words, lacking either

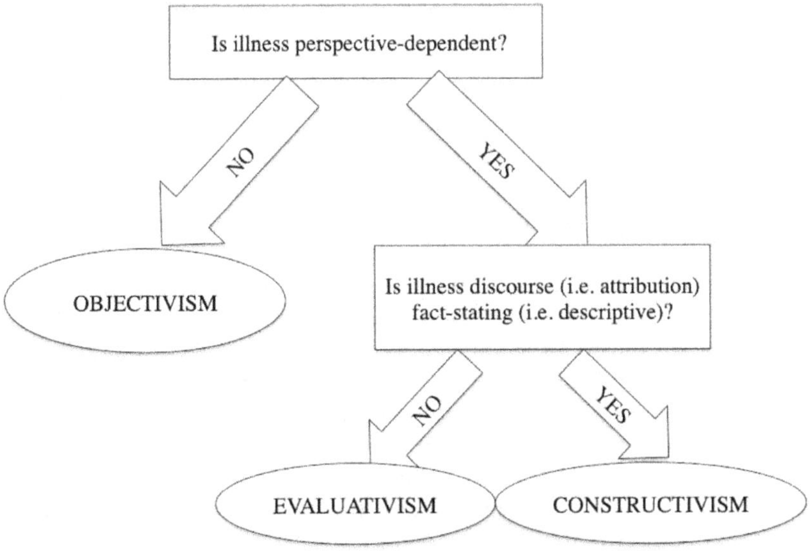

Figure 2.1 A flowchart representing different philosophical accounts of illness

would mean that you are not truly ill, but having both would alone suffice for you to count as being ill. At the end, we look at arguments against the idea that there should be any objective criterion at all. The resulting views are compatible with constructivist and evaluativist positions.

2.2.2 How viable is objectivism?

Before answering this, it is worth briefly mentioning why an objectivist, value-neutral account of illness might be desirable, let alone viable.

The main motivation behind objectivism is the avoidance of an unacceptable kind of relativism in our thinking about illness. To echo Glackin (2019), it seems unacceptable that someone should ever be able to cure themselves simply by crossing a border! Another closely related motivation is that we don't want our medical practices to be biased by political considerations (broadly construed). These motivations for objectivism can also be marshalled against values-based accounts, and we will also see potential responses to these concerns.

Undoubtedly the most well-known and influential statement of objectivism is to be found in the work of Christopher Boorse (1977, 1997). In an early paper, he states the following.

> The medical conception of health as absence of disease is a value-free theoretical notion [. . . whose] main elements are biological function and statistical normality. . . . Diseases are internal states that depress a functional ability below species-typical levels.
>
> (Boorse 1977, p. 542)

One first thing to note is that, unlike many theorists, Boorse distinguishes disease from illness (something that, as we saw earlier, I would commend). According to him, the former is the "theoretical" notion of ill health, the latter the "practical" one. However, disease is *necessary* for illness. Correctly attributing illness involves the practical, normative judgment that a disease is undesirable. But if disease is not present, then neither can illness be. In other words, Boorse distinguishes illness from disease and focuses his philosophical account on the latter. But the former, practical notion is reliant on the latter, theoretical notion. You should retrospectively revise your attribution of illness upon the discovery (which should be a scientific one) that someone doesn't have a disease.

So, what is disease for Boorse? A key component is biological malfunction, where biological proper function is seen as a "species-typical" contribution to survival and reproduction (1976, pp. 62–3). This notion of "species-typicality" is a purely statistical notion, which is defined in terms of being a certain distance from the population mean. Other factors are taken into account, such as the age of the organism. An octogenarian should not be assessed by the same standards as a 20-year-old. The set of individuals, the population, that you should be compared to, is called "the reference class."

There are, broadly speaking, two kinds of critique of Boorse's objectivism. First, there are ones that take issue with his notion (or more generally *the* notion) of biological proper function. Second, there are those who don't take issue with this but who claim that, nevertheless, it can't do the work that he needs it to in defining disease. Let's examine these in turn.

What's wrong with Boorse's notion of biological function? There are three quite different concerns here. The first is about where you draw the line between proper function and malfunction. How far must the departure be from the mean? How is this quantified? How is the reference class delineated? Such judgment calls begin to quickly look arbitrary and hence look less and less objective.

The second concern is about whether function is best thought of statistically. Some things that we might think of as malfunctions are widespread (e.g. tooth decay) and some that we wouldn't think of as being malfunctions might be in a minority. Indeed, some may be adaptive for certain purposes, like how being double-jointed in both arms and legs reportedly helped Michael Phelps excel at swimming. There are, thankfully, other accounts of biological function that don't rely on statistical prevalence (e.g. Millikan 1989; Neander 1991). Karen Neander gives the following definition:

> It is a/the proper function of an item (X) of an organism (O) to do that which items of X's type did to contribute to the inclusive fitness of O's ancestors, and which caused the genotype, of which X is the phenotypic expression, to be selected by natural selection.
>
> (Neander 1991, p. 74)

Something like tooth decay (or respiratory problems in highly polluted cities) can be statistically prevalent but still go against proper function (and hence be pathological) to the extent that they exhibit features that go against what, for example, the tooth or the lung contributed to the fitness of the organism. Conversely, a statistically minoritarian trait might be adaptive (biological fitness enhancing) and so wouldn't be a malfunction.

The third concern is a very general concern with not just Boorse's notion of proper function but the very notion of normality in general (Amundson 2000). Amundson has three main points:

i. Normal function is at odds with heritable variation (needed for Darwinian evolution).
ii. Normal function is at odds with developmental plasticity.
iii. Normal function is not (as Boorse claims) implicit in biomedical science, which actually allows for benign variations in physiology and anatomy.

All three points require technical elaboration that I can't go into here. Nevertheless, this last point relates nicely to one argument motivating the second type of worry. Recall that the second type of worry is: suppose that we accept a notion of biological normality (it needn't be Boorse's). Is it enough to define disease? As Bolton (2008) argues, it is not clear that Boorse has the resources to distinguish difference from disease (this is very similar to Amundson's point about benign variation).

But this can be seen to amount to a deeper worry. It is not just that Boorse's objectivism is wrong in its details. Rather, the worry goes, you can never ensure that something different counts as pathological, without making that further step towards value judgement. Indeed, this sort of worry is in the same conceptual vicinity as ideas that have a long and distinguished philosophical pedigree, namely, Hume's law. Hume's law is often stated as "You cannot derive an ought from an is." But some might argue that this is precisely what Boorse is doing. In saying that something is pathological (whether cast in terms of illness or disease), we are in the business of saying that things aren't as they ought to be. And yet, by talking about statistical norms, we are simply stating how things are.

2.2.3 Hybrid accounts

An influential and popular view accepts the unavoidably evaluative nature of pathology but accepts that something objective must be at its core. This effectively *anchors* pathology in the objective, rendering an objective condition necessary for illness, but not sufficient. An additional evaluative component is what is required.

By far the most well-known of these is Jerome Wakefield's (see Glackin 2019 for a similar view, but from a very different starting point). Here is his definition of a disorder (which for Wakefield is synonymous with disease, illness, pathology, etc.).[2]

> A condition is a disorder if and only if (a) the condition causes some harm or deprivation of benefit to the person as judged by the standards of the person's culture (the value criterion), and (b) the condition results from the inability of some internal mechanism to perform its natural function, wherein a natural function is an effect that is part of the evolutionary explanation of the existence and structure of the mechanism (the explanatory criterion).
>
> (Wakefield 1992, p. 384)

Such hybrid views, in tying illness to an objective underlying condition, put them more in the objectivist camp than the values-based camp, since, arguably, the central claim of values-based accounts (of constructivist or evaluativist varieties) is that an underlying, definable objective condition is not even necessary, let alone sufficient, for someone to genuinely count as ill.

Notice that you can end up as a hybrid theorist, coming either from the starting point of objectivism or from the starting point of anti-objectivism. Wakefield does the former, saying, in effect: "There must be an objective criterion, of course, but it's not enough: there must be an irreducibly evaluative judgement (namely, of harmfulness)." Glackin is another hybrid theorist who comes at this from the other direction. He is saying in effect: "Of course illness is evaluative, but it must be grounded in something objective." Notice how both of these views can be seen as closer to the objectivist than to the values-based theorist, to the extent that an objective criterion remains necessary, even if not sufficient. To reiterate: the whole point of the values-based account is that there need not be any objective grounding for someone to genuinely count as ill. In other words, illness is values-based in the sense of being values-*grounded*, rather than merely values-*embellished*.

What is wrong with hybrid accounts? Logically speaking, while they avoid the worry that it will be impossible to find any objective criterion (or criteria) that will be *sufficient* for something to count as illness, they do, by their nature, commit themselves to the requirement that it be at least necessary. In other words, on pain of becoming values-based, the hybrid views need to say that if someone doesn't fulfil the relevant objective criteria, then they cannot be ill. This is seen as a desirable position, since is "grounds" illness in objective reality. This in turn avoids the sort of relativism that means that an "ill patient can be made healthy just by crossing a border" (Glackin 2019, p. 271).

Nevertheless, this does mean that the hybrid accounts will clash with values-based theorists over whether certain conditions that lack clearly definable objective criteria are in fact illnesses. For example if certain forms of trauma or depression lead to profound personal dysfunction but don't have the right kind of objective underlying condition, then the hybrid theorist has to say that we're not talking about "real" illness. To sum up, then, while the objectivist is committed to objective criteria being both necessary and sufficient for illness/disease, hybrid accounts think that the relevant societal and/or normative response (or similar) is needed to give the underlying objective criterion the status of an illness/disease. This of course means that though not sufficient, an objective criterion remains *necessary*. According to one way of seeing the argumentative landscape, values-based accounts deny this. An objective criterion is not even necessary, let alone sufficient.

2.2.4 Values-based accounts

There are a number of different values-based theories. One influential expression is to be found in the work of R. M. Hare in his classic paper "Health" (1986). Hare was nearing the end of a long and illustrious career as a moral philosopher (indeed, from 1966 to 1983, he held the White's Chair of Moral Philosophy at Oxford), when he turned his attention to the concept of health in this paper (he also gave a lecture in 1978 entitled "What can Philosophy do for Medicine?"). All of this is by way of saying that one species of values-based account is motivated by parallels between ethics and medicine. A view, which Hare adheres to, but traces back to Plato and Aristotle, is that "expressions such as "good man" behave in some ways like the expression "healthy man"" (p. 174). Nevertheless, drawing this similarity still leaves open the question of how it is that these expressions do indeed "behave." In Hare's paper, he distinguishes between those who adhere to "descriptivist ethical theories (i.e. those which assimilate moral words . . . to standardly descriptive words like 'red' or 'triangular')" (ibid.) and those who reject this. Those who reject this think that moral words are not in the business of describing reality. Since claims that describe reality are seen as representing reality (and are hence capable of misrepresenting it if they get it wrong), and that what is represented is often seen as something that we derive from the cognitive state of the speaker (most commonly thought to be "factual belief"), the broad family of views that takes moral discourse to *not* be in the business of representing reality in this way is often called *non-cognitivism*. This, of course, is a negative claim and would usually be accompanied by a positive one, in other words, one that answers: if it is not in the business of describing reality, what *is* it in the business of doing? Non-cognitivists often have accounted for moral discourse by saying that it involves the *expression* of something other than factual belief. As such this slightly narrower family of views is called *expressivism*. Varieties of expressivism are then generated depending on what mental state is expressed. Emotivists like A. J. Ayer (1952) thought it was emotions, rather than beliefs, that were expressed by moral claims. Prescriptivists, like Hare himself (1952), thought moral claims expressed universal prescriptions (imperatives or directives). Although Hare's prescriptivism began in ethics, we can see that he applied this to the medical terms of illness and health.

It is worth mentioning, however, that a number of options are left open in this meta-ethically inspired analysis of the concept of illness. In principle, you could explore whether other non-descriptivist expressivist meta-ethical positions fit illness discourse. These include, most prominently, emotivism (which we've just encountered), projectivism and norm-expressivism.

Tristram Engelhardt is another theorist who has prescriptivist leanings. He felt that "the concept of disease," which he doesn't distinguish from illness, "acts not only to describe and explain, but also to enjoin to action. It indicates a state of affairs as undesirable and to be overcome" (Engelhardt 1975, p. 127).

A very prominent and influential version of a values-based account is to be found in the work of Hare's one-time student, Bill Fulford (1989). "The conceptual structure of medicine is essentially evaluative (rather than factual)," Fulford argues (1989, p. 260). Like Boorse, Fulford distinguishes illness from disease and (in contrast to him) focuses on the former, which he takes to be of greater importance (at least to clinical practice – which is not necessarily something that Boorse would dispute either). As we mentioned earlier, illness afflicts entire people, and Fulford very much agrees. Fulford, however, is more specific: illness is intimately bound up with problems with "intentional doings."

Fulford's central insight is that people tend to seek medical attention when their ordinary "intentional doings" are interfered with or going wrong. Of course, the medical professional might find that such problems are underpinned by underlying disorder or disease of tissues or organs, but it doesn't *have* to be. These two scenarios call for different courses of action: the former would call for targeting the underlying disease or disorder in the hope that it will remedy the situation, the latter would not. The problem is that the former course of action has become dominant (for a number of reasons), and this has promoted the false and pernicious conclusion that complaints that are not suited to this course of action are somehow less real or even fictitious. Fulford hopes to correct this by emphasising the primacy of illness over disease (what he calls "the reverse view" since it goes against the mainstream view that disease is primary).

Another, somewhat different version of a values-based view appears in Reznek (1987). According to Reznek, disease (which, unlike Boorse, Fulford and myself, he takes to be interchangeable with illness) is inextricably tied to the notion of harm (in this Reznek resembles Wakefield), which is an irreducibly normative, values-based notion. Biological malfunction, Reznek acknowledges, is objective and value-free, but it is neither necessary, nor sufficient, for the presence of disease (and in this Reznek and Wakefield part company).

All of these variants of values-based accounts are based on the idea that illness discourse is, at least partly, evaluative. As there are other forms of values-based accounts (as per Figure 2.1, which we might alternatively call "relativist" accounts), we might call this more specific subset of views, evaluativist views, since they take illness talk (or illness and disease talk) to be profoundly evaluative, where the evaluative is then typically seen in a non-descriptivist, non-cognitivist manner.

Other accounts are values-based in a more indirect way. Illness discourse is fact-stating through and through and is the expression of factual belief. But it is not robustly, objectively real either. On this view, illness and disease discourse aren't expressive but descriptive, although the facts described are *constructed.* This means that they are brought into existence and supported by societal and/or institutional structures. Money is a classic example of a social construct. National borders are another. Claims about money and boarders are not expressive, but descriptive, and yet the facts described aren't objective in that, in a possible world where human societies don't exist, these facts don't obtain, and where the societies are relevantly different, these facts will differ accordingly. However, they aren't *subjective* either – they don't rely on individual, or even group, perspective but are underpinned by complex institutional structures, agreements, laws, etc. A recent constructivist is Alex Broadbent (2019), who takes his position to be both value-independent, but constructivist. This occupies that slice of logical space we saw earlier, namely, of being anti-objectivist while not being directly values-based either.

What's wrong with anti-objectivist accounts more generally? There are two main arguments against them, the first of which is somewhat question-begging, whereas the second poses a genuine challenge. They are as follows:

Objection 1. Illness *must* have some kind of objective grounding, otherwise there is too much modal fragility to the concept.

Objection 2. You need an account of how to distinguish medical disvalue from other kinds of disvalue.

Let's address these in turn. Objection 1 expresses the concern that if things had turned out slightly differently, then illness/disease would look different, or indeed not exist at all. Furthermore, its supposed cultural variance raises the concern we'd already encountered, that someone could be cured by moving from one culture to another.

There are two responses to Objection 1. The first response is that it is question-begging in that the anti-objectivist would simply insist on this fragility and variability. The second response involves the anti-objectivist clarifying two points that makes their position more palatable. First, it needn't be that the concepts themselves are fragile, any more than the concept of the morally wrong is: the variance (modal fragility) comes in at the level of what is deemed to *fall under* the concept, not the concept itself. If a different culture has morals radically different to us, and they deem radically different things to be morally right and wrong, it is precisely because they have the *same* concept of right and wrong that there is genuine conflict in our morals. It is precisely because they mean the same thing by "right" and "wrong" that we are not talking past each other. The second, related, clarification is that, as Roberts, Krueger and Glackin (2019) point out, the status granted to the phenomena may vary in line with values, but the phenomena themselves (which may or may not attract that status) are very much objective. There is only potentially a change to how they are evaluated across cultures and value systems: there are no broken bones magically mending as you cross a border.

Objection 2 presents the following, genuinely quite thorny challenge. The anti-objectivist will (to varying degrees of directness) want to tie illness (and perhaps also disease) to an undesirable or disvalued condition. However, we routinely distinguish different forms of undesirable condition; poverty, ugliness, unhappiness and so on. Giving an account of what specifically *medically* disvaluing amounts to is the nub of the challenge. This is especially the case with psychiatric illness: when does someone need antidepressants, and when do they need a better job, nicer accommodation, a happier marriage, and more buoyant bank balance? In stark contrast, the objectivist can meet the challenge relatively straightforwardly. What distinguishes medical from other forms of disvalue is that in the medical case, there is something actually, objectively, medically wrong with you. This brings us smoothly into the next section, for reasons that will become clear as we progress.

2.3 Is mental illness a myth?

As we said at the start of this chapter, "What is mental illness?" divides into questions about illness as opposed to good health and about *mental* illness as opposed to *physical* illness. We turn to the latter issue now.

In a landmark paper (1960) and subsequent book (1962), both entitled "The Myth of Mental Illness," Thomas Szasz argued that mental illness is a myth. What did he mean by this? On one interpretation, to call something a myth is simply a somewhat elaborate way of saying that it doesn't exist. On another interpretation, however, the claim might not so much focus on the *negative* fact that the stuff of myth doesn't exist but rather on the *positive* fact that myths can play a certain role. Myths, and narratives more generally, are things that we humans live by. It is clear, once you look at what is being argued for, that Szasz's main preoccupation is with the first interpretation: that mental illness doesn't exist or, somewhat more accurately, that the concept of mental illness is confused. However at times, especially in the book, Szasz elaborates on this notion of myth in the richer sense.

2.3.1 Szasz's argument against the existence of mental illness

Szasz's argument is, at its core, beautifully simple.

1. Illness is an objective concept based on the notion of a physical abnormality (e.g. a "lesion" 1960, p. 113).
2. Talk of mental illness wrongly extends this concept to the mental realm, where it is not appropriate.
3. What are called "mental illnesses" aren't illnesses at all, they are "problems in living" (Szasz 1960, p. 114).
4. Where there is genuine illness, it is physical. It is a brain disease with behavioural symptoms. (Ultimately, you need a neurologist, not a psychiatrist.)

Although Szasz was a psychiatrist by profession, his conceptual sensitivity is philosophical. His reflection on what *kind* of concept a given concept is and what kinds of things are appropriately given a particular status is very welcome and important. There have been a number of criticisms of Szasz's argument; so many, in fact, that I can't begin to even try to catalogue them (although see Schaler 2004). As a result, I shall have to be selective.

One is that he builds his argument on some questionable assumptions. First, in spite of his radical reputation (as a founder of anti-psychiatry, no less), he seems to be assuming an unquestioning medical objectivism. Physical illness is a legitimate concept, but it cannot be applied to the mental. But as we have just seen in some detail, a great many theorists want to deny that illness/disease, physical or otherwise, really is tied this closely to physical, objective reality. For example some tie it to notions of being disvalued in a particular way, and it might well be that such reactions of disvalue can be towards physical and mental phenomena. More generally, there is a certain complacency surrounding the supposed simplicity of the idea of physical/somatic illness.

In short, Szasz claims that when there is something medically wrong with you, you are ill, but there are many ways of things being wrong that we wouldn't naturally call illness or being ill. This is very nicely illustrated in a rich, if somewhat eccentric, paper by T. E. Champlin (1981). I hope the reader will permit a lengthy, and amusing, quote.

> Imagine that you wake up one morning to find yourself in unfamiliar surroundings. You are lying in one of a row of beds in what appears to be a

> hospital ward. The person in the next bed assures you that this is indeed a hospital and that each patient is ill. He offers to whisper to you the name of the illness from which each inmate is suffering. 'He is very overweight; he is blind; he is senile; he is deaf; he has a broken arm; his illness is not so bad – he has toothache.' With each observation your astonishment increases. 'But being overweight is not itself an illness', you protest. 'Not one of the things you mention is an illness. What is true is that each person has something physically wrong with him. A man who is blind or who has toothache may also be ill, but neither blindness nor toothache are themselves illnesses. You are muddling up all sorts of different things, some of which are causes of or caused by illness, and then calling them all illnesses, which they are not!' . . . 'I know what illness is; you are ill when you have measles, chicken-pox, mumps, pneumonia and 'flu, etc. Noses run, chests wheeze, heads throb and stomachs gripe; there is fever, aches and pains in the limbs, vomiting and diarrhoea; you feel awful and just want to go to bed and lie down.' At this point you wake up, relieved to find that the whole episode has been a nightmare. My hope is that for a moment you strongly identified with the dreamer's denial that what the man in his dream kept calling illnesses could possibly be illnesses.
>
> (p. 469)

Champlin goes on to argue that, whereas mental *illness* is neither a myth, nor a metaphor, mental *disease* is. And this leads us on to another apt criticism of Szasz's argument and that is that he uses "illness" and "disease" interchangeably, indeed, synonymously. He argues that *disease* means *physical disease*, but then uses that as a premise to argue for the Myth of Mental *Illness*. And yet, as I and several others (Champlin, Fulford, Cassell and even Boorse) have suggested, one can and should usefully distinguish illness and disease. As a result, those of us who draw this distinction would disagree with the first premise as cast in terms of illness. *Illness* is not an objective concept identified with physical abnormality (Ruse 1988 makes precisely this point). But *disease* might well be. If we make this distinction between disease and illness, then we can rephrase Szasz's central claim to express something that Champlin, Fulford etc. agree with: "Mental *disease* is conceptually confused, since something mental cannot be diseased." This seems correct but is far less radical. Indeed, one might even say it is obvious.

Furthermore, it leaves the viability of mental illness as a concept intact. What the critic would say (echoing Champlin) is that it is neither bodies, nor brains, nor minds that are ill. It is *people*. Sometimes the illness has a bodily nature, at other times it has a more mental nature. Note how "nature" here does not mean cause, since surely many of us think that in a trivial and uninteresting sense, all causes are physical, and so surely the causes of mental illness are physical too. But to the extent that we talk of the mind at all (and Szasz clearly is), we can talk of our ailments as being mental (anxiety, depression, psychosis) or somatic (pain, weakness, fatigue, etc.).

2.3.2 *What does the myth achieve?*

So much for the narrow and negative interpretation of the "myth claim," according to which to call mental illness a myth is simply to claim that it doesn't exist. In what way might we look at mental illness as a myth in a more positive, constructive, sense?

The features of myths that are relevant to mental illness are as follows. In addition to being, strictly speaking, factually inaccurate, myths are (the subset of stories that are) a profoundly embedded part of a culture. They also often serve a particular purpose. They might illustrate a point, hold up certain exemplars (both positive and negative) and in this manner (or others) provide guidance or instruction. On this view, mental illness doesn't objectively exist, but it serves a purpose, which both enables and is enabled by its deep cultural embedding. All of this, of course, leaves open whether the presence of the myth in cultural circulation, so to speak, is a good or bad thing. To say, for example, that it serves a purpose is not to say that that purpose is a beneficial one.

In this instance, Szasz thinks it is a bad thing, not least because it is based on, and proliferates, conceptual confusion. But Szasz, like Foucault, also thinks that it embodies a pernicious power grab: an attempt to crush, silence and sweep away difficult problems in life by calling them medical. Nevertheless, we can put aside such socio-political considerations and evaluations, in part because it is beyond the scope of what we're doing here, but also because one might argue that the response to the myth being used for ill is not to jettison it altogether and to divide all current psychiatric case-work into neurology on the one hand, and social work on the other, but rather to use the myth properly, as a force for good.

We saw, when looking at values-based theories, that there are interesting comparisons to be made between psychiatric and moral discourse, in particular within expressivist traditions. Recently, I have explored comparisons with a position known as moral fictionalism (Wilkinson 2022). Moral fictionalism agrees with forms of non-cognitivism, like expressivism, that moral facts aren't robustly real. However, it differs from these view in insisting that moral discourse is descriptive (i.e. not expressive). In other words, it describes facts that don't robustly exist. What happens when we engage in moral discourse is that, in some sense, we engage in a useful, perhaps even indispensable, fiction. Fictionalism has been applied beyond the moral, to mathematics (Yablo 2002), as well as to the mental more generally (Wallace 2022). It takes a number of different forms depending on how the precise nature of the "fictional engagement" is cashed out. It would be too much of a digression here to go into these details, but how might fictionalism about mental illness work?

Roughly, the basic idea is that, whereas mental illness is not objectively real (because neither is illness, nor indeed perhaps is the mental), it is an indispensable part of our narrative practices. In particular, there are two important aspects to this.

First of all, the narrative practices, like myths, suffuse various aspects of our society and culture. Thus, deeming someone to suffer from mental illness has multi-track consequences for how they are to be treated in relevant circumstances by loved ones, friends, the medical profession, the legal system (most prominently but not exclusively the criminal justice system), the educational system, the world of insurance and so on and so forth. So the narratives of mental illness fit with other narratives, some of which are quite "weighty," like moral narratives of blame and responsibility. As Craig Edwards very neatly puts it:

> Mental illness is not a label that picks out a set of consistent qualities in a mental condition; it is a label that stipulates how people *should* respond to the condition, and in particular whether they should respond *morally* or *medically*. . . . To diagnose someone as mentally ill is to declare that the person is entitled to adopt the sick role and that we should respond as though the person is a passive victim of the condition.
>
> (Edwards 2009, p. 75).[3]

Second, and in a related manner, this means that, like the expressivist (e.g. prescriptivist), the attribution of illness has certain motivational consequences. However, for the fictionalist, these claims are still *externally* related to the ensuing action, rather than *internally* related (as they are for the expressivist). What do I mean by this?

To understand this, it is worth mentioning (without digressing too far) one main argument in favour of expressivism about moral discourse. That motivation comes from the idea that moral discourse (e.g. "Murder is wrong") is somehow intrinsically ("internally") related to motivational states (e.g. an all-other-things-being-equal reluctance to murder). Someone who said "Murder is wrong, but that in no way disinclines me to do it" is contradicting themselves in a way. Hence, the argument goes, "Murder is wrong" (and moral claims in general) cannot express a factual belief since factual beliefs don't motivate in and of themselves.

An expressivist about illness, or mental illness, would have to make a similar claim. And what the fictionalist would argue is that there may be something unusual, but not contradictory, about someone saying "This person is ill, but I have no inclination to . . . [help them, treat them, cut them the appropriate sort of moral slack etc.]." This issue, it seems to me, is where the crux of the matter would lie in terms of deciding between fictionalism and expressivism.

2.3.3 Taking stock: is mental illness real?

It might be tempting to think that only objectivists are committed to the reality of mental illness. But this would be overly simplistic. After all, this depends on what you mean by "real." If by "real" you mean that the thing in question is consequential, important, makes a difference, etc., then constructs have a good claim to counting as real: they constitute constructed reality, social reality.

Indeed, part of why many constructs come into existence is in order to perform certain important roles. As we've discussed, money is a construct, but it is highly consequential. The borders between sovereign states are another example of constructs of extreme consequence. The same, it seems, can easily be said about mental illness, in general, as a construct, or specific mental illness categories as constructs (more on this in the next chapter). What could be more consequential than being deemed mentally ill? Or of receiving a particular diagnosis?

One might, however, think that those who are furthest removed from objectivist, the evaluativist, are going to embrace the idea that mental illness isn't real. This may well be the case, but it is important to see that it doesn't necessarily follow. Indeed, in meta-ethics, the "quasi-realism" of Simon Blackburn (1993) might provide a relevant comparison. The quasi-realist project is to find a way for something that is not "robustly real" in any standard metaphysical sense (like a chunk of matter), to *earn the right* to be spoken about with the sort of seriousness that it is associated with reality.

In light of this, those who want to deny the reality of mental illness aren't so much those who are opposed to objectivism but rather those who want to dispense, whether now or ultimately, with mental illness discourse, as a serious and consequential form of discourse.

A final point of clarification is in order. Not only have we been careful in separating questions about what makes mental illness *illness* from questions about what makes mental illness *mental*, we must also be mindful of the fact that the positions presented here are concerning mental illness in general. They are not about specific diagnostic categories. That is the topic of the next chapter, but it is worth mentioning that the position that one adopts with regard to the one needn't transfer without question to the other. In other words, one could be an objectivist or expressivist about mental illness but a constructivist about, say, schizophrenia or major depressive disorder or vice versa.

Summary

- Debates about the nature of mental illness divide into debates about the nature of illness more generally, and debates about whether and how we can apply illness to the mental.
- We get better purchase on the target phenomenon of illness by contrasting it with neighbouring concepts, including disease and disability.
- Philosophical views on the nature of illness divide roughly into objectivists and values-based theorists (there are also hybrid accounts that use components of both).
- Values-based theorists further subdivide into constructivists and evaluativists, where the former take mental illness discourse to be descriptive and fact-stating, whereas the evaluativists take it to be doing something else altogether.
- Some theorists, more famously Thomas Szasz, have denied that illness can coherently apply to the mental, since illness is a physical condition.
- However, if we reflect on differences between illness and disease, it is not so simple as simply saying that illness is when something is wrong with you.
- There are other more positive ways of interpreting the claim that mental illness is a myth that are worthy of consideration.

Questions

- What sort of conditions do you think of when you think of illness? How about mental illness?
- Can you think of conditions that are difficult to be neatly categorised as either illness or some other disvalued condition or misfortune?
- Can you think of conditions that are difficult to be neatly categorised as mental or physical? Are they combinations of mental and physical elements?

Notes

1. In French, for example, it would be normal to use the word "*maladie*" to refer to both illness and disease.
2. Since Boorse, unlike Wakefield, distinguishes illness from disease, he is a hybrid theorist about illness but not disease. This is often overlooked.
3. Note the telling use of "as though".

Further reading

Boorse, C. 1975. On the Distinction Between Disease and Illness. *Philosophy and Public Affairs* 5:49–68.
Carel, H. 2007. Can I Be Ill and Happy? *Philosophia* 35:95–110.
Champlin, T.S. 1981. The Reality of Mental Illness. *Philosophy* 56 (218):467–87.
Hare, R.M. 1986. Health. *Journal of Medical Ethics* 12 (4):174–81.
Szasz, T. 1960. The Myth of Mental Illness. *American Psychologist* 15 (2):113–18.

3 Psychiatric diagnosis and the medical model

Overview

One significant debate in philosophy of psychiatry centres around the issue of whether psychiatry ought to be seen as a branch of medicine, where this is often stated in terms of subscribing to "the medical model." One key pillar of this model is that psychiatry ought to operate via procedures of diagnosis. Key questions then become: What is diagnosis? And do we want our psychiatry to be heavily reliant on diagnostic practices? We discuss different varieties of medical model and then examine several issues that arise when thinking about psychiatric diagnosis. We end by looking at challenges and alternatives to diagnosis and, by extension, to the medical model.

3.1 Diagnosis and the medical model

One central debate in philosophy of psychiatry revolves around the question of whether psychiatry ought to adhere to "the medical model." Unfortunately, this term gets used in different ways by different people. One problem here is that, if two people use the term "the medical model" differently, it is not clear how to adjudicate between them (assuming that both are tolerably sensible). It is a term

DOI: 10.4324/9781315450124-4

of art and doesn't wear its meaning on its sleeve. Although I lay no special claim to the term, I want to present a way of thinking about this that is useful, that centres around a genuinely substantive issue and is about how many theorists in this debate use the terms. This is the issue of whether psychiatry ought to base itself around *diagnosis*.

In this section, I start by explaining why I choose to understand "the medical model" in this way. Then I present motivations for the medical model and what is meant by "diagnosis." Most importantly, diagnosis is far more than mere taxonomy. Then I end by presenting different "strengths" of medical model, based on these ideas.

3.1.1 The relationship between diagnosis and the medical model

To subscribe to the medical model for psychiatry is, at a first pass, to say that psychiatry ought to be a branch of medicine. But what does *this* mean? One major proponent of the medical model, describes it as follows:

> This [namely, the medical model] means simply that the concepts, strategies, and jargon of general medicine are applied to psychiatric disorders: diagnosis, differential diagnosis, etiology, pathogenesis, treatment, natural history, epidemiology, complications, and so on.
>
> (Guze 1992, p. 4)

In this sense, Western psychiatry is broadly where these theorists want it to be: it already adheres to the medical model, and the trajectory taken by those who become psychiatrists abundantly reflects that (although we will see existing alternatives later in this chapter). Our question is: *Should* it? And if not, why not?

A certain assumption reappears here. As we saw with the last chapter, many (but not all) theorists are tempted to make the following move: "we have a good grasp of somatic illness, but does it apply to the mental?" However, as we saw, the assumption that somatic illness is well understood needs to be questioned. Similarly, here theorists say: "Somatic medical practice ("general medicine") is unproblematic, but does psychiatric practice follow suit?" But do we fully grasp medicine, somatic or otherwise?

As Lisa Bortolotti (2020) nicely points out, there is an assumption that doctors treat disorders. But a moment's reflection betrays that this is both too narrow and too simplistic. First of all, there are things that might be wrong with you that fall short of disease/disorder/illness: a broken leg, a sprained ankle or high blood pressure. Second, we call upon the wonders of modern medicine when there is nothing wrong with us – for example in childbirth or when getting a vaccine. One might retort that these are preventative of disease, to some extent, but this shows that diagnosis and treatment of a disease/disorder/illness are not necessary to everything that we call medicine. Nevertheless, diagnosis is absolutely central, if not entirely (although nearly) ubiquitous, to our overall conception of Western medicine.

We can, as Guze does, list concepts and practices other than diagnosis. However, it is worth noting that all of them can be linked, more or less directly, to diagnosis. For example as mentioned, "differential diagnosis" is logically implied by diagnosis in general and becomes explicitly salient when two or more candidate diagnoses present themselves in light of the symptoms/observable data, hence prompting further exploration (e.g. a scan, a blood test). Another example is the concept of "disease," which again can be derived from diagnosis since it is that which you diagnose someone *with*. "Treatment," in the relevant medical sense, comes downstream from diagnosis, and treatments are selected on the basis of them, at least in the first instance. In short, diagnosis has a certain centrality within the medical model, and I find it useful to tie the viability of the medical model to the viability of psychiatric diagnosis, specifically as a species of medical diagnosis (as opposed to diagnosis in a more generic sense, i.e. as applies to informatics, engineering or economics).

At this point, I'd like to present and put aside some unhelpful uses of the term "medical model." To begin with, some people think that the medical model means that mental illnesses are treated as if they were physical/somatic. This is too strong a position, and an unfairly easy critical target. In any case, many who are self-professed proponents of the medical model (Guze 1992, being one of them) will deny this.

Similarly, there are some who think that "the medical model" implies that mental illness should be, where possible, treated pharmacologically. This, again, is too strong and implausible a view. This view of the medical model probably comes from the fact that within the medical model, psychiatrists are doctors who can prescribe drugs, whereas other mental health professionals (clinical psychologists, psychotherapists etc.) are not and typically cannot. But these institutional constraints say nothing about which conditions should be treated with drugs. Most, probably all, of those who subscribe to the medical model will accept that there are effective and appropriate non-pharmacological treatments (exposure therapy for phobias, cognitive behavioural therapy for depression and anxiety, etc.). This doesn't conflict with the medical model, not least because the same applies to somatic illness. A low-sodium diet helps with hypertension, cardiovascular exercise with obesity and so on.

Finally, there are others who think that the medical model simply means that it is scientific. This, in contrast, is too weak a position. Many who take themselves to be virulently opposed to the medical model do so in the name of science. No – to my mind, the most useful way of thinking of the medical model is as follows. Wanting psychiatry to subscribe to the medical model is to say that it should make central, pivotal, use of diagnosis.

3.1.2 Motivations for the medical model

Why would anyone be tempted to subscribe to the medical model? We can separate motivations into two categories: institutional ones and theoretical ones. We take these in turn.

Institutionally speaking, psychiatry, in tying itself to medicine, inherits the credibility and authority that come with that. A corollary of this is that mental illness gets taken as seriously as somatic illness, and this, in itself, seems like a very good thing (although the demand for parity between mental and physical illness is at times a double-edged sword, as we will see at several junctures).

It isn't just credibility (and prestige) that the medical model acquires for psychiatry. It also opens up access to medical funding for psychiatric research. But this also shapes the nature of psychiatric research. As a result, it is heavily driven towards following the model for somatic medicine: the gold standard is the randomised control trial (RCT) of which the simplest and most prominent type is the drug trial (with placebo control condition). Whether the trial is for depression or asthma makes little difference as to the design of the trial and the evidential threshold. This equivalence between psychiatric and somatic medicine has been criticised, not least because outcome measures for mental illness tend to be harder to pin down (e.g. Ross 1995). The rate at which someone's depression is improving is harder to both ascertain and quantify than the rate at which someone's asthma improves, tumour shrinks or viral load decreases. Furthermore, the extent to which adequate controls can be in place, and confounding factors eliminated, for something as environmentally embedded and multi-factorial as mental illness is highly debatable.

This institutional motivation can be cast in a slightly cynical light. However, coupled with robust theoretical considerations, it becomes more justifiable. Indeed, a fairly mainstream view (and it is certainly reflected in the structure of Western healthcare institutions) is that medicine quite rightly has credibility and prestige and that psychiatry, when practiced well, is rightly considered a branch of medicine and inherits that prestige and scientific rigour. Dissenting voices, in contrast, will either dispute (at least to some degree) the scientific status of medicine (e.g. Engelhardt 1975; Saunders 2001) or the justifiability of counting psychiatry among its ranks (e.g. Szasz 1962).

The theoretical motivations for adopting the medical model are comparatively simple. The concepts, procedures and standards of general medicine are, the claim goes, helpfully and/or accurately applied to mental illness. Of course, the helpful and the accurate are not always the same, and, as we will see, we get different varieties of medical model depending on whether we focus on utility or accuracy. Nevertheless, the central principle that the high evidentiary standards of scientific medicine should be applied to furthering the understanding and treatment of mental illness is something that is hard to dispute. Some might even say that, due to its intangible nature, mental health practice is particularly susceptible to hacks and charlatans, and so the medical model is all the more important as a protective framework.

Another boon for the medical model is that it structures scientific research and links it up geographically and across time. The aim is that, just like medical researchers working on HIV or rheumatoid arthritis can build on work from past researchers (and also replicate findings) as well as share their work across the globe, so can, for example, depression or schizophrenia researchers do the same. This is enabled by the fact that these researchers in different times and different places are working on the same phenomena. This requires a degree of stability in both the categories and the phenomena themselves. The same applies not only to the sharing and advancing of research, but to diagnosis and clinical practice as well. People in different times and places can get the same diagnosis, and therefore similar treatments with demonstrated efficacy, because of this same sort of stability. We will see that many theorists are critical of this picture of stability for psychiatric categories and diagnoses.

As I've already said, a cornerstone of the medical model, of whatever variety, is the concept and practice of *diagnosis*. Part of what it means to say that psychiatry is a branch of medicine is to say that psychiatrists not only pronounce on whether people are ill as opposed to well (and are in need of medical help, as opposed to not), but also have one particular illness as opposed to another, and that various solid inferences can be drawn on that basis (e.g. the course of the illness will run as follows; the patient will need a particular course of treatment as opposed to another; these factors will aggravate, these will ameliorate, and so on).

3.1.3 What is diagnosis?

Diagnosis is central to medicine, and to psychiatry namely viewed as a branch of medicine. But what is diagnosis, and is it appropriately applied to mental illness? We will address this, first, by reflecting on the concept of diagnosis in general and medical diagnosis in particular and, second, by reflecting on mainstream existing practices of psychiatric diagnosis, viewed as a subspecies of medical diagnosis.

Diagnosis is often associated with medicine, but, of course, it has a totally general meaning that gives it applicability to a wide variety of areas (this is in keeping with the fact that "medical diagnosis" is not pleonastic). These areas tend to be those where there is an underlying cause of a phenomenon of interest, and the ascertaining of that cause is a non-trivial epistemic achievement (i.e. discovery). In this respect, diagnosis is more specific than mere classification or taxonomy. It doesn't simply divide things into categories: it involves a commitment to things beyond both the obviously visible and the here and now. It not only delves beneath the surface presentation of a phenomenon of interest to something underlying that presentation, but also delves into the future (and often the past) to generate a prognosis (or indeed an aetiological mechanism). There is, for example, talk of diagnosis in the context of informatics and computing. (One might wonder whether this is literal or metaphorical, but, if it ever began as a metaphor, it certainly seems to be used very literally.)

The phenomenon of interest tends to be of interest because it is a *problem*. The purpose of diagnosis is, if possible, remediation. Indeed, one might think that it is more than a mere tendency that diagnosis tends to point to a problem. It does after all sound distinctly strange to talk about diagnosing (or searching for a diagnosis of) something that is not problematic. If, with no knowledge of context, you overheard someone ask "So, what's the diagnosis?" you would assume that there was some sort of problem (they might be talking to a doctor, an IT specialist, a car mechanic). You might hear talk about diagnosing an economic recession, or a drop in share price, but it would be odd to talk of diagnosing an economic boom or share price rise. This is not of course to say that these things wouldn't have underlying causes; it's rather that "diagnosis" has built-in implications of negativity and (which might come to the same thing) a need for remediation.

From this preliminary reflection, we can derive a rough, working definition of diagnosis:

> *Diagnosis:* The identification of an underlying cause of a phenomenon, where that phenomenon is problematic to some degree.

But what about specifically *medical* diagnosis? A medical diagnosis is the identification of an underlying condition that causally accounts for a person's problems. The underlying condition is sometimes thought of as a disease, and the problems are traditionally thought of as symptoms or signs *of* that *disease*. This idea feeds into our standard ways of talking and does so in a way that strays from other domains where diagnosis is used. In medicine, we talk about diagnosing *somebody* with a disease (or condition). We don't usually diagnose a computer with a problem, we diagnose the problem. An obvious reason for this is that medicine is patient-orientated. The person being served (well or ill) by the diagnosis, and potential downstream remediation, is the person. In computing, it's not the computer, but the computer's owner/user.

So, tied very closely to the notion of medical diagnosis is the notion of a disease or condition (underlying and explanatory cause) and of symptom/sign. A canonical instance of diagnosis is: diagnosing someone with a disease, as a way of explaining symptoms. Sometimes, the diagnosis, the identification of the disease, will be based on symptoms. At other times, it will be based on some other diagnostic procedure (e.g. a blood test) in which case the disease can be identified prior to, or in the absence of, any symptoms. One example of this might be the detection of a virus, which would lead to diagnosis before the onset of any symptoms.

Of course, when you diagnose someone with a disease, you are implicitly diagnosing them with one disease as opposed to another. Two things should be said about this. First, this means that a central part of diagnostic practice is the process of *differential diagnosis*, namely, distinguishing between two diseases that share similar symptoms (e.g. Is this dry cough a symptom of COVID-19 or pollen allergies?). Second, although (as I said earlier) diagnosis is a narrower, more specialised, concept than mere taxonomy, it inevitably involves taxonomy. To state the relationship of logical entailment: all diagnosis involves taxonomy (i.e. distinctions between diagnostic categories), but, rather obviously, not all taxonomy involves diagnosis. Taxonomising types of cloud or plant species is all-too-obviously not diagnosis! Less obviously, distinguishing types of cough (e.g. dry versus productive coughs) is taxonomy but not (not yet, at least) diagnosis.

3.1.4 *Current practices of Western psychiatric diagnosis: the DSM and the ICD*

The two diagnostic manuals that carry the most authority in Western psychiatry are the American Psychiatric Association's "Diagnostic and Statistical Manual of Mental Disorders" (or DSM for short) and the World Health Organization's International Classification of Diseases (or ICD for short). Both manuals have undergone significant revisions over several decades. We are currently (in 2021) at the DSM-5 (2013) and the ICD-11 (2019). Since the ICD is about health problems in general, the DSM is what is most commonly held up as "the psychiatric bible" (especially in the United States), and it is this that has tended to be the focus for philosophers (see especially Cooper 2004, 2018).

The oft-used term "bible" is hyperbolic but justified. The far-reaching power and authority wielded by the DSM is really quite striking. It doesn't just guide clinical diagnosis, it structures (with potentially distorting effects) mental health research of all kinds. Participant inclusion criteria will frequently be based on whether individuals have DSM diagnoses or not. Furthermore (something that UK readers will be less aware of than US readers), it is the benchmark for health insurance rulings. Changes to the DSM could in principle change your health insurance premiums or indeed whether you get health insurance pay-outs or cover at all, as well as whether you get access to forms of welfare, social care, sick leave, etc. The DSM could not be further from a sterile academic treatise: it is of the utmost practical consequence.

With each new edition, the DSM changes significantly. Not only do the diagnostic categories themselves change in name, criteria and how they are organised and clustered within the manual, but overarching conceptual changes are present too. This includes changes to what the DSM takes itself to be doing, fundamentally. There are two big issues here.

The first centres around whether the DSM is committed to its diagnostic categories being "natural kinds." There are several different ways of interpreting the natural kind claim, but, roughly, to say that something is a natural kind is to say that it has an objective nature or essence and it is a real thing (kind of thing) in the world, like water or gold. This then gets unpacked in different ways (e.g. natural kind terms have been widely thought (at least since Putnam 1975)) to function like proper names and be *rigid designators*, namely, refer directly, rather than descriptively, and pick out the same type of thing in every possible world where that thing exists. However it is interpreted, to accept the natural kind claim would be ultimately to claim that the DSM subscribes to a *strong* version of the medical model (more on this shortly). Cooper (2004) adds to this worry that, even if there were natural kinds to be unearthed in psychiatry, it is unlikely that the institutional and financial forces that are tied up with the DSM's conception would allow them to be accurately captured by DSM categorisation.

The second issue is whether the DSM is as categorical as it first appears or whether it incorporates dimensions. On this latter point, it is widely accepted that the shift from the DSM-IV to the DSM-5 (apart from the dropping of Roman numerals) represents a shift towards a more dimensional approach. Although many have hailed this as progress, it is worth mentioning that, since natural kind membership doesn't come in degrees, this shift also appears to signal a shift away from a natural kind interpretation of the DSM.

As mentioned, we are now at DSM-5. Although it is historically interesting to look back to the very earliest transitions, going back to the First Edition in 1952, we need to prioritise what is most relevant for our enquiry here. In this vein, it is instructive to examine the transition from DSM-IV (1994) (with a revised version coming out in 2000) to DSM-5 (2013)).

Although there are widely accepted improvements in the DSM-5 (see Regier, Kuhl and Kupfer 2013), there are still institutional sticking points that prevent changes from taking place that are unimpeded attempts to follow evidence. As Rachel Cooper writes:

> Workgroups of specialists worked on the various sections of the manual (mood disorders, substance-related disorders, and so on). Each workgroup reviewed relevant literature published since the publication of the DSM-IV, drafted proposals for revisions to diagnostic criteria in their area, and then solicited and responded to feedback on their proposals. Throughout the revision period, mental health conferences and journals were full of discussion about proposals for revision. Draft versions of the DSM-5 were periodically posted online, and anyone who wished could contact the workgroups with their ideas and suggestions. Field trials, where clinicians used draft criteria to check they could be reliably applied, were used to test some draft diagnostic criteria. *Behind the scenes, various interest groups (representing different types of therapist, and patient, but also business groups) lobbied the APA in attempts to get the classification to fit their interests.*
>
> (Cooper 2018, p. 50, emphasis added)

What Cooper goes on to very convincingly argue is that, even when this lengthy and rigorous process has been completed and evidence emerges in favour of certain changes, there are still various institutional obstacles to such changes being implemented. This is no doubt a bad thing, but it may, practically speaking, be unavoidable.

3.1.5 Strong versus minimal medical models

So far, I have referred to "*the* medical model" as if it were a single, unified position. However, nothing could be further from the truth. Many theorists and practitioners working within psychiatry as a branch of medicine, and thereby embodying the medical model to some extent, will differ in how they think about mental illness in general and psychiatric diagnostic categories (psychiatric nosology, taxonomy, etc.) in particular. We will see more of these issues in the next section, but, here, I want to point to a very broad distinction (made, e.g., by Murphy 2013) between "strong" and "minimal" forms of medical model.

A proponent of a *minimal medical model* takes medical concepts like "disease" and "diagnosis" to apply to mental illness but does not take this to entail specific commitments about the underlying nature of the disease. Nevertheless, the minimal medical model can't be *too* minimal on pain of no longer being counted as a medical model at all. The view is that diagnostic categories provide an indispensable framework for predicting and understanding mental illness.

A quintessential proponent of the minimal model is Emil Kraepelin. Kraepelin is seen by many as one of the founders of modern, scientific psychiatry. Nevertheless, his approach was fundamentally pragmatic. As Kendler and Jablensky (2011) point out, "Kraepelin saw psychiatric disorders as multifactorial, arising from the difficult to untangle action and interaction of internal and external causes." As a result, psychiatric diagnostic categories were to be built on careful observation and longitudinal following up of cases. There was perhaps a hope that, someday, underlying "biomarkers" of specific mental illnesses might be found. Nevertheless, this would have felt such a distant prospect in the late nineteenth and early twentieth centuries that it didn't feed into diagnostic practice. One interesting question is whether Kraepelin would remain a minimalist today. Minimalism as a theoretical position shouldn't really be about technological limitations. It should be about the fundamental nature of mental illness in general and psychiatric categories in particular, not about our epistemic access to them.

More recently, Guze (1992) and McHugh and Slavney (1998) probably count as minimalists in a similar sense. They take psychiatric categories very seriously, but the psychiatric categories do not even attempt to pick out underlying conditions or things unified by these conditions. They are constructs that usefully bind symptoms into clusters that enable clinical and scientific endeavours to flourish.

In contrast, the strong medical model (Andreasen 1997) takes the correct application of a valid diagnostic category to go beyond mere observable regularity and predictive success: it implies the presence of an underlying condition (state or process), and two separate applications of the same category (at least disease category) must involve the same underlying condition (if they are to both be correct, that is). As things currently stand, however, this distinction between strong and minimal forms of medical models remains vague and ambiguous, not least because it's not always clear what is meant by "underlying condition" or "underlying disorder." At this point, it might be useful to run through more specific distinctions and dichotomies and see to what extent they might contribute to more or less minimal versions of medical model.

3.2 Issues in psychiatric diagnosis

We have seen, very roughly, how mainstream psychiatric practice operates via diagnosis, as enshrined in the DSM, which in turn fits in all manner of aspects of life beyond treatment, from insurance, to benefits, to the criminal justice system and so on. Putting the DSM aside, what are the philosophical issues surrounding diagnosis? These could initially be divided into political and theoretical issues, but here we will focus on the latter. The former we will examine as part of the next section, when we come to look at challenges and alternatives to diagnosis.

3.2.1 Causalism versus descriptivism

In Chapter 2, we were interested (at least in part) in what makes someone ill as opposed to not ill. Here, we are concerned (at least in part) with the conditions under which, given that someone is ill, they have one illness or disease as opposed to other. In other words, on what basis should we categorise disease/illness? Whereas *causalism* categorises on the basis of *causes*, *descriptivism* categorises on the basis of what can be observed and *described*, namely, clinically relevant characteristics (see Zachar and Kendler 2007, p. 557, for a nice discussion of this distinction).

As Zachar and Kendler explain, there are at least three separate positions one can take towards this issue. The first, which they call "temporising," begrudgingly accepts that "we will have to settle for descriptive approaches until we understand the real causes" of psychiatric illnesses (ibid., p. 557). The second, "robust descriptivism," rejects causalism in principle: it is not something that we should aspire to. It is, quite simply, inappropriate for psychiatry. Mental disorders, unlike infectious diseases, are far too complex in their nature to be identified by a unique underlying cause. The third is an intermediary position. We can grant this complexity, but it might still be useful, for practical purposes, to prioritise a particular class of causes (e.g. genetics or brain structure).

Although I agree that these are three available positions, I'd like to add two points at this stage that add a bit of further complexity. The first is that a rejection of causalism needn't be based on complexity. It could be based on concerns about the kinds of concepts at play. In other words, the sort of thing that a mental illness is makes causalism fundamentally misguided, regardless of the complexity involved. My second point seems like a pernickety linguistic point but is actually of profound importance. It is vital that we distinguish between, and avoid conflating, "identified with" and "identified by." To identify X with Y is to say that X and Y are the same. Thus, I might identify Batman *with* Bruce Wayne. To identify X by Y is to use Y as evidence that something is X. Thus, I might identify Bruce Wayne by a distinctive birthmark (or his finger prints, or his dental records, or whatever). I am obviously not saying that Bruce Wayne is his birthmark!

What relevance does this have to how we should think of causalism about disease? If we are identifying the disease *with* the cause, then we are saying that the cause *is* the disease. This seems both implausible and conceptually confused. First, something cannot be its own cause. Second, it seems like a category mistake to say that, for example, a virus, or anatomical abnormality, is the disease. It's not even the right *kind* of thing. What is clearly the charitable and correct interpretation is to say that we are identifying the disease *by* the cause. This seems relatively plausible, but it means that the cause plays the role of an *identifying* piece of information. Although plausible, this is a relatively weak claim for two reasons. First, like an individual's identifying birthmark, the identifying piece of information ("biomarker") can in principle be an identifier, all while telling us very little about the nature of the disease. This means we can know such an identifying piece of information while knowing remarkably little about the disease itself. This is potentially problematic, since, if we know so little about the disease (or condition) then how can we know that this is genuinely an identifier? There is a danger that this starts to look more like a stipulation than a discovery. Second, and in a related way, the status of this piece of information as identifying is open to revision in light of further investigation (just like someone might turn out to have a birthmark exactly similar to somebody else's, rendering it no longer identifying of that individual).

3.2.2 Essentialism versus pragmatism

This issue is very closely related to two things we have already mentioned, namely, strong versus minimal forms of medical models and also the natural kind claim.

Essentialists think that specific categories of disorder or illness are (or should be) defined by their *essences*, their underlying nature. If this is what you think about natural kinds, then essentialism amounts to subscribing to the natural kind claim (but not all philosophers think of natural kinds in this way). For *pragmatists*, in stark contrast, the categories are (or should be) defined in practical terms: they are based on usefulness and are used, and deemed good or bad, based on how well they perform certain useful functions. Zachar and Kendler (2007) call this pragmatism "nominalism." Not much hangs on this terminological choice, but I prefer "pragmatism" because of its connotations of practicality. I also associate "nominalism" with certain debates in metaphysics about whether certain things exist, which to me seems antithetical to pragmatism. This debate is not really about whether certain things exist but about how we carve up what does exist. In a related manner, if "nominalism" is applied to this question, it suggests that a category is brought into being in virtue of being named, but it says nothing about whether our naming practices are guided by practicality or utility.

Pragmatism comes in varying degrees of both *strength* and *scope*. Starting with strength, an extreme pragmatist will think that psychiatric categories should be purely practical, either with no hope or expectation of an underlying nature, or even with the claim that the very idea of an underlying nature is unintelligible. A more moderate pragmatist will simply say that there is an underlying nature but our practical concerns shape our categorisations, but that this isn't a bias to be corrected, but rather something positive (or necessary) that should be embraced.

Now let us look at how scope varies. One could be a *local pragmatist*, namely, only a pragmatist about psychiatric categories (or other things that are relevantly similar), while being an essentialist (and objectivist) about other things like, for example, chemical elements. Alternatively, one could be a *global pragmatist*, and therefore be a psychiatric pragmatist, on pain of inconsistency, as a natural consequence of a much wider all-encompassing view. In other words, if tables, chairs, elements, atoms, quarks, are all practical constructs, then *of course* schizophrenia is.

One might think that global (wide-scope) pragmatism will come hand in hand with high-strength pragmatism, but that's not necessarily the case. For example a narrow-scope, high-strength pragmatist might think that mental illness is special, both in how intangible and elusive an underlying essence might be, and in how bound up it is in deep, important practical concerns at a societal level, such that it is *deeply* constructed, but the same does not apply to the vast majority of other things that fail, for example, to have these interactions with morality, responsibility, autonomy, etc. In contra-distinction, a wide-scope, low-strength pragmatist might think that, like so much else (chairs, corkscrews, electrons), mental illness categories will have an underlying nature, but the ways and places where we draw the lines will be shaped by practicalities. Such a view entails that, as Zachar and Kendler (2007) elegantly put it, "categorization requires decision as well as discovery" (p. 558).

3.2.3 Objectivism versus evaluativism

Is deeming something to be a mental illness or disorder a purely factual matter or is it (at least in part) irreducibly evaluative? We saw in Chapter 2, how this can be applied to mental illness in general, but it also applies to specific categories. It is useful to see that a pluralistic picture could emerge here in a number of ways.

One could be an objectivist about specific disorders (e.g. schizophrenia) but an evaluativist about the application of mental illness status. Again, the status/constitution distinction could help here (see Roberts, Krueger and Glackin 2019). What *constitutes* a particular condition is something objective, but giving that condition the status of illness is evaluative. Also, one could be an objectivist about some categories (e.g. schizophrenia) but an evaluativist about others (e.g. delusion or addiction). What's more, there might be some patterns here: purportedly scientific concepts (like schizophrenia) are intended as value-free and hence might suit an objectivist view. But concepts with more of a folksy or common-sense provenance, which emerge from social interaction about what you ought to believe or how you ought not to behave, like delusion or addiction, might suit an evaluativist view (see Wilkinson 2020 for an evaluativist approach to delusion).

3.2.4 Internalism versus externalism

Again, this is a distinction we find in Zachar and Kendler (2007), and it is presented as follows. "Should psychiatric disorders be defined solely by processes that occur inside the body (internalism) or can events outside the skin also play an important (or exclusive) defining role (externalism)?" (p. 558). There is a big and important ambiguity here, similar to the one we saw earlier with "identify by" and "identify with," and that is this notion of "defining role."

Zachar and Kendler (2007) rightly point out that psychiatric orthodoxy has internalistic tendencies at least in so far as what is deemed to be wrong is usually something with the sufferer's mind, which is in turn taken to be underpinned by the nervous system. Thus, many of the most reputable psychiatric treatments, building on this, but also on the ease and authority of RCTs for pharmacological interventions, tend to be pharmacological. In stark contrast, a quintessentially externalistic intervention would be, for want of a better term, ergonomic, namely, changing the agent's physical or social environment. Having said this, there are a number of conceptual distinctions that are important to make here, which show how one can be an externalist and, at times, advocate pharmacological treatment and also be an internalist and advocate an ergonomic or "externalist" intervention. I'd like to focus on two, although there are more.

The first is borrowed from discussion about externalism about the mind in general and, in particular, the Extended Mind Hypothesis (EMH) (Clark and Chalmers 1998) and is the cause/constitution distinction. This is most commonly known in this literature as the "coupling-constitution confusion" (Adams and Aizawa 2001) in the context of criticism of the EMH. The second is our old friend, the status/constitution distinction. In both of these distinctions, "constitution" means the same thing but with inevitably different emphasis. Let's look at them in turn.

The EMH (Clark and Chalmers 1998) is the claim that the mind isn't bound by "skin and skull" but literally extends into the world. The basic relevance of this for our purposes is the idea that, if the mind extends "beyond skin and skull," then so must mental illness (Davies 2016). The original example of mind extension is of Otto and his notebook. Clark and Chalmers compare two individuals, Otto and Inga. Inga knows her way to the MOMA, whereas Otto uses his notebook to get there. Clark and Chalmers (1998) contend that to the extent that the information in the notebook functions exactly like Inga's biological, brain-bound memory, it should count as much a part of his mind as Inga's memory is. Therefore we really should be saying that Otto knows his way to the MOMA after all!

There is a vast and long-running debate about this, and I can't do it justice here (see Gallagher 2018 for an excellent review of the state of play), but I'll illustrate the point that I need. One point of debate centres around the distinction between cause (or "causal coupling" to narrow it down to a smooth, reciprocal causal relationship) and constitution. In other words: does Otto's notebook make a constitutive contribution to his mind, or merely a causal, enabling, one? Applied to mental illness then, we need to be very careful not to attribute a deeply implausible position to our argumentative opponents. Even the most staunch internalist isn't going to deny that environmental factors play a *causal* role. The question is whether they are *constitutive* of the condition. To take an example Zachar and Kendler use to illustrate externalism, you cannot suffer from PTSD unless something has traumatised you. But is the traumatic event a cause of the PTSD, or a part of it?

Now for the second distinction, which I feel may help with the first, between status and constitution. Remember that the status question asks: "what is it about a particular condition that gives it a particular status?" Whereas the constitution question asks: "what constitutes this condition (given that it has that status)?" Now consider an example: a mosquito bite. We can surely only answer the status question externalistically. A mosquito bite has to be made by a mosquito. However, you could in principle recreate an exact replica of a mosquito bite, that has no causal or historical connection to mosquitos (or to biting for that matter). This condition (what we call it) can be understood internalistically. It seems that a similar move can be made by the internalist as a way of defusing the externalist's objections. *Of course* PTSD requires trauma, otherwise we wouldn't call it PTSD. But, once that status is granted, the constitution of the condition itself can be understood without appeal to that which both caused it and gave it its status. It is this latter claim that the externalist has to target. To be an externalist about a particular condition is not simply to say that something external needs to be appealed to in order to get both the cause of the condition and its status: the external factors need to be part of the condition. (One side observation worth making is that, if these factors are literally part of the condition, then they cannot also be its cause, since something cannot cause itself.)

3.2.5 Categories versus dimensions

Are the boundaries between disorders clear-cut? Or are these conditions on a continuum along different dimensions? This issue is closely related to the essentialism versus pragmatism debate, although the dichotomies do not match up entirely. If you are an essentialist about mental disorders, then you will think that they constitute something like a natural kind. Essentialism will typically entail a categorical approach. However, you could be a pragmatist who defends the practical validity of categories over dimensions.

Another approach, which is neither obviously essentialist nor pragmatist, is to say that different disorders are amenable to different approaches. We see this in somatic medicine, with hypertension being a clear continuum, where furthermore what has been deemed "of clinical severity" has shifted over the years, whereas being infected by a virus (although viral load may vary) is a categorical state. This approach is very nicely exemplified by Haslam's "Kinds of kinds" approach (2002). He presents the flowchart as given in Figure 3.1.

As you can see from the flowchart, the idea is that different conditions will result in different answers to each questions. Haslam's own examples are as follows:

Non-kind: Neuroticism
Practical kind: Depression
Fuzzy kind: Borderline Personality Disorder
Discrete kind: Melancholia
Natural kind: Williams Syndrome

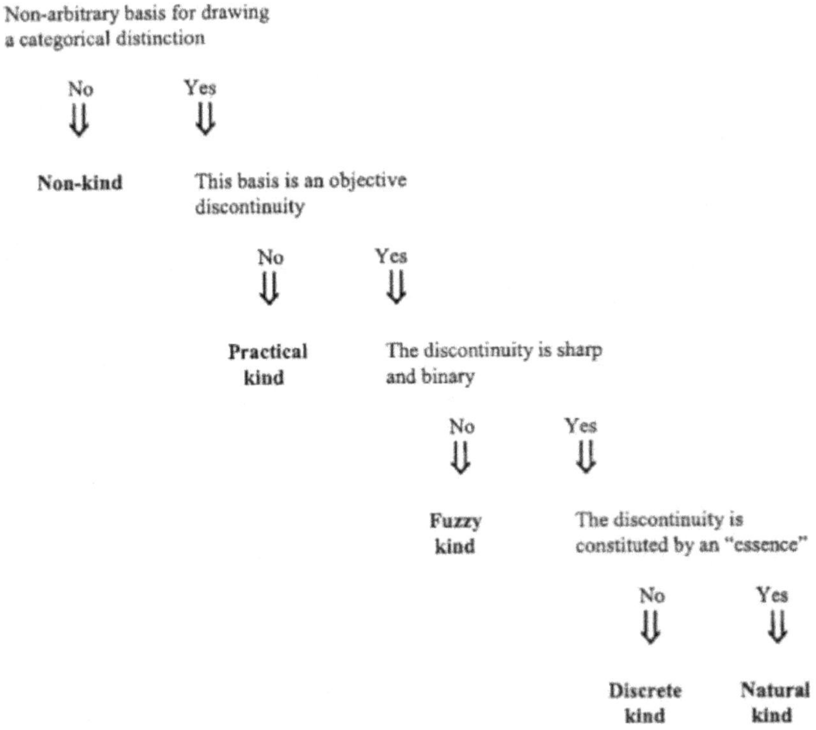

Figure 3.1 Haslam's proposed taxonomy of psychiatric kinds (Haslam 2002)

Let's walk through these examples and the rationale for placing them in each category. Neuroticism is a spectrum, and no place along the spectrum is more suited than any other for separating "neurotics" from "normals." Depression is also a spectrum (Haslam appeals to Ruscio and Ruscio 2000, 2002, to support this claim), but there are practical reasons for deeming some depression to be of clinical severity. In this respect, it is like high blood pressure. You draw the line based on practical judgements (that are liable to shift subtly across contexts) about risks to health and well-being. Borderline Personality Disorder is a *fuzzy kind* because although it is deemed to be objectively discontinuous with "normality," the discontinuity isn't sharp. As Haslam puts it, "the discontinuity is less a 'joint' than it is a bend or graded inflection" (p. 208). Although this is a matter of significant dispute, "melancholia" (as presented by Haslam) is intended to point to a particular depressive variant. This is particularly severe and exhibits a sharp discontinuity. And yet, this discontinuity is not associated with an underlying essence but rather with a threshold (Schotte et al. 1997). Finally, Williams Syndrome is a natural kind because "it results invariably from the submicroscopic deletion of chromosome 7" (p. 213). If anything counts as an essence, and hence a natural kind, surely this does.

Although this pluralism seems attractive, two things are to be noted. First of all, the example of a natural kind that Haslam uses in his taxonomy of *psychiatric* kinds is not easily categorised as a mental disorder or psychiatric condition. It is odd to say that someone with Williams Syndrome has a psychiatric diagnosis. Williams Syndrome (like Down Syndrome) is a genetic neurodevelopmental condition. It is not some affliction affecting the person: it is a part of who they are. Now this is not me just being pernickety: it may actually be revealing of our concept of mental disorder that it is hard to think of a properly psychiatric natural kind. Second, as Zachar writes in his response to Haslam (Zachar 2002), a thoroughgoing pragmatism allows for the sort of pluralistic kinds of kinds that Haslam presents but along pragmatic lines.

In other words, for Zachar, all psychiatric classifications (and other classifications too) are grounded in practical considerations, even the so-called natural kinds. As Zachar (somewhat controversially) writes:

> [A]lthough we may have *discovered* that everything we call *hydrogen* has one proton in its nucleus, we *decided* that having one proton is the essence of hydrogen. . . . [S]cientists might just as well decide to nominate an isotope number to be the essence of hydrogen.
>
> (Zachar 2002, p. 220)

Note that, if this is correct, you could replace "hydrogen" with "Williams syndrome" and "one proton" with "deletion of chromosome 7," and the same point could be made. *If* there were those who presented just like Williams Syndrome individuals, but lack the chromosomal deletion, we could *decide* to make the presentation and not the underlying chromosomal state, the thing that the label "Williams Syndrome" tracks. To our knowledge there aren't those individuals, so it makes *practical* sense for us to make deletion of chromosome 7 the essence of Williams Syndrome.

One criticism of this general pragmatic view is that it extends the domain of the "practical" to encompass everything, including the sorts of things that are typically distinguished from it. One might think that this renders it too broad, and therefore trivial. As Zachar admits:

> Deciding what counts as practical is complicated. With respect to categorizing psychiatric disorders, we should consider many things, including but not limited to available treatments; potential management strategies; the effects of labeling; maximization of true positives and true negatives in identification; establishing within-category homogeneity for creating groups in experimental research; uncovering etiologic scenarios (especially for spectrum disorders); mapping time courses; predicting prognosis; achieving coherence with basic science in genetics, physiology, and psychology; being both clinically informative and easy to use; and meeting psychometric standards such as reliability and validity. On some level, these are all practices. We do things with category members. We

> interact with them and based on that interaction we learn how to think about (or use) the category. These practices have to be balanced vis-a-vis each other as well.
>
> (ibid., pp. 221–2)

3.2.7 *Taking stock*

There are a number of different issues with regard to psychiatric diagnosis or even simply taxonomy more generally. I have, for simplicity (and I don't think at the expense of too much accuracy), presented these in terms of oppositions or distinctions (X versus Y). In thinking about these, it is important not to conflate them, or to think that there are strict entailments between them when in fact there are not. How do these oppositions relate to each other? Yes, one can think of causalism, essentialism, objectivism, internalism and categorical approaches to all be conducive to one another, but one can mix and match, too. For example you can be an externalist and an objectivist; a causalist and a pragmatist; a categorical evaluativist and so on. Some of these permutations may be less *plausible*, or less well motivated, than others, but they are not *incoherent*. As a result, I would warn against any temptations to make simplifying assumptions about what goes with what.

Another interesting question worth asking is: how many of these apply to somatic medicine too? One notable example of this, with regard to the internalism versus externalism opposition, can be found in Shane Glackin's response to Will Davies. Davies (2016) claims that in various ways, psychiatric medicine distinguishes itself from somatic medicine with this externalism. In response, Glackin (2017) points out that all of Davies's examples have parallels in somatic medicine too. We also saw how Haslam used hypertension as well as depression to highlight practical kinds along a continuum. Again, as a general rule, it is important to avoid an overly simplified and epistemically optimistic view of somatic medicine. Medicine, in general, and somatic illness, in particular, are not simply about viruses and vaccines, bacterial infections and antibiotics; they too can be messy and conceptually difficult.

3.3 Challenges and alternatives to diagnosis

Why might someone challenge psychiatric diagnosis? It is extremely important to divide these challenges into two very different kinds, with very different targets. First, there are challenges to the *practice* of psychiatric diagnosis. These focus on the negative effects of diagnosis with a particular focus on unwanted effects on society or the well-being of the individual being diagnosed. Second, and in principle entirely independently, there are challenges to the scientific validity of psychiatric diagnosis and diagnostic categories. We take these in turn.

3.3.1 *Concerns about the impact of psychiatric diagnosis*

Putting aside the theoretical viability of diagnosis in general, and specific diagnostic categories in particular, there is a significant literature on the negative effects of psychiatric diagnosis, and in particular, the negative effects on the person being diagnosed. The literature on this is vast, and much of it is in the social sciences rather than in philosophy. Nevertheless, I'd like to focus on two particular themes that I take to have significant philosophical importance. The first is the theme of *labelling and stigma*. The second is the theme of *medicalization*.

Many critics of psychiatric diagnosis focus on the pernicious use of psychiatric labels. Giving people a diagnosis has a tendency to label them and hence encourage various negative behaviours towards them on the part of society, family, friends, even themselves, which are prominently examined under the umbrella term *stigma* (an important subspecies of which is *self-stigma*). Stigma roughly refers to the (implicit or explicit) negative attitude or evaluation that comes as a result of having been labelled or categorised in a certain way. As Hayward and Bright (1997) put it:

> [I]f a person threatens his neighbours he will not be well thought of, but many argue that he will incur an additional negative opinion from those neighbours if they believe that his behaviour is caused by mental illness; we might call this the "stigma of mental illness."
>
> (p. 346)

Two things are worth mentioning at this point. The first, as I hinted at, is that the negativity of the opinion or attitude towards the stigmatised individual need not be conscious or explicit. It can be a mild discomfort or aversion towards an individual. The second is that stigma is more a feature of public perception than it is of the realities of the purported individual being labelled. This latter point has led some to think that jettisoning diagnostic labels altogether would be to throw the baby out with the bath water. What we really should do is try to remove the stigma from diagnosis by better informing the public and by discouraging certain practices that promote stigma. Others, however, are rather more pessimistic: you can take steps to reduce stigma, but you cannot eliminate it. Even if you change your labels from ones that are pejorative to ones that are more neutral, you'll still find yourself objectifying individuals, flagging them as individuals who lack autonomy, control and who, as a result, cannot be trusted by society and who shouldn't have faith or optimism in themselves.

Medicalization refers to the encroachment of medical diagnosis into areas where it shouldn't be (Conrad and Schneider 1980). Indeed, the term is often used pejoratively to signal that this is not merely (or not even) a factual error, but something more vicious like an abuse of power, an expression of institutional laziness or social control (Conrad 1992). Two grounds where this particular battle has been prominently fought are, first, in the context of depression and its medicalising of sadness (Horwitz and Wakefield 2007) and, second, ADHD (attention-deficit/hyperactivity disorder) and the perception that this (often? always?) involves the deeming of children who are naughty or otherwise behaviourally challenging, as having a *medical* problem that they do not in fact have (Donzelot 1997; Malacrida 2003).

The reasons for this focus are relatively obvious: each "condition" has a readily available non-pathological analogue (roughly, sadness for depression, "naughtiness" for ADHD). However, charges of medicalisation can also apply to more severe illnesses that lack such mundane analogues. For example one can talk about the medicalisation of voice hearing in the context of thinking that these experiences aren't simply malfunctions to be quashed with antipsychotics but rather meaningful responses to life events that need to be understood, or emotionally confronted, dealt with, etc. Note, in this vein, an accusation of medicalisation is not synonymous with an accusation that something is deemed more severe than it should be and more of a problem that it actually is. Often, in fact, it is quite the contrary: it is an unwillingness to face deep and difficult, but non-medical, problems – *life* problems. Many of the opponents of medicalisation fully accept that the conditions for whose de-medicalisation they are campaigning are extremely serious. Rather, it is not about severity but about the type of problems that they are deemed to be, namely, medical problems in need of *diagnosis*, and the relationship of this framework to pharmacological treatment, via the view that RCTs (randomised control trials), which prioritise pharmacology in their structural design, are the gold standard. Ultimately, it is a great deal easier to put someone on medication than it is to delve into every aspect of their lives in search of a solution.

Medicalisation, stated simply as wrongly deeming something to be an illness (or more generally a medical issue) when it is not, looks different depending on your views of illness and mental illness (see Chapter 2). For example an objectivist will tend to take it to rest on a factual issue about, say, whether or not depression and ADHD are in fact illnesses. An evaluativist will take it to be an encroachment of one normative landscape upon another. It is not about objective things being biologically wrong with these individuals: this is a red herring. Rather, we end up "treating," and depriving of autonomy, individuals that we should be either blaming or helping in other ways that acknowledge rather than remove their autonomy.

3.3.2 Theoretical concerns with diagnosis

Quite aside from diagnostic practice, and the specifics of different diagnostic manuals, with their different categories and outlooks, what might be wrong with the theoretical underpinnings of diagnosis in general?

When two people get the same diagnosis, the medical establishment will treat them in the same way. This is because, to the extent that we are grouping them together via diagnosis, this amounts to saying that they have the same underlying condition. Two problems for this are as follows.

First, unlike in many instances of somatic illness, there is no *distinctness* between underlying condition and symptomatic presentation. Second, unlike in some canonical instances of somatic illness, *disunity* is rife. In other words, whereas there is relatively robust uniformity across all cases of, say, Ebola, there is relatively little uniformity across all cases that are diagnosed with, say, schizophrenia. Furthermore, in various cases of somatic illness, variations in symptomatic presentation are put down to different ways in which the same pathogen, or pathogenic process, can operate upon the body. The same degree of understanding cannot be said about mental illness.

Problem 1: lack of distinctness between illness and symptom

In somatic medicine, there are objective measures that can be used to make an accurate diagnosis. You might have an X-ray, and, despite your protestations, your doctor can say: "I can see that your ankle is not fractured: it's just a bad sprain." Similarly, we might use a blood test to detect a disease before the patient exhibits any symptoms. The same, the critics will say, does not apply to mental illness. If you are depressed, or anxious or hearing voices, the doctor cannot do a blood test or a brain scan to tell you otherwise. In short, in mental illness, we are not talking about symptoms *of* illness: the so-called symptoms *are* the illness. I say so-called, because a symptom is always a symptom *of* something, that's just part of what symptom means. If someone says to you, "You have a symptom," the natural follow-up question is "Symptom of *what*?" This is why Bentall (2003) suggests that we talk of *complaints* rather than either symptoms or diseases in mental health contexts.

One response to this is to take a tactic we took in the previous chapter, namely, distinguishing disease from illness. A disease is an underlying pathogenic condition. In contrast, an illness is the medical problem that afflicts the person as a whole. Someone can be ill because they have a disease, but this is not always the case. Mental illnesses are perfectly legitimate illnesses, but they should not be thought of as diseases. When you use a blood test to detect the presence of a virus, you are diagnosing someone with a disease, but until there are symptoms, they are not ill.

Problem 2: disunity within a diagnostic category

A diagnostic category is supposed to group together individuals who have the same condition (illness, disease, disorder, etc.) with a view to informing treatment, course and prognosis. The validity of diagnosis gets undermined when it looks like people who are getting the same diagnoses have quite different conditions. Quite independently of examining clinical realities, simply looking at the DSM can raise questions. As we will see abundantly later in this book (e.g. Chapter 7 on schizophrenia), many diagnostic categories have *disjunctive* criteria, namely, they will show, for example, two or more of any five symptoms, thereby enabling the possibility that two individuals with no symptomatic overlap will get the same diagnosis.

As Taylor (2016) points out, psychiatrists don't actually blindly follow diagnostic categories when deciding on a course of treatment but take a more nuanced and symptom-based approach.

> Psychiatry is different. The biochemical or pathological basis of most mental health conditions is, at best, poorly understood. Our knowledge of the precise actions of effective drugs is likewise incomplete. Diagnostic criteria shift and sway like in no other area of medicine and named mental health conditions appear and disappear with a disturbing frequency. Alongside these factors is the troubling awareness of the frequent overlap of symptoms across formal diagnostic entities. It seems trite to point out, for example, that people with depression sometimes hear voices, or that people with bipolar disorder may suffer from anxiety.
>
> (Taylor 2016, p. 224)

Taylor goes on to reflect on the significance of this for the prescribing of drugs and in particular the effect it has on "off-label prescribing," namely, prescribing drugs for things that they are not licensed for:

> When I first worked in psychiatry in the 1980s, I was struck by the extent to which drugs were prescribed outside their license. . . . Treatment of choice for alcohol withdrawal was thioridazine, a drug previously only known to me as an antipsychotic. This was also a time when antipsychotics were prescribed to almost everyone with behavioral disturbances associated with dementia.
>
> Over the years since then, my experience and observations of prescribing decision-making have been fairly uniform. Rarely is a firm diagnosis arrived at before prescribing and, perhaps more tellingly, rarely do prescribers know the precise licensed indications for every drug they prescribe. So, antipsychotics tend to be prescribed for psychotic and manic symptoms, and antidepressants tend to be prescribed for symptoms of low mood. Lithium is prescribed for mood fluctuations. Benzodiazepines are prescribed for pretty much everything. Never, in over 25 years in psychiatry, have I observed prescribers make any noticeable attempt to match a diagnosis with the licensed indication of a drug.
>
> (ibid.)

Many individuals defend the practice of psychiatric diagnosis on the basis that, like in general medicine, it structures treatment, including the prescription of drugs. This assessment certainly casts doubt on that.

A reply is clearly available to the defendant of psychiatric diagnosis. These observations are perfectly consistent with the epistemic point (which Zachar and Kendler 2007 call "temporizing") that we are, quite rightly, aiming at bringing psychiatry in line with the rest of medicine, and it's just that, as things currently stand, our knowledge and understanding are limited.

A more damaging version of Taylor's lament would take it further to the claim that we have no reason to think that clear-cut diagnostic categories are ever attainable in psychiatry. This is the view of a number of critics of the medical model, including, prominently, Bentall (2003).

3.3.3 Alternatives to diagnosis

One may think that diagnosis in general is indispensable. First of all, one might think that it is vitally important to allow clinicians to build on pre-existing knowledge: you cannot treat each patient from scratch. Categorising patients is an absolute practical necessity. Second (as Rachel Cooper has been at pains to point out), these psychiatric diagnostic systems are intertwined with many other institutions and systems, and so they are "locked in." Changing, or jettisoning, psychiatric diagnosing would have knock-on effects for benefits, insurance, sick leave, legal proceedings, etc. Nevertheless, this has a flipside. Some people who are very unwell may fail to get the institutional response that they need or deserve (whether it be insurance, healthcare, benefits or the legal system) because they "fall through the gaps" between diagnostic categories. In any case, some have felt that psychiatric diagnosis is either so harmful or so theoretically misguided that they have proposed alternatives. Before ending this chapter, I want to present two such alternatives.

Alternative 1: psychological formulation

The first alternative to psychiatric diagnosis is psychological formulation. There are in fact different ways of integrating it into clinical practice. One is to reject psychiatric diagnosis altogether and to use formulation as an *alternative* to diagnosis. The other is to incorporate formulation into mental health practice alongside psychiatric diagnosis. The latter resembles in some respects the multi-axial approach promoted in DSM-IV. Since psychological formulation is most clearly presented in contrast to psychiatric diagnosis, I will present it in the first way: as an alternative. Then the question arises later as to whether aspects of diagnosis can and should be retained.

Psychological formulation is a practice that is officially supported by the British Psychological Society (BPS). Some consider it to be a (perhaps *the*) central skill of clinical psychology in the UK. It is also a central part of practice for counselling, health and forensic psychologists (Health Professions Council 2009). As Johnstone (2018) puts it:

> [Psychological formulation] can be defined as the process of co-constructing a hypothesis . . . about the origins of a person's difficulties in the contexts of their relationships, social circumstances, life events, and the sense that they have made of them.
>
> (p. 32)

This "co-construction" takes place between service users and clinicians. A formulation is not about an "underlying condition," or even underlying, explanatorily relevant deficits. As a result, it does not involve an expert judgement on the part of the clinician in the face of a passive patient. In contrast to psychiatric diagnosis, the formulation requires the clinician to find out a great deal about the service user's life, and the formulation is seen as collaborative.

This may all sound rather woolly at this stage, so what might a formulation look like?

Here is an example, again, from Johnstone (2018):

> You have told me that you had a happy childhood until your father died when you were aged 8. We have also discussed how, as a child, you felt very responsible for your mother's happiness and pushed your own grief away. Later, your mother remarried and when your stepfather started to abuse you, you did not feel able to confide in anyone without risking the break-up of the marriage. The only escape was to leave home as soon as you could, and you found an apartment and a job in a store. However, you found it increasingly hard to deal with your boss, whose bullying and controlling ways reminded you of your stepfather. You gave up the job, but long days at home in your apartment made it hard to push your buried feelings aside any more. One day, after your stepfather made some threatening comments on the phone, you started to hear a hostile male voice telling you that you were "dirty" and "evil." This seemed to express how the abuse had made you feel, and it also reminded you of things that your stepfather had said to you when you were a child. You found day-to-day life increasingly difficult as past events caught up with you and many feelings came to the surface. Eventually, your mother persuaded you to contact the mental health services, and for the first time you have been able to tell someone what you went through as a child. Despite your difficult experiences, you have many strengths, including intelligence, determination, and self-awareness, all of which have helped you to survive. You recognize the need to revisit some of the unresolved feelings from the past, and we have started to do this work together.
>
> (p. 33)

The first thing to note is that, viewed from a different, psychiatric perspective, this individual could, on the basis of their voice hearing, quite easily attract a diagnosis of schizophrenia (or similar psychotic disorder). Putting aside whether this is accurate or helpful, formulation presents a radically different approach. What typically follows a formulation is some kind of therapeutic plan. It could be any kind of therapy, the most well-known of which is CBT (cognitive behavioural therapy).

This may trigger the following observation: there is a sense in which formulation is, in a broad sense, a diagnosis, in the sense that it is about isolating what has "gone wrong" and trying to improve matter. However, it is not a medical diagnosis, and it is not even diagnosis in the sense it is used in informatics or mechanics. It is not a medical diagnosis, because it emphatically does not use diagnostic categories (still less, diagnostic *disease* categories), and it is not a diagnosis in the mechanical sense, since it is not isolating what is wrong with the person. It is about what has gone wrong in the person's life. There is no implication of pathology or anything being broken in, or wrong with, the person themselves. It is about their life and their trajectory through it. On the contrary, it often comes with the reassurance that someone's problems stem from normal reactions to challenging life events.

As a result, formulation is deeply personal. Unlike a psychiatric diagnosis, whose whole point is to group people together as being the same in relevant respects, no two formulations are the same. In a related manner, the process is centred around sense-making: it is around beliefs, emotions, reactions. It is not centred around a causal, mechanistic approach (e.g. neurotransmission).

Formulation is not without its challenges and criticisms. Because it is so deeply personal, it's very time-consuming for both parties involved, and hence very resource-intensive (whether the service is publicly or privately funded). Also, many will worry that it doesn't do much real explanatory work. And just because an explanation satisfies the service user, it doesn't mean it's a good explanation (scientifically speaking) (see Chapter 6 for discussion of these issues). In the example given earlier, the "and then you started hearing voices" comes abruptly and is unexplained (and yet it is the first robustly psychiatric symptom).

Another clear limitation of formulation is that there are cases of acute mental illness that may not suit a formulation. For example someone who is deeply psychotic wouldn't be able to partake in the conversation that would yield a formulation. Clearly a natural response to this would be to suggest a hybrid model. Some patients need psychiatric diagnosis and pharmacological intervention. We mustn't lose sight of the fact that in many cases, pharmacology can improve a very desperate situation. In other cases, something more discursive will be appropriate. As it happens, mental health services within the NHS strive towards this judicious hybrid use of clinical psychology and psychiatry.

Alternative 2: network theory

Network theory (NT) is not motivated by the sociological, practical or political motivations that gave rise to psychological formulation and related approaches. It is not what we might think of as a soft, humanistic approach: in fact, it is uncompromisingly scientific. Nevertheless, it differs markedly from standard psychiatric diagnosis in the following way. It questions whether the things that we group together as specific mental illnesses or disorders (i) share "pathogenic pathways," (ii) are the causal effects of the same underlying condition or "disease" and (iii) are indeed even the effects of a common cause. Nevertheless, grouping patients and symptomatic presentations in a legitimate way, the Network Theorist claims, is indispensable for progressing with both treatment and research.

The solution proposed by NT is not so much to view psychiatric symptoms as causes of an underlying condition (or common pathogenic causal pathway) but rather to see symptoms as causes of *each other*. As a leading Network Theorist puts it:

> For instance, if one thinks that other people can read one's mind (delusion), this may generate extreme suspicion (paranoia); this paranoia can lead one

> to avoid other people (social isolation), which, because one is no longer exposed to corrective actions of the social environment, may serve to sustain and exacerbate the relevant delusions.
>
> (Borsboom 2017, p. 5)

On this view, disorder categories are not based on hypothesised-yet-elusive underlying conditions but are reconceptualised as *symptom networks*. A number of points of clarification are in order.

First, NT is very *open-minded about how these causal relations are instantiated*:

> Direct causal connections between symptoms may be grounded in basic biological (e.g., insomnia/fatigue) or psychological (e.g., loss of interest/feelings of guilt) processes, in homeostatic couplings (e.g., appetite and sleep both interact with the biological clock, so that when one is disturbed, the other is likely to be disturbed as well), in societal norms (e.g., dependence on heroin increases the probability of contact with law enforcement agencies in countries where it is prohibited by law), or in still other processes.
>
> (Borsboom 2017, p. 6)

This agnosticism is grounded in a relaxed acceptance of an interventionist theory of causation, according to which, within a system, A is causally connected to B, if an intervention on A brings about a change in B. Note, however, that the lack of sensitivity to the sorts things we are talking about (as values for A and B) could be seen as grounds for criticism. (Are we, for example, talking about objects, states, events, processes, phenomena?)

Second, we need to distinguish the *symptom network* from the *external field*. The external field refers to conditions outside the symptom network that can have a causal effect on the network. A canonical example might be an adverse life event (e.g. trauma or bereavement). However, this notion of "external" should not be taken too literally. A factor can be inside the organism but count as external to the symptom network (an example of this might be inflammation). Conversely, a symptom "within" the network may well be environmentally and socially distributed (e.g. social isolation). (Note how this fits nicely with the externalism discussed in Section 3.2.4 (see also de Boer et al. 2021 for a detailed discussion).)

A third crucial component of the theory requires an introduction of the notion of *hysteresis* (Cramer et al. 2010). Hysteresis is not simply a feature of psychiatric theorising but is a potential property of any complex system. Roughly, the idea is that, whereas the presence of an external trigger (external in the sense just sketched) can *activate* a symptom network, its subsequent absence does not *de-activate* it. In other words, the triggering event sets the network in motion, and the network proves itself to be *self-sustaining*.

This rings true of a number of examples of mental illness. It also sheds light on why it is that certain symptoms can be recreated in certain contexts, but until they prove themselves to be self-sustaining, we are not talking about mental illness. For example hallucinations under LSD or delusions under Ketamine are present for as long as the external factor, namely the drug, is impacting on the system. To take a very different example, someone may exhibit many of the symptoms associated with depression in the context of bereavement, and this may be long lived, but to the extent that the bereavement (and the usual psychological processes associated with it) is considered a still-present external factor, it lacks the self-sustaining nature associated with a mental-illness-like depression.

The notion of hysteresis can also be used to make some very general suggestions. For example the notion of mental health (i.e. *good* health) can be defined as *the stable state of a weakly connected network* (Borsboom, p. 9). Note that this is very different from a definition based on "absence of symptoms." Rather, mental health is defined as an "equilibrium state, to which a healthy system returns if perturbed" (ibid.). Thus, symptoms can be present in a healthy system, the symptom network, because it is weakly connected and hence doesn't reach hysteresis.

In contrast, mental illness/disorder can then be defined as *the stable (alternative) state of a strongly connected network*. And conversely there can be periods where the person is asymptomatic (e.g. because of medication), but the network is strongly connected and liable to self-sustaining symptom networks. (I note in passing that this has affinities with the "Flexibility Hypothesis of Healing" (Hinton and Kirmayer 2017) according to which "many forms of healing and psychotherapy may work by inducing positive psychological states marked by flexibility" (p. 3).)

In a related manner, this can also be used to characterise the two key dispositional states of *resilience* and *vulnerability*:

> The concept of *resilience* can be defined as the disposition of weakly connected networks to quickly return to their stable state of mental health, and the concept of *vulnerability* as the disposition of strongly connected networks to transition into a state of disorder upon a perturbation in the external field.
>
> (ibid.)

There are many further ways in which NT can be developed. Crucially, this general approach allows powerful statistical tools to be used and applied to a variety of constructs, including depression, anxiety disorders, post-traumatic stress, psychosis, etc. Promising though NT may appear, there are a number of criticisms.

First, a hard-nosed proponent of the medical model might just dig in their heels and claim that the key motivating fact behind NT, that common pathogenic pathways (underlying "diseases") have not been discovered, is just a reflection of our sorry epistemic state of affairs. It is not to say that there are no such common underlying causes. Furthermore, drawing on some of the ideas floated in Chapter 2, they might say that in the situations where there is no underlying condition, we are talking about "problems of living." Admittedly, Borsboom does at times talk about mental disorders in these terms, and this is further corroborated by the fact that NT gets applied to quality of life, personality traits, etc. But this may concede too much. It is ultimately just a way of modelling patterns of human behaviour and activity: it is not specifically a theory of mental illness. In a related manner: why use the word "symptom"? A symptom is always a symptom *of* something, and yet in NT we have nothing beyond (viz. underlying) the symptom.

NT is very open and relaxed about the ontology of all of these things: the causal networks and the things that take part in these network can be almost any kind of thing. This gives the framework a high level of flexibility and applicability, but at what cost? In particular, one might ask: how do you delineate a symptom/problem? Problems come in all shapes and sizes. Is it really satisfactory to put social isolation on a par with hallucinations?

Taking a very different critical approach, one might say that, regardless of NT showing promise on theoretical grounds, the view could be criticised on the same practical and sociological grounds as the standard medical model has been. Viewing personal problems as nodes on a network is a far cry from the sort of personalised patient-centred approach promoted in psychological formulation.

3.3.4 To diagnose or not to diagnose?

One interesting observation one could make is that the theoretical concerns against diagnosis can largely be accommodated by certain minimal forms of medical model, especially ones that have tendencies towards descriptivism and pragmatism. Indeed, they might even constitute motivations for being a minimalist as opposed to subscribing to the strong medical model. Then the argument about whether we should diagnose becomes a profoundly practical one: is it beneficial?

Looking at practicalities, however, we may accept that, given unlimited resources, we should adopt a deeply personalised view: no two individuals are the same, no two lives (past and present) are the same. But we haven't got unlimited resources and, in any case, research needs statistically significant sample sizes, and that, in turn, requires people to be clustered into meaningful sets that stand out from "control" groups.

This seems to me to be a sensible and moderate position: some diagnoses structure research and clinical practice in a helpful way. But we should be careful about our diagnostic categories and make sure that they do not overreach. We should also make sure that they do not promote labels that stigmatise or have other unhelpful consequences. We should also be wary of the idea that finding an underlying disease is our aim in scientific psychiatry. And this is not necessarily because we will not find any, but perhaps because the act of seeking will systematically mislead us and hamper our understanding. What this amounts to, however, is a view where psychiatry may (with care) taxonomise, but it shouldn't diagnose, in the strict medical sense.

Summary

- Proponents of the medical model for psychiatry take the central concepts and practices of general medicine to be applicable to mental illness.
- One such notion is diagnosis, in large part, because it is intrinsically connected to other medical notions (such as disease, which is usually taken to be the entity that one is diagnosed *with*).
- There are, roughly, two varieties of medical model: the minimal medical model, which rests on the practical utility of diagnostic categorisation, and the strong medical model which aims to capture objective and underlying disease conditions.
- Among those who discuss psychiatric diagnostic categories, there are a number of different debates and oppositions that we summarised.
- Two different concerns for psychiatric diagnosis, and by extension the medical model, are that diagnosis is problematic in practice and in theory.
- Concerns about diagnostic practice include specific concerns about the DSM (the main diagnostic manual in psychiatry) but also general concerns that the diagnostic labelling of individuals is unhelpful and sociologically pernicious (e.g. promotes social stigma).
- Theoretical concerns include (i) that there are no underlying conditions in psychiatry, which are separable from the symptomatic presentation of the illness and (ii) that there is no unity within the diagnostic categories.
- We end by exploring alternatives to psychiatric diagnosis, including categorising on the basis of symptoms without any commitment to underlying disease and approaches that eschew categorisation altogether.

Questions

- Imagine that, in spite of yourself, you cannot socialise in the ways that you would like, because you have crippling social anxiety. Does it make sense that you should see a doctor (a psychiatrist), or should you see a psychologist?
- How do you (or would you) react to someone telling you that they have a psychiatric diagnosis (like schizophrenia or bipolar disorder)? What does this mean to you, and how might it impact how you interact with this individual?
- Would it bother you if you received the same diagnosis (and hence treatment plan) as someone else who seemed to have very different symptoms/complaints? If so, why?
- Can mental illness *literally* (even if only partially) reside in the patient's environment?

Further reading

Cooper, R. 2014. *Diagnosing the Diagnostic and Statistical Manual of Mental Disorders*, London: Karnac.

Drayson, Z. 2009. Embodied Cognitive Science and Its Implications for Psychopathology. *Philosophy, Psychiatry, and Psychology* 16 (4):329–40.

Kendell, R.E. 1975. *The Role of Diagnosis in Psychiatry*, Oxford: Blackwell Scientific Publications.

Roberts, T., Krueger, J. & Glackin, S. 2019. Psychiatry Beyond the Brain: Externalism, Mental Health, and Autistic Spectrum Disorder. *Philosophy, Psychiatry, and Psychology* 26:E51–68.

Zachar, P. 2002. The Practical Kinds Model as a Pragmatist Theory of Classification. *Philosophy, Psychiatry, and Psychology* 9 (3):219–27.

4 Mental illness, moral responsibility and the boundaries of the person

Overview

What is the relationship between mental illness and moral responsibility? One common way of thinking about this is as follows. When we think of someone's behaviour as attributable to mental illness, the behaviour should not be as strictly morally evaluated, if indeed it should be morally evaluated at all. What does this mean, and why might this be the case?

In this chapter, we start by examining intuitions about the relationship between mental illness and moral responsibility. We then explore different accounts of responsibility and, more specifically, the different requirements suggested by theorists for an individual to count as morally responsible for a given action. We then examine how these accounts relate to views explored in Chapter 2 concerning the nature of mental illness in general, and also to narratives of illness, responsibility and personhood.

4.1 Introducing the issue

It is widely acknowledged that there is, at the very least, *some* sort of relationship between the attribution of mental illness on the one hand and the attribution

DOI: 10.4324/9781315450124-5

of moral responsibility on the other. Many have noticed that mental illness has a tendency to reduce moral responsibility, at least in cases where the nature of the mental illness is relevant to the action that is a candidate for moral evaluation. If we agree that there is this relationship, we might ask: *Why* is there this relationship, and how strong is it? For example does (relevant) mental illness *always* reduce moral responsibility? And is the reduction of responsibility, when it happens, full or partial, binary or gradual?

Jerome Wakefield nicely articulates this issue in the following way:

> Why is it that when we classify a mental condition as a mental disorder, that tends to take the condition out of the sphere of moral responsibility or virtuousness of character? For example, it is not uncommon when someone commits suicide and the question is raised as to why he or she did it, that instead of a reason or character trait that would place the act within a moral calculus, one is offered the explanation "she was clinically depressed" or "he was bipolar," and this tends to terminate discussion and make reasons and moral evaluation superfluous. Why does such a comment have that effect?
> (Wakefield 2009, p. 91)

The idea here is that the attribution of depression doesn't *quantitatively* reduce moral responsibility. It takes us *qualitatively* into a different sphere, outside a "moral calculus." One could question whether this is in fact the case, but even if it is, it is not entirely obvious why.

4.1.1 Responsibility and related notions

For now, let's take a step back. What is responsibility? In the context of this topic, when we talk about responsibility, we are typically talking about *moral* responsibility, but, of course, not all responsibility is moral (otherwise "moral responsibility" would be pleonastic). At first approximation, responsibility, to say that someone is responsible, means that something can be attributed to someone. Sometimes, our concerns when making such an attribution are moral; we want to know whether and whom to morally blame or praise. But other non-moral concerns elicit attributions of responsibility; for example in artistic agency, we similarly want to know whether and whom to aesthetically praise or critique. "Who is responsible for this wonderful painting?" a prospective Renaissance benefactor might ask on a visit to Florentine studio.

Although moral and aesthetic evaluations are important and involve high-stakes applications of attributions of responsibility, there is a coherent sense in which someone can be held responsible for something in an entirely zero-stakes, and hence non-evaluative, context. It sounds odd, because one wouldn't usually bother to explicitly say so (the stakes are too low), but, yes, I was responsible for opening the fridge to get milk to put on my cereal this morning. It was neither good, bad, beautiful, nor ugly, but I did it. It was fully attributable to me. What made it attributable to me?

Let us look again at the aforementioned Wakefield quotation. He interestingly suggests that a "moral calculus" operates on "reasons" and "character traits." These are two quite different things, but they seem to apply in this mundane case: my behaviour was (somewhat trivially) "in character" (and certainly wasn't out of character), and it was motivated by reasons. In more high-stakes, moral cases, we may negatively evaluate character traits and/or the reasons that generate certain actions. We will see more on this later.

What is the relationship between responsibility and these evaluations, when we make them? It seems at first glance that you need to be responsible for something in order to be the target of these kinds of evaluations of praise or condemnation (moral, aesthetic, etc.). What are these evaluations anyway? Some have thought of them in terms of "reactive attitudes" (Strawson 1962), and there is a highly influential view (which we will examine very closely in Chapter 10) according to which, in some cases of mental illness, you should hold the individual responsible for their actions, but not engage in such attitudes, specifically *blame* (see Pickard 2011, 2017).

Another interesting question is: what is the relationship between responsibility and related concepts like autonomy? Sometimes, the very same issues are viewed through the lens of autonomy, rather than responsibility (see Radoilska 2013). Autonomy and responsibility are closely related, but the former is a complex, compound concept that is often taken to relate closely to legal notions like "the right to self-determination." There is a significant variation in the sorts of things we call autonomous. We can talk about autonomous actions. But we can also talk about autonomous individuals. This is similar to the distinction between reasons and character traits, where the former attaches to certain actions, while the latter attaches to certain people. It also relates to another distinction in thinking about responsibility, and it is as follows.

When we talk about responsibility, we can be talking about something *capacity-based*, applicable to an individual generally and over a long time, or we can be talking about something that we attribute in the context of a specific action, at a specific time. Thus, we might say that a young child lacks responsibility, not in the sense that any particular action carries reduced responsibility, but rather that, in general, this individual lacks the capacity to act in ways that they can be held (fully) morally responsible for. This contrasts starkly with the more specific, *action-based* use. We can say of someone who has the general capacity for actions for which they can be held responsible (e.g. they are a "responsible adult"), that, on a given occasion, a given action has certain features that reduce responsibility or blameworthiness. There is perhaps more focus in this debate on the latter, namely, on the attribution of responsibility for specific actions, rather than the general capacity. That said, the action-based and capacity-based uses are clearly related since someone lacking the capacity to be held responsible whatever the occasion won't be held responsible when a particular action is under consideration. Conversely, though, somebody, who is in the capacity-based sense responsible, might act in ignorance or under coercion on a given occasion in a way that reduces responsibility in the action-based sense.

Let us start our exploration by looking at a variety of concrete cases where we might be tempted to say that responsibility for action is reduced.

4.1.2 A variety of concrete examples

The example from Wakefield, discussed earlier, is a case of suicide, and there are a number of factors that make suicide unlike other potentially morally reprehensible acts (e.g. the harm to others, though often emotionally or financially severe, is indirect). But the moral exculpation provided by the recognition of mental illness extends well beyond this. For starters, it isn't only suicide for which responsibility and blame are diminished in the context of depression. Far less harmful behaviours, such as absenteeism (e.g. not turning up to birthday drinks), carry reduced responsibility when we learn the individual in question is clinically depressed. Indeed, the case of absenteeism shows that when we talk about reduced responsibility, we aren't necessarily talking about the realm of the morally wrong. The breaking of "softer" social norms and the coming across as weird or rude are also the sorts of things that should be excused and revised in light of finding out that someone has a mental illness. For example someone failing to excuse themselves for missing a commitment is (or *should be*) no longer considered rude in light of discovering that the individual was depressed.

Going beyond depression, but back to questions of moral judgement, there are many instances where mental illness is directly involved in the harm to those other than the ill subject. Since these sorts of cases are what much of the theorising turns on, let us look at some concrete examples. The examples are invented but certainly representative of real-life cases.

Example 1: delusional misidentification

> Mrs A. killed her husband. She was convinced at the time that that the man that she killed was not in fact her husband but a malevolent android, programmed by the Russians to spy on her. Now that she is aware that this is not the case, she is devastated by her actions, which resulted in the death of a man she loved.

There are real-life cases like this (see, e.g. Casu, Cascella and Maggini 1994), and, as we will see, the tendency in criminal law is to take the psychiatric circumstances to be significantly attenuating.[1] In other words, the level of responsibility attributed is reduced and punishment is either lessened or the legal course of action is not deemed to be punishment at all (but rather care and treatment).

In this example, Mrs A. becomes aware of what she did and is deeply saddened and remorseful. Nevertheless, reduced responsibility would still very much apply if her delusion remained (although she would almost certainly not be deemed fit to stand trial).

Why are we so universally tempted to say that Mrs A. is not guilty of the act that "on the surface" she appears to have committed (namely, the murder of her husband)? One way of answering this is to first rephrase the question in Wakefield's terms: "What is it about Mrs A.'s delusion that interferes with our 'moral calculus'?" And to answer this, we need to think about a canonical instance of someone acting in a way that we *would* deem open to face-value moral evaluation. This requires us to think about the nature of action and motivation (we will revisit some of these issues in Chapters 10 and 11).

A mainstream view (in both philosophy and decision theory) is that action is both caused and explained by beliefs and desires, where this is not intended in the everyday sense of belief or desire, but rather where "belief" is the basic informational state and "desire" the basic motivational state. In other words, belief is about how you take things to be in the world, and desires are about how you want things to be in the world. Obviously, the former constrains the latter. It is against the backdrop of you taking the world to be a certain way that you have certain desires, namely, you typically will not want that which is (believed to be) already achieved (and, although this is more contentious, you will typically not want that which is so unachievable as to be motivationally irrelevant). On a standard account, beliefs and desires come together to form intentions, which give rise to actions. To take a mundane example, I'm hungry and *want* to ease my pangs of hungers. I *believe* that eating food will do that, and that there is food in the fridge. I therefore form the intention to go to the fridge to get food. When I go to the fridge, I act in accordance with that intention, and we have a canonical case of rational, autonomous, responsible action.

A less mundane example can be constructed in terms of someone hating her husband and wanting him dead and thereby forming the intention to kill him. This, however, is not what happens in the case of Mrs A. This is because Mrs A. is profoundly misinformed about the state of the world and, by extension, misinformed about what she is actually doing. You cannot hold someone morally responsible for something that they are doing unknowingly. You might want to hold her responsible for what she *thought* she was doing (namely, killing a malevolent android), but she didn't succeed in doing that.

In other words, Mrs A. is seen to be less responsible for her actions, because (on the occasion of that action) her delusional misidentification was leading her astray. She was misinformed about the state of the world and misinformed about what she was doing. Had she known what she was doing, she would certainly not have done it. We have here an *informational problem*.

Notice, of course, that such informational impediments to responsibility are not specific to mental illness or even to unusual psychological states (e.g. delirium or psychedelic states). People often act under ignorance or misinformation. Oedipus married Jocasta but did not know that she was his mother. He was responsible for what he did under the description of marrying "Jocasta," or "that woman," but not under the description of marrying "his mother." In a more specialised sense, he didn't know what he was doing either!

Now let's look at a very different case.

Example 2: addiction

> Mr B. is a gambling addict. He has been fighting his addiction for several years. He has a regression, and, having spent all of his own money, breaks open his 10 year-old son's piggy bank. Fully aware of his actions, he is wracked with shame and remorse.

This is a rather common example (and we'll see much more on this in Chapter 11, when we focus more closely on addiction). One of the things that make addiction so unpleasant and harmful is that it leads us to behave in ways that seemingly conflict with our better judgment, namely, what we deem to be morally right, or even best for us. Mr B. is deeply regretful, and (and we'll see in Chapter 11) our standard conceptions of addiction (at least should) accommodate the idea that "deep down," he didn't want to break open his son's piggy bank.

At one level, Mrs A. and Mr B. are in a similar predicament: Mrs A. didn't really want to kill her husband, and Mr B. didn't really want to steal from his son. Neither was forced (at least not overtly/visibly) to act in those ways from the outside. And yet the case of Mr B. is very different. The relevant impairment for understanding the action is not *informational.* Mr B. knows exactly what he is doing; he just doesn't really want to be doing it. It is a *motivational problem.* One might say that in the one case we have *problematic belief*, while in the other we have *problematic desire*. (Whether depression can be conceptualised in either way is an important debate we examine closely in Chapter 10).

The case of Mrs A. is relatively straightforwardly made sense of in the terms of belief–desire psychology: she was simply profoundly misled by her delusion. But if Mr B. didn't want to break open his son's piggy bank, then why did he *do that*? We'll see much more on this in Chapter 11, which is dedicated to addiction. A very natural response is to say that there is some kind of *internal conflict.* There are, for example, strong urges that he has no direct control over, and that all things considered he would rather not have, and these pull him in a direction that conflict with his more reasoned, long-term policies and plans. Mr B.'s addiction pushes him in a direction that he doesn't really want to go.

Again, as with Example 1, the criminal justice system will typically take addiction to be an attenuating circumstance (albeit usually only moderately). The agent is forced into acting in ways that he otherwise wouldn't. The intuition here is that you typically do not hold people responsible for things that they were coerced into doing, regardless of the source of that coercion (be it external or internal; someone else or your own nervous system). The appropriate course of action, one might think, would simply be that the person needs help with their addiction, rather than punishment for what they did.

Examples 1 and 2 are relatively uncontroversial. Although there are differences in the precise details, the fact that these examples, at least to some extent, reduce responsibility, and the reasons why there is such a reduction are widely accepted. You cannot be held fully responsible for doing something that you didn't know you were doing (Example 1), and you can't be held fully responsible for acting under coercion (Example 2). Example 3 is one that, in contrast, seems to divide opinion.

Example 3: Foetal Alcohol Spectrum Disorder

> Mr C. gets easily angry, is impatient, impulsive and lacks resilience. One day, Mr C. assaults his wife. It emerges that Mr C.'s mother drank heavily while she was pregnant with him. Evidence strongly suggests that the character traits that have led to Mr C.'s actions are associated with Foetal Alcohol Spectrum Disorder (FASD).

First of all, it is worth mentioning that it is not at all clear that FASD counts as a mental illness (for the sorts of reasons that we saw in Chapters 2 and 3). However, whether it does or not, it still functions as an interesting case when reflecting on the nature of responsibility and conditions of exculpation.

In stark contrast to the first two examples, here the individual action (the assault) is not one that can be explained in terms of either misleading informational states or coercive motivational states. What we seem to have here is somebody behaving in a particular way because he is a particular sort of person. However, it is not his fault that he is that sort of person. Of course, an obvious retort is that the same can be said of anyone. No one controls his or her DNA or upbringing. Nevertheless, many people feel that some special allowance should be made for a case like this one. Mr C.'s FASD certainly does not seem *irrelevant* when we are evaluating him morally (Carel 2016). This is a case that raises significantly different issues from the first two, and we will examine it in more detail later in this chapter.

At this point, however, it is worth mentioning that Example 3 raises two significantly different issues that are sometimes conflated, and perhaps because both are capacity-based rather than action-based. One of these issues concerns the level of control that the person has over character traits. Mr C. is a victim of FASD, which resulted from his mother's behaviour while he was *in utero*, and that makes him irritable and lacking in self-control. Another issue concerns the understanding required in order to be bound by the relevant norms. This latter point applies to FASD only to the extent that we feel that there are cognitive (i.e. intellectual) differences rather than simply emotional or temperamental ones. More canonical cases of these sorts of grounds for exculpation involve examples of intellectual disability, and, here, the diminished responsibility is often presented independently of volatile character traits. Although these are even more clearly not cases of mental illness, they are worth considering for theoretical illustration. With that, let's consider the following.

Example 4: intellectual or learning disability

> Mr D. committed a serious crime. It transpires that Mr D. has a significant learning disability as evidenced by a number of factors, including a significantly lower than average IQ.

In Example 4, the rationale behind Mr D. being exculpated, at least to some extent, is somewhat akin to the rationale behind exculpating young children. What is at issue is whether Mr D. has the cognitive ability to understand what is expected of him as a member of society. But there are other conditions that raise similar issues, but where partaking in the normative economy is impaired, but not for reasons of impaired intellect. Indeed, in the cases we are about to examine, above average intellect (at least as measured by IQ) can be reported.

Example 5: psychopathy

> Mr E. has committed a number of violent and reprehensible acts. He is extremely intelligent, manipulative, but simply doesn't seem to feel bound by the same moral boundaries as the rest of society. He feels no remorse and doesn't seem to understand why he shouldn't have acted in the way he did.

This example suggests that a degree of appropriate emotional reactivity is required in order to be bound by or *feel* the pull of the moral norms of society. Exculpation for psychopathy is highly contentious, and they are often the sorts of people who attract the extreme label of moral revulsion: "evil." In fact, it is an open question whether their neurological differences (whether innate or acquired through trauma) make them *breakers* of moral norms or simply *non-partakers*. In the former case, they would exhibit moral wrongdoing (immorality). In the latter case, they are morally alien (amoral) and, one might argue, less deserving of moral disapproval.

Let us for now sum up these five cases since they illustrate five ways in which the "moral calculus" might be disrupted, leading to potentially reduced responsibility and exculpation:

- Example 1 → Misleading informational state (delusion) and hence not knowing what she was doing.
- Example 2 → Coercive motivational state (addiction) hence not really wanting to do what he did.
- Example 3 → Pathological changes to behavioural tendencies, of which the subject is a victim.
- Example 4 → Intellectual disability leading to a lack of understanding about the world in general, but (most relevantly) legal and moral norms in particular.
- Example 5 → A lack of ability to partake in the emotional community that grounds the normative landscape of society.

Before moving on, I'd like to make a few observations about these five examples. Example 1 and Example 2 are the only relatively uncontentious cases of mental illness. They are also the only purely action-based cases. There is no suggestion that Mrs A. and Mr B. should not be held responsible for anything they ever do. Exculpation should only be considered for the specific actions that are of direct relevance to the pathological informational and motivational states. Examples 3, 4 and 5 are more capacity based. The exculpation is not about the precise circumstances surrounding the target action, but rather about the kinds of people we are presented with: people with problematic character traits, cognitive and emotional capacities and dispositions that they carry with them all the time.

4.2 Approaches to mental illness and moral responsibility

The most institutionally tangible context in which questions of mental illness and moral responsibility arise is in the legal system, and criminal justice in particular. Legal responsibility and moral responsibility don't necessarily coincide, but there is something refreshing about the codification that legal contexts bring. More importantly, however, reflections on the legal case can feed into our thinking about the moral case.

4.2.1 Mental illness and legal responsibility: "the insanity defence"

So far, we have been talking about moral responsibility. However, the legal system is supposed to codify and reflect our moral intuitions and guide legal action (in this case, most relevantly, criminal justice). The intuitions we have about mental illness and moral responsibility are reflected in what is often referred to (perhaps suboptimally) as "the insanity defence." (I say "suboptimally" not only on the theoretical grounds that the term is only very loosely defined, but also because it is stigmatising.)

As Malatesti, Jurjako and Meynen (2020) point out, the insanity defence tends to involve two parts, which they name *the incapacity clause* and *the mental illness clause*. The incapacity clause "concerns the presence of certain mental incapacities at the moment of committing the act" (Malatesti, Jurjako and Meynen 2020, p. 1). The mental illness clause "is the requirement that these incapacities should be attributable to a mental illness" (ibid., p. 1). What is philosophically interesting about the insanity defence, presented thus, is that we can scrutinise how necessary each clause is. For example Matthews (2004) argues that our practices of exculpation are only essentially about moral agency, not about "insanity" (mental illness) *per se*. The reason why we might deem a young child to have reduced responsibility is not fundamentally different to the reason why we might deem someone who has a mental illness to have reduced responsibility. To rephrase this in terms of the two clauses, Matthews' suggestion is that we should jettison the mental illness clause, since it is not strictly relevant (and by extension the insanity defence) and replace it with a generalised defence purely in terms of the incapacity clause. As Malatesti, Jurjako and Meynen (2020) put it: "it is not mental illness that reduces responsibility, it is the incapacity to conduct oneself in accordance with the law, regardless of the cause." The same point is made by Nicole Vincent (2008) with regard to psychopathy and its classification as a disorder. As she puts it, regardless of whether we choose to think of psychopathy as a disorder (or illness),

> our judgments about psychopaths' responsibility should not be affected, because responsibility hinges not on whether a particular difference is (referred to as) a disorder or not, but on how that difference affects the mental capacities required for moral agency.
>
> (Vincent 2008, p. 199)

This, on reflection, might seem like a rather plausible, indeed obvious, point. So what logical work is the mental illness clause supposed to be doing? What difference does it make if incapacity is brought about by mental illness? Malatesti, Jurjako and Meynen (2020) consider three answers in response to this.

The first answer is that incapacity brought about by mental illness undermines responsibility in a way that some other forms of incapacity do not. A particularly salient example of the latter is incapacity brought about by an earlier decision for which you are accountable, such as driving under the influence of drink or drugs. This kind of incapacity does not exculpate to the same extent, or even at all. In contrast, those with mental health conditions are victims of, not agents responsible for, the conditions that result in their incapacity.

The second answer is that the attribution of mental illness can play a practical or epistemic role (Malatesti, Jurjako and Meynen 2020, p. 2) in helping us establish "whether the agent at the time of the act had relevant incapacities." This is not as substantive as the first answer. The incapacity is still what is doing the exculpatory work, and it is simply that "insanity" provides a stronger evidential footing for that lack of capacity.

The third and closely related answer is also practical: "it enables an interactive relationship between the legal practice and its conceptual apparatus, on the one hand, with clinical advancements and its scientific conceptual apparatus, on the other" (Malatesti, Jurjako and Meynen 2020, p. 2).

The third answer, like the second, acknowledges (along with Matthews and Vincent) that the mental illness clause does no exculpatory work *per se*. It is merely practically useful in the absence of certain pieces of information or for fostering potentially enlightening interdisciplinary dialogue. It is only the first answer that is based on the acceptance that mental illness makes and concrete difference in and of itself (rather than merely causally contributing to the thing that really matters).

This legal issue, concerning the centrality or otherwise of mental illness, closely mirrors debates in the moral, rather than legal, realm. In other words, there is a question about whether mental illness plays a direct, conceptual role in our judgements about moral responsibility, or whether some of the conditions that we happen to deem to be cases of mental illness sometimes play a causal role in undermining responsibility, which, crucially, is a role that could be played just as well by something that we don't deem to be mental illness. We examine this distinction now.

4.2.2 Incidental versus intrinsic accounts

There are several different accounts of the relationship between mental illness and moral responsibility, and it is useful to divide them, at least initially, into what I call "intrinsic accounts" and "incidental accounts."

Intrinsic accounts, as the name suggests, take the relationship between mental illness and moral responsibility to be very close and, in fact, *intrinsic*. The idea here is that it is somehow packed into the *concept* of mental illness that the agent's moral responsibility is reduced. There are different ways of putting this tight conceptual connection (in part because there are different ways of thinking about concepts and meaning in general), but one way is in terms of the function of mental illness attribution. In other words, by saying that someone is mentally ill, we are preparing ourselves (and others) for moral leniency if any relevant untoward behaviour takes place. The intrinsic account is nicely expressed in the following quote we saw in Chapter 2.

> Mental illness is not a label that picks out a set of consistent qualities in a mental condition; it is a label that stipulates how people *should* respond to the condition, and in particular whether they should respond *morally* or *medically*
>
> (Edwards 2009, pp. 74–5)

You will no doubt have noticed parallels between this view and some forms of values-based approach to mental illness, namely evaluativism. It is hardly surprising that intrinsic views will elucidate the relationship between mental illness and moral responsibility by unpacking the former notion (mental illness) in terms of the latter (moral responsibility). In short, according to this view, the relationship falls out of an accurate treatment of the concept of mental illness.

One obvious initial worry for this view is that many mental health conditions seem morally irrelevant. How might someone's anxiety disorder lead them to commit face-value morally questionable actions? And even with psychosis, many people will never do anything of relevance to moral disapproval. This worry misunderstands the intrinsic view but provides us with a useful opportunity to clarify matters. The misunderstanding in question has two interrelated aspects.

First of all, when we talk about "responding morally," we mean something all pervasive and subtle: not something that is only relevant to moral wrongdoing. When we observe or interact with others, we subtly evaluate their behaviour in all sorts of ways that fall short of full-blown moral judgement. There are general social norms that invisibly guide our reactions to the behaviour of others. Etiquette, politeness, subtle and not-so-subtle aspects of interpersonal dynamics might be negatively evaluated in all manner of ways that fall well short of the moral realm. For example it might just "rub us up the wrong way," feel socially odd or unacceptable and make us like someone a bit less. Depression or social anxiety might lead someone to exhibit socially unacceptable levels of absenteeism. Knowing the individual has a mental health condition for which this is causally responsible means that we should make allowances for that, and we shouldn't think less of the individual; we should override our tendency to judge them and encourage a more compassionate response. We recognise that these behavioural tendencies are not reflections of the individual "deep down." Cure them of their anxiety or depression, and they won't be like this.

The second misunderstanding concerns something we have already mentioned, namely, the two senses of responsibility attributions: one to do with general capacities and the other with specific actions. The proponent of an intrinsic account can insist that, even if the individual doesn't actually perform any morally reprehensible actions, if they *were* to, they would not be held morally responsible in the same way. However, the (actual and hypothetical) action-spaces need to be relevant to the condition in question. A depressed person may be deemed less responsible for failing to come to a birthday party or to adequately take care of a child, for example, but not for any wrongdoing whatsoever (e.g. murder).

Incidental accounts, as the name suggests, take the relationship between mental illness and moral responsibility to be *incidental*. This is very closely related to those critics of the insanity defence who, like Matthews (2004) and Vincent (2008), take the mental illness clause to be irrelevant to establishing culpability. On these incidental accounts, mental illnesses are the kinds of thing that *tend* to negatively impact on, or interfere with, capacities or features required for someone to count as fully morally responsible. It is perfectly possible that some mental illnesses have no impact on moral responsibility at all, and, conversely, many things that do similarly have that impact aren't mental illnesses. Since the impact that mental illness has on moral responsibility is incidental, in stark contrast to the intrinsic view, the account of the relationship does not fall out of our understanding of the general concept of mental illness. The focus instead is on (i) what is required for someone to count as morally responsible, and (ii) how some of the specific conditions, that we happen to deem to be mental illnesses, interfere with those requirements. It is central to the logic of incidental accounts, in contrast to intrinsic accounts, that this interference has no direct bearing on whether that condition *counts as* mental illness.

Different incidental accounts ultimately stem from differences in the postulated requirements for moral responsibility. With this in mind, let's put intrinsic accounts aside for now and focus on different accounts of the requirements for responsibility. It is worth bearing in mind that these accounts can be examined entirely separately from any considerations of mental illness. These requirements are entirely general and could be seen through the lens of, for example, establishing whether children of a certain age should be deemed morally responsible, etc.

4.3 Requirements for responsibility

Giving an account of the requirements for responsibility is particularly pressing for someone who adopts an incidental view of the relationship between mental illness and moral responsibility. This is because, if their claim is that some mental illnesses incidentally (i.e. happen to) interfere with the requirements for responsibility, then they need to provide an account of those requirements. In other words, there is an explanatory burden on proponents of the incidental view to elucidate the requirements in the good case (i.e. where someone counts as morally responsible) and then explain how certain forms of mental illness can, at times, disrupt the fulfilment of those requirements or prevent them from being met.

So, putting mental illness to one side for now, what are the requirements for responsibility more generally? What does it take for you to be morally responsible for your actions?

4.3.1 A simple view: information and control

One simple view that was brought out by some of our examples is that your actions need to be based on accurate information, and they need to be under your control (these conditions are what David Shoemaker (Shoemaker 2007) calls the epistemic and motivational conditions). Delusion violates the former, addiction the latter. Notice, however, that they are not *special* in how or why they reduce responsibility. According to this simple view, delusion reduces responsibility in much the same way as someone lying to you might (or you making an honest mistake leading to you being informationally misled). Addiction reduces responsibility in much the same way as someone coercing you might. That's perfectly fine for the proponent of what I'm calling the incidental view. Mental illnesses aren't special: they just happen to, sometimes, interfere with the conditions of responsibility. What is more problematic for this view is that, while misinformation and coercion are sufficient for reduced responsibility, they don't seem necessary for it. In other words, there are cases where people are not misinformed, nor are they coerced, and they still lack responsibility. For example young children, or non-human animals, aren't responsible, even though (or even *when*) they aren't obviously misinformed or out of control. Some have thought that responsibility runs deeper and that a better account is available that explains the cases the simple view does in a more satisfactory way, and succeeds in accounting for those it can't accommodate at all.

One now-classic paper on this topic is Susan Wolf's "Sanity and the Metaphysics of Responsibility" (Wolf 1987). In this paper, she presents a view, associated with several other philosophers before her, that she calls "The Deep Self View," which she criticises, and presents an improved view, which she calls "The Sane Deep Self View." Because these are two very important views, I will present them here. However, it is worth noting that Wolf's use of the word "sane" is somewhat non-standard and carries no implication of mental health issues (as will become clear).

4.3.2 The Deep Self View

The Deep Self View (sometimes called the "Reflective Self View" or "Hierarchical View") starts with the observation that to deem someone responsible for an action is to attribute it to them. In the examples given before, or at least certainly the first two (Mrs A and Mr B), we aren't comfortable saying that the actions in question, as they appear on the surface, come from, or are a reflection of, the "*real them*." As Will Cartwright (2006) puts it:

> The familiar excuses of accident, ignorance, coercion, and so on, are to be explained on this view as cases where the action does not reflect, and so cannot be attributed to, the agent's real self.
>
> (p. 145)

This immediately raises the question: when *are* actions attributable to the *real self*? On the Deep Self View, they are if they flow from desires that have been endorsed by the deep self. But what is the deep self? Perhaps the best-known presentation of this view is found in Harry Frankfurt's (1971) classic paper, *Freedom of the Will and the Concept of a Person*, although relevantly similar views are also found in the work of Garry Watson (1975) and Charles Taylor (1976). What they all have in common is that, in order to be responsible agents, our actions cannot be the product of desires that are just found "*in* us, but expressions of characters that come *from* us, or that, at any rate are acknowledged and affirmed *by* us" (Wolf 1987, p. 365). What does this mean?

Frankfurt lays the groundwork for the view as follows:

> Besides wanting and choosing and being moved to do this or that, [persons] may also want to have (or not to have) certain desires and motives. They are capable of wanting to be different, in their preferences and purposes, from what they are. Many animals appear to have the capacity for what I shall call "first-order desires" or "desires of the first order," which are simply desires to do or not to do one thing or another. No animal other than man, however, appears to have the capacity for reflective self-evaluation that is manifested in the formation of second-order desires.
>
> (1971, p. 7)

A cornerstone of Frankfurt's position is this introduction of a hierarchy of motivational states. Human beings have the capacity to not only want certain things, but also want to want certain things. For example, I might want to eat the bacon sandwich I can tantalisingly smell, but want to not want to, because I'm a vegetarian. Conversely, I might not have an appreciation of classical music, but wish that I did have such an appreciation of and hence appetite for it, because I feel that it might enrich my life.

With this distinction between first-order and second-order motivational states, we get a clear account of the addict's behaviour. Mr B. has a very strong first-order desire to gamble, and the drive towards fulfilling that desire leads him to break his son's piggy bank. However, Mr B. has a second-order desire: a desire to not desire to gamble, a desire to not be a gambling addict and all of the negative consequences that ensue.

Note that in Example 1, the delusional misidentification case with Mrs A., if the desire was endorsed by a second-order desire, it was not the desire to kill her husband, but to protect herself and her family from a malevolent android that was endorsed.

The Deep Self View also has the resources to explain a general capacity-based lack of responsibility. Just as Frankfurt highlights that non-human animals may lack "the capacity for reflective self-evaluation that is manifested in the formation of second-order desires," the same might be said of some humans. It would be an empirical issue as to precisely where this line may be drawn, but obvious candidates are young children and those with intellectual disabilities. They may for example lack the capacity to reflect upon overarching long-term plans and policies that regulate, endorse and reject first-order impulse.

One obvious problem with the Deep Self View is that the explicit endorsement that the view describes seems to happen rather rarely. This would make responsibility a "rarer thing than we suppose it to be" (Cartwright 2006, p. 146). There is, however, a quick response to this concern, which is simply that second-order endorsement needn't be an explicit and conscious *act* of endorsement. Indeed, there is a perfectly legitimate sense of endorsement that is dispositional. When you ask someone "Do you endorse this?," you are often not asking them to form the endorsement on the spot; you are often asking them to reflect on whether they have been endorsing it all along. On a related note, the vast majority of our actions are so deliberatively mundane (getting dressed, walking down the stairs etc.) that, although we would endorse them if we were to reflect on them, in practice we never do. That doesn't, however, mean that we're not responsible for these actions. Of course we are.

This response is especially clear in Watson's (1975) version of the Deep Self View, which emphasises the distinction between "mere" desires that I find myself "stuck with" and desires that are expressions of my values, things that I want because I deem them to be good. The relevance here is that I can deem something to be good to flow in accordance with my values, without doing so through an explicit, reflective psychological event.

Another worry for the Deep Self View is that it paves the way for a potential infinite regress of successively deeper selves (Wolf 1987). Your first order desire is legitimate and endorsed because it is in accord with a second-order desire. But what are we to say about that second-order desire? Where does *that* come from? Who endorsed *that*?

One relatively attractive way of blocking the infinite regress is not only to distinguish between first and second-order desires, but also to distinguish between different kinds of motivational state. This line of thinking goes back at least to Aristotle and is also quite explicit in Watson's view. For example you might say that there are relatively abstract, long-term desires that have as their object things that are spatiotemporally distant. These "plans" or "policies" or "reasoned desires" have various distinguishing features. For example they might be relatively emotionally cold, and they might not physically drive the agent in quite the same way as the other kind of desire. This other kind we might call "urges." These are emotionally charged, visceral, behaviour-driving, and, above all, they are the kind of thing that you "just find yourself with" (and cannot reason yourself out of).

In light of this distinction between these cold, reasoned, long-term, long-distance desires and "hot," short-term, short-distance ("stimulus-bound") urges, one might then suggest the following constraints. Reasoned desires can be first-order, world-directed or second-order urge-directed in the human psyche. I might have the overarching desire to get healthier (first-order). I might (in the service of that) want to not crave unhealthy foods (second-order). Urges, on the other hand, can only occupy first-order position. They just aren't the right kind of thing to occupy second-order position. You cannot have the urge to not have an urge. That's for a number of reasons, one of which being that urges target worldly stimuli and drive behaviour towards those stimuli. Mental states can't play the role of the worldly stimuli that urges latch onto. Now comes the constraint that blocks the regress: reasoned desires can only occupy second-order position if their targets are urges, not if they are reasoned desires. You cannot have the reasoned desire to have a reasoned desire since that would simply collapse in the first-order reasoned desire. If that's true, then you cannot have third-, fourth- and fifth-order desires ad infinitum.

Whether or not this way of blocking the regress works, it could be argued that it misses the full force of the criticism. One way of interpreting the criticism is not so much in terms of the psychological mechanism involved in embedding motivational states, but simply as a concern about the origin of the "deep" or "true" self and the somewhat arbitrary nature of "stopping the buck" there.

To appreciate this version of the worry, consider the case that Cartwright describes, namely, the case of the murderer Robert Harris. He might, Cartwright argues, have reflected on his desires and actions and endorsed them, but if that endorsement itself is simply a feature of Harris's deeply troubled childhood, what control is there over this endorsement? Proponents of the Deep Self View might simply bite the bullet and say that Harris is fully responsible. That simply *is* what it is to be a bad person. Regrettably, there are many things that are ultimately out of our control, but the buck has to stop somewhere. A troubled childhood might contribute to the sort of person you are, but so might your DNA or, as we saw in Example 3, whether your mother drank while you were in utero.

However, some have found this an unpalatable response and added a further condition on responsibility, and this is where the second view emerges: the *Sane* Deep Self View (in a rather specialised sense of "sane").

4.3.3 *The Sane Deep Self View*

The "Sane Deep Self View" (Wolf 1987) takes the Deep Self View to attribute the full force of moral responsibility to cases where it should not apply. In particular, these are cases, like Harris', where there is endorsement of action, but we still feel the pull of exculpatory intuitions. The rationale behind this is elegantly expressed by Cartwright, paraphrasing Wolf's position: "responsibility requires more than just control [viz. endorsement as per the Deep Self View]; it requires both cognitive and moral understanding" (Cartwright 2006, p. 149). Wolf's own case, unlike Cartwright's use of Harris, is not a case of childhood trauma but an example of cultural variance (which makes Wolf's choice of the term "sane" perhaps even more surprising). Here is the well-known (imagined) case of JoJo:

> JoJo is the favorite son of Jo the First, an evil and sadistic dictator of a small, undeveloped country. Because of his father's special feelings for the boy, JoJo is given a special education and is allowed to accompany his father and observe his daily routine. In light of this treatment, it is not surprising that little JoJo takes his father as a role model and develops values very much like Dad's. As an adult, he does many of the same sorts of things his father did, including sending people to prison or to death or to torture chambers on the basis of whim. He is not coerced to do these things, he acts according to his own desires. Moreover, these are desires he wholly wants to have. When he steps back and asks, "Do I really want to be this sort of person?" his answer is resoundingly "Yes," for this way of life expresses a crazy sort of power that forms part of his deepest ideal.
>
> (Wolf 1987, pp. 367–8)

Wolf then goes on to claim that our intuitions pull us in the direction of denying that JoJo is responsible for his actions. This is because it "is unclear whether anyone with a childhood such as his could have developed into anything but the twisted and perverse sort of person that he has become" (ibid., p. 368). She rehearses two explanations for this reduced responsibility – the first she rejects in favour of the second.

The first is that JoJo had no say in who his deep self might become. Quite rightly, Wolf dismisses this explanation: viewed "from the outside" none of us have! Indeed, as Wolf memorably puts it, "self-creation is not just empirically but logically impossible" (p. 368). The second explanation is in terms of what Wolf calls "sanity" (in an admittedly "special sense" (p. 369, p. 370)). An agent is "sane," in this sense, if she is able "cognitively and normatively to understand and appreciate the world for what it is" (1987, 387).

This adds yet another sense in which someone might be excused for "not knowing what they are doing." We have seen how someone might literally not know the behaviour that they are performing (e.g. they might be sleep-walking) or they might know what they are doing in a basic sense, but not know either its true consequences, or facts relevant to how the action should be described (e.g. marrying a woman, not knowing that she is your mother; killing an android, not knowing it's your husband). In this case, in contrast, JoJo knows what he is doing both physically and cognitively – he just doesn't *know that it's wrong*. In other words, this notion of understanding the world for what it is both factually and normatively opens up the possibility that two forms of exculpation are the result of agents "not knowing what they do," but in different senses. Some agents don't know what they are doing in a factual sense, either in the direct sense like Mrs A., or because they are ignorant (like Oedipus not knowing that Jocasta was his mother), or indeed they don't appreciate the consequences of their actions (like not knowing someone's fatal nut allergy). Other agents don't know what they are doing in the normative sense, namely, although they know what they are doing, they don't know that what they are doing is wrong. It is in this latter sense that JoJo "knows not what he does" in that he is a victim of unavoidable moral ignorance.

So, to sum up then, JoJo is let off the hook, so to speak, because he is "insane," in the sense of not knowing right from wrong. To paraphrase Wolf, JoJo's actions are driven by mistaken values, values that he unavoidably has. In general, then, the Sane Deep Self View adds another condition to the Deep Self View: yes, your actions need to be endorsed by your Deep Self, *but* (as the JoJo case is supposed to illustrate) that Deep Self has to be "sane," in the aforementioned sense.

The Sane Deep Self View has two advantages. It inherits the advantages of the standard Deep Self View in allowing us to differentiate between good cases of full responsibility from cases of addiction, delusion, brainwashing, posthypnotic suggestion, etc. Second, it "deals with the case of Jojo [the son of an evil and sadistic dictator] and related cases of deprived childhood victims in ways that match our pretheoretical intuitions" (Wolf 1987, p. 382). In other words, endorsement, though perhaps necessary for moral responsibility, appears not to be sufficient, as cases like JoJo or Harris seem to demonstrate.

What are the problems with the Sane Deep Self View? At this point, I will present two potential criticisms and leave it to the reader to evaluate them.

The first criticism might simply disagree with the intuitions that it is built on. Is it clear that JoJo should be let off the hook? Sure, we cannot expect him to have turned out differently, but is that really central to the issue of moral responsibility? Viewed from a particular perspective, might that not be true of everyone?

The second problem with the Sane Deep Self View is that it relies on at least some degree of moral realism. The reason for this is that Wolf relies on the notion of moral knowledge, of seeing the world normatively, for how it is. And that requires some acceptance of moral fact, of moral truth. And that is a highly contentious issue. Indeed, it is arguably the central dispute in meta-ethics. Meta-ethics is (roughly) not concerned with the question of what actions are morally right and wrong – that's ethics – but rather concerned with what makes claims and judgments of rightness and wrongness in general the sorts of claims and judgments that they are (see Chrisman 2017 for an excellent textbook on this). There are many different positions that are built on an anti-realist view. I can't go into details, but many theorists think that morality emerges to regulate society, and it involves approving or disapproving of, encouraging or discouraging, certain actions. As such, there is no factual domain that could be the target of moral knowledge.

4.3.4 *Taking stock*

So, assuming an incidental view about the relationship between mental illness and moral responsibility, the strategy is as follows. You ask yourself what the conditions are for someone to count as morally responsible and then ask how this can be interfered with by any number of things, some of which may be classified as mental illnesses. In light of this, we examined two views of moral responsibility, namely, the Deep Self View and the logically more demanding (but morally more forgiving) Sane Deep Self View. The former is about endorsing (in an explicit or dispositional sense), via second order desires, one's first-order desires. The latter builds on this but adds the further requirement of "sanity," roughly understood to be knowing right from wrong. This means that, like JoJo, someone might fully endorse their own abhorrent actions but still not be fully responsible for them because they are acting, and indeed endorsing, in a state of moral ignorance.

In spite of the use of the term "sanity," it is not at all obvious that JoJo (or Robert Harris, or indeed any "psychopath") is someone that we would think of as having a mental illness. So, moving from questions about responsibility in general to the issue we are interested in here, we need to ask: Under what circumstances does mental illness in particular interfere with moral responsibility? This depends on what we think mental illness is.

4.4 Mental illness revisited

How might reflections on responsibility tie in with notions of mental illness? I now recap some of the positions from Chapter 2 on the nature of mental illness, but through the lens of its impact on moral responsibility. We will see that interesting consequences emerge, not only in the sense that our accounts of mental illness impact on the relationship between mental illness and moral responsibility, but also in the sense that reflection on moral responsibility can influence our thinking about mental illness.

4.4.1 Objectivism and the incidental view

If you are an objectivist, then it is not clear why the sorts of criteria that define illness in general, and/or mental illness in particular, should bear any sort of close or necessary relationship to moral responsibility. This point stands on its own but is further bolstered by the observation that a central feature of objectivism is that it hopes to make illness attribution factual, descriptive and value-free. In stark contrast, if any domain of discourse is to be non-factual, non-descriptive and value-laden (*pace* robust moral realists), it is morality and the related notion of responsibility (another strong candidate, of course, is aesthetic discourse).

This suggests that objectivism is conducive to (although it doesn't quite strictly *entail*) an incidental view. If you think that being ill (mentally or physically) involves being in some objective state of dysfunction, then the easiest way of accounting for exculpation, when it happens, is as follows. Some of the things that are mental illnesses interfere with the conditions of responsibility. Addictions coerce, psychoses mislead, but it is the fact that they coerce and mislead, and not the fact that they are mental illnesses, that explains the exculpation. Indeed, other mundane forms of coercion or misinformation will exculpate in the same way, and mental illnesses that lack these features won't exculpate at all.

Granted, an intrinsic objectivism would be coherent, but would have to postulate a coincidence, namely, that all relevant dysfunctions that lead to mental illness also *happen to be* responsibility reducing.[2]

4.4.2 Values-based accounts and the intrinsic view

In Chapter 2 I chose, first, to mark a coarse-grained distinction between objectivist and non-objectivist views (calling the latter values-based views). Then, within the values-based camp, there is a further division between those who take mental illness attribution to be descriptive, and those who take it to be more irreducibly evaluative. In other words, according to the former view, which I call "constructivism," saying that someone has a mental illness is like saying that someone has brown hair, namely, it is (like objectivism) attributing a property, but only that (unlike objectivism) that property is "constructed" or "a construct," which is to say that it varies (or even ceases to be) across cultures. According to the latter, which I call "evaluativism," you are not (or not only) attributing a property, but (also) evaluating the individual. What does this mean?

Perhaps the simplest way to think of these irreducibly evaluative statements is to think of them in terms of expressing evaluative states of mind and that these states of mind are fundamentally unlike descriptive states of mind. In turn, these evaluative states of mind are thought to be more directly action-driving than descriptive ones. What all of this has to do with our discussion here is that another area, beyond mental illness and illness discourse that get this kind of evaluativist treatment, is moral discourse (usually under the label "non-cognitivism" or "expressivism").

What this allows, then, is the idea that these two intrinsically action-driving domains of evaluative discourse are *intrinsically related*. In other words, this is one way of understanding the (otherwise somewhat perplexing) intrinsic view. The intrinsic relationship between the moral and the medical is that the courses of action with which those discourses are tightly (even constitutively) bound pull in opposite directions. In other words, the enjoinment on society to (medically/clinically) help someone typically suggests that they are not at fault and should not be punished. Conversely, the attribution of responsibility typically enjoins punishment rather than medical/clinical help.

Two things should be noted here, which complicate matters significantly. First of all, the word "typically" suggests that we are talking about a tendency that can be overcome (and perhaps it sometimes ought to be overcome (see Pickard 2011)) rather than a strict logical entailment. Second, there are many non-medical, and non-clinical, ways of helping people. One of the challenges that the objectivist lays at the feet of values-based theorists, and evaluativists in particular, is that of distinguishing between different kinds of problem and the different kinds of help they require/request/trigger. I am not saying the challenge cannot be met, but it is a difficult one. If the evaluativist looks to define illness in terms of help-seeking/help-needing, but then distinguishes the appropriate type of help (e.g. medical versus social, financial, what have you) in terms of illness, then this is circular. There is, however, another form of intrinsic view worth considering, namely, one that is built around constructivism rather than evaluativism.

4.4.3 Fictionalism about mental illness and morality

Just as evaluativism about mental illness has a parallel in meta-ethical expressivism, so does constructivism find a parallel in meta-ethical constructivism. Having said that, constructivism in meta-ethics takes various forms, many of which have a more realist flavour than the constructivists about mental illness would accept. As a result, the closer parallel is actually with what is called moral fictionalism (e.g. Joyce 2001; Kalderon 2005), rather than moral constructivism (e.g. Rawls 1980; Korsgaard 1983). To make this clear, let us call a particular subset of constructivism about mental illness, psychiatric fictionalism (Wilkinson 2022).

What is fictionalism more generally? To be a fictionalist about a domain of discourse is to say that, while that which is being talked about does not literally exist, claims made in that discourse are not straightforwardly false or meaningless, and that is because they are part of a useful, perhaps indispensable, fiction. To get a quick sense of where this is coming from, consider the fact that "Sherlock Holmes lives on Baker Street" is not false, or at least not false in the way that "Lyon is the Capital of France" or "Dogs are felines" is. Indeed, if you were in a pub quiz and the question was "On what street does Sherlock Holmes live?," if you answered "Nowhere – Sherlock Holmes is fictional," you wouldn't be awarded the point (even though you are saying something literally true).

Debates surrounding fictionalism, its varieties and the viability of those varieties are highly complex and nuanced, and I can't do justice to them here. However, I can quickly sketch why fictionalism about mental illness might combine with moral fictionalism to form a kind of intrinsic view.

Fictionalism has seemed appealing to philosophers for various domains of discourse. These include, in addition to morality, mathematics (Yablo 2000), the mental (Toon 2016) and also mental illness (Pickering 2005; Wilkinson 2022). Fictionalism is, in essence, a coupling of a negative ontological claim, namely, that the stuff talked about doesn't really exist, and a positive linguistic claim, namely, that the way the stuff is talked about is somehow interesting, important, or even indispensable. Varieties of fictionalism then emerge from different fine-grained accounts of this so-called fictional discourse. This importance or indispensability is usually thought of in terms of its practical and theoretical utility.

This leads us on to one central argument for fictionalism known as "The Oracle." This thought experiment appears in Yablo (2000) as part of an argument for mathematical fictionalism. A fully trustworthy oracle comes to you and says, "Psst . . . numbers don't exist." Would you stop talking about numbers, using numbers etc.? Of course not. We need talk of numbers to do a vast array of things. Now suppose an Oracle said "Psst . . . mental illness doesn't exist," would you stop talking in terms of mental illness? One might similarly think that the answer, like in the mathematical case, is also "No," because of the ways in which it is embedded in our social and institutional practices. But this invites the question: "which social and institutional practices?" One might be tempted to respond to this that, among other things, our practices of exculpation, and attributing diminished agency, autonomy and responsibility. And here we get an elaboration and explanation of the intrinsic view presented by Edwards, and that several others, too, have felt tempted by.

4.4.4 *Narratives of illness, responsibility and personhood*

Here is one counter-intuitive implication of fictionalism. Although it is motivated by light-touch metaphysical commitments for the domains of discourse, which are deemed in need of the fictionalist treatment, this implies that they take the robust objective reality of other domains of discourse seriously. In contrast, there are those who are more broadly sceptical about objective ontologies and focus instead on many different discourses and narratives and their power. Nevertheless, the work that is done within these traditions (associated perhaps with so-called continental philosophy – not in the phenomenological tradition, but more in the tradition of postmodernism and social critique) dovetails nicely with fictionalism and the intrinsic view and can be used to embellish it nicely.

On such a view, the relationship between mental illness and moral responsibility is not that some mental illnesses get in the way of the conditions required to be seen as culpable. Not only, one might think, do we not have a good account of what those conditions are; we have no clear picture of what makes something a case of mental illness either. Instead, the story goes, we have all sorts of ways of using various different narratives to shape and control group behaviour by endorsing/rewarding/praising or forbidding/punishing certain ways of behaving. There are informal ways, expressed in our moral reactions, and more codified versions, as expressed in law. Yet another way is through psychiatry and narratives of mental illness. But these are not to be viewed in an objectivist or essentialist light: they are simply another way of reacting to, representing, evaluating people. When we decide to treat rather than punish, say, the troublesome teenager, on account of their difficult upbringing, and give them a diagnosis (it may be "antisocial personality disorder"), we are "not carving nature at its joints." We have not detected some medical fact about this person. We are simply deciding to take, as a society, a particular course of action towards this person, one that we may well evaluate as more humane, sensible and progressive. But it is not "accurate" or "inaccurate." Needless to say, these arguments are more applicable to diagnostic categories where we are dealing with narrow boundaries between medicalising and moralising narratives. In other words, I am talking about cases where the very same person can be seen through the lens of delinquency, naughtiness, poor aptitude, on the one hand, or personality disorder, attention hyperactivity deficit disorder, dyslexia on the other. At what point, for example, is it a primary school child's fault that they can't concentrate? Or at what point is it just a faultless part of who they are? Is this a factual issue? One might be tempted to say that the answer is No.

This issue is deeply embedded in questions of personality and personhood. It is standard in moral philosophy to think of actions as the primary targets of moral evaluations. But we also target persons and personalities. In our moral narratives, there aren't only bad *acts*, but also bad *people*.

According to one way of thinking about the relationship between mental illness and moral responsibility, it is not so much that mental illness interferes with the ways in which responsible *actions* are executed but rather interferes with how *people* can show their true, authentic personalities. This generates similar accounts for examples we saw earlier: the depressed, addicted or delusional individuals are not behaving in accordance with their true selves. The similarities are unsurprising; after all, the accounts of responsibility we examined refer to "Deep Selves" and "Sane Deep Selves." Having said that, the focus on the person rather than the action, rather than offering a conflicting account, casts things in a different light.

To see this, let's return to Example 3, namely, Foetal Alcohol Spectrum Disorder (FASD). As Carel (2016) points out, the symptoms of FASD look just like character flaws, just the sorts of things that we talk about in the context of negative personality evaluations. However, since we know that these characteristics are the result of alcohol poisoning while they were in utero, we are encouraged to temper, or even eliminate, our moral disapproval and adopt a compassionate and helpful stance. In other words, it is not their fault that they are like this. Indeed, it is somebody else's: this was done to them.

This intuition seems solid, but it raises some very thorny issues indeed. As we've already mentioned, there are all manner of things that we have no control over and that aren't our fault (our genes, education, upbringing, etc.). Surely we don't exculpate on the basis of these things, otherwise we might never negatively morally evaluate anyone ever again! No, are we not rather tempted to say that these factors are best thought of as causal contributors to who you are and have become, rather than factors that interfere with your true self? This is really how we see things, and this is why we don't typically allow people to protest: "Don't blame me; it's not my fault that I'm a bad person!"

So why exactly are our intuitions about FASD (and, for example, traumatic childhoods) different? Perhaps the more fruitful question to ask is not so much: "are these intuitions justified?" But rather: "where do they come from?" A number of things feed into this. First of all, the person's "symptoms" have to be severe enough to be noteworthy, to be problematic in daily life. Somewhat counter-intuitively, the more problematic someone's personality is, the more likely we are to cross a threshold, at least partially, out of a moralising sphere. Then there are the narratives about whether the person could have done otherwise. We are not talking about metaphysical determinism here. In a more metaphysically soft sense, there are various things that (whether determinism is true or not) we have no control over (genetics, upbringing). However, our psychological tendency to exculpate the individual gets stronger the more that can be attributed to another's agency. It is not a parent's fault that they pass on their genes to their children, but it is (or so the narrative goes) that they drink while pregnant or abuse their children. In those settings, we are presented with vivid contrast cases. What if this person's mother hadn't drunk heavily while pregnant? What if this person's father hadn't brutally abused him? We think about a less dysfunctional, less problematic human being, and we feel compassion towards the person who could have been so different in a nearby possible world, had things turned out differently, and had their parents acted differently. This compassionate stance is also compounded by imaginative episodes. We imagine the poor, helpless, poisoned foetus. We imagine the vulnerable, impressionable toddler. The question "What if this person had different genes?" does not present us with the same contrast-cases. Indeed, we may be tempted to say: "well, he wouldn't be a better version of himself: he wouldn't be himself at all!"

This brings us on to another narrative landscape, namely, the narratives of personhood and identity. The simplest way of presenting this issue is via generalised anecdote. Imagine you are a loving, conscientious, engaged parent. Your son, who has always been happy and has flourished, had lots of friends, has a nice partner, likes sport and music and does well at school. Suddenly, your son, who is now 17 years old, starts to act strangely. First, he finds various basic things profoundly significant and meaningful. Then, he starts to become paranoid, and, over a few weeks, he becomes floridly psychotic. Whether there is something you can point to is irrelevant: your son has been afflicted by something, and you want him cured, you want him back. The same applies to a descent into depression, or addiction, or an anxiety disorder, etc. What these all have in common (at least in the narratives, if not in the objective reality of the matter, whatever that may be) is the notion that there is the person, the intact person that you know and love, and outside of them, but worryingly close *to* them, is a condition that is afflicting them, that they need to shake off, be rid off. You want them back. And often they themselves want to be back to their true selves.

It is against this backdrop that it is helpful to think about conditions that feature in the DSM, but that aren't like this. One such condition that has drawn a significant amount of controversy, often through the lens of the neurodiversity movement, is so-called "Autism Spectrum Disorder." Central to the claim of neurodiversity, or simply diversity, rather than disorder, is that people with autism are not ill, not disordered, but simply *different*. There is a spectrum of human diversity, and they are somewhere on that spectrum. It is a matter of *mere* difference (and some would say even *good* difference) rather than *bad* difference. This way of looking at the issue focuses on the extent of difference and whether a valence can be placed on it. However, there is a different way of looking at it, and that is, as we have been discussing here, about notions of personhood and identity. This goes as follows. When, for example, a parent of someone with autism insists that their child isn't ill or disordered, they are not only (sometimes not even) drawing attention to the extent of the child's difference of their functionality, but also talking about the fact that, whatever the extent of the difference (and the extent to which it may present challenges), it is a part of *who the child is*. Unlike the descent into depression or psychosis, there is no coherent sense in which the child can be cured of their autism. Indeed, the mother may insist that, on the contrary, they wouldn't get their child back by getting rid of their autism, they would get rid of an integral part of their child. Note that (unlike the claim based on the extent and nature of the difference) such a claim can be made in the face of quite severe autism (conversely, such a claim might be withheld or disputed in the face of relatively mild presentations).[3]

Summary

- A starting point for debates in this area is that mental illness, at least sometimes, reduces an individual's capacity for moral responsibility.
- The relationship between mental illness and moral responsibility can be seen as intrinsic (namely, part of the concept of mental illness, or the act of attributing mental illness, involves exculpation) or incidental (namely, some of the things that we call mental illness *happen* to interfere with the prerequisites for moral responsibility).
- There are a number of different views of what these requirements of moral responsibility are. They all have problems, which might make one think that the intrinsic view is on the right track.
- The intrinsic view encourages us to think about mental illness and what the attribution of mental illness does.
- It also encourages us to reflect on the boundaries of personality and personhood: when is somebody's behavioural profile a part of who they are (and hence fair game for ethical judgement) and when is it the product of something that afflicts them?

Questions

- Can you think of any cases where someone might behave in a way that, on the surface, appears to be morally reprehensible, but which is in fact excusable in light of knowing that the person has mental health issues?
- Can you explain why you might be tempted to excuse this person? Does the rationale make it a matter of degree or is it all-or-nothing?
- Is there something special about mental illness in terms of its power to exculpate? If there is, what is it? If not, what else has similar properties?
- When is something that looks like a personality trait in fact a symptom of a mental health condition?

Notes

1. In UK criminal law, a judgement of "diminished responsibility" reduces a sentence from murder to manslaughter.

2. Indeed, Wakefield's (2009) critique of Edwards (2009) highlights this implausible coincidence. This is unsurprising given that Wakefield is a (moderate) objectivist, and Edwards is a values-based theorist.
3. It is worth noting that a relatively small subset of individuals develop typically in early childhood (typically 15–30 months) and then have an onset of autism, at times very "low-functioning." In this kind of "regressive autism" (or "setback-type autism") (Stefanatos 2008), the child may acquire language and a typical degree of sociality, which they then go on to lose. I bring up this example to show how sometimes part of the biographical trajectory makes diversity narratives more or less fitting. In this case, one can more easily imagine a parent wanting to cure their child, rather than simply accepting that this is part of human diversity.

Further reading

Broome, M., Bortolotti, L. & Mameli, M. 2010. Moral Responsibility and Mental Illness: A Case Study. *Cambridge Quarterly of Healthcare Ethics* 2 (19):179–87.

Ciurria, M. 2014. Moral Responsibility and Mental Health: Applying the Standard of the Reasonable Person. *Philosophy, Psychiatry, and Psychology* 21 (1):1–12.

Perring, C. 2009. The Place of Moral Responsibility and Mental Illness. *American Journal of Bioethics* 9 (9):32–3.

Pickard, H. 2013. Responsibility Without Blame: Philosophical Reflections on Clinical Practice. In B. Fulford, M. Davies, R. Gipps, G. Graham, J. Sadler, G. Stanghellini & T. Thornton (eds.), *The Oxford Handbook of Philosophy and Psychiatry*, Oxford: Oxford University Press.

5 Religion, culture and pathology

Overview

What is the influence of culture on psychiatric phenomena? This can be seen as dividing into two separate questions. First, what is the influence of culture on the phenomena themselves? Second, what is the influence of culture on how the phenomena are received, evaluated and judged by a particular community? One interesting place to start examining these questions is in the relationship between mental illness and religion, which is a widespread, stable and yet extremely varied cultural phenomenon (indeed one might think of religion as the cultural phenomenon *par excellence*).

One noteworthy appearance that religion makes in psychiatry is in the well-known but controversial exemption of religious belief from delusional status in the DSM. There are two ways of understanding the rationale behind this exemption. One is to do with the exemption of widespread belief, of which religious belief is a canonical example (please excuse the pun). This is the explicitly stated rationale in the DSM. However, there is another path to the exemption of religion, which is not about how widespread beliefs are, but about a certain special status that religious phenomena, in particular, seem to have. The special status can be seen from not only a values-based perspective (that focuses on the role of religion in well-being and meaning-making), but also an objectivist

DOI: 10.4324/9781315450124-6

one (that focuses on biological adaptivity). Finally, we ask ourselves about situations where mental illness has religious themes and conclude with general reflections on the relationship between cultural norms and mental illness.

5.1 The controversial exemption of religious belief

What is the influence of culture on psychiatry? A good place to start is with the controversial exemption of religious belief from delusional status. I begin with an excellent illustration of this issue by Ross and MacKay (2017).

> Some people have rather striking beliefs. Consider the following examples:
>
> (1) Arnold believes he has two heads, the second of which belongs to his late wife's gynaecologist.
> (2) Simone believes that each Sunday she drinks the blood of a long-since-murdered man whose mother was a virgin and whose father was the creator of the universe.
>
> Arnold suffers from a somatic delusion (a singular condition labeled "perceptual delusional bicephaly"; see Ames 1984). He is hospitalized and prescribed antipsychotic medication. Simone, in contrast, escapes medical attention and lives a happy and fulfilling life. Why should this be? After all, both Arnold and Simone hold beliefs that, on the face of it at least, are bizarrely implausible, violating widely established biological and physical principles.
>
> (Ross and MacKay 2017, p. 316)

One relatively uninteresting answer to the question of why there is this disparity between our treatment to these two beliefs is "because the DSM says so." More specifically, if you look at the definition of delusion in the glossary of the DSM-5, you will find the following:

> A false belief based on incorrect inference about external reality that is firmly held despite what almost everyone else believes and despite what constitutes incontrovertible and obvious proof or evidence to the contrary. *The belief is not ordinarily accepted by other members of the person's culture or subculture (e.g., it is not an article of religious faith).*
>
> (APA 2013, p. 819, my emphasis)

The exemption expressed by this final sentence is highly controversial. It directly contradicts, for example, the sentiment behind the title of Richard Dawkins's book, *The God Delusion* (Dawkins 2006). Several theorists side with Dawkins in finding this exemption unjustified. Among other things, they argue, it is a cowardly and *ad hoc* attempt to avoid hurting the feelings of religious believers, who constitute a very substantial section of the human population. But this is not a good reason for letting religious belief off the hook. Truth, they argue, should not be swayed by such diplomatic and cowardly concerns! Furthermore, they will continue, religious belief is a significant source of conflict and harm in the world.

Note, however, that if you look closely, you will see that the exemption is not *directly* about religious faith. It is simply about the idea that beliefs that are sufficiently widely held, *of which one canonical example* is religious belief, should be exempt from being attributed to delusional status.

Is this idea right? Is the psychiatric exemption of widely held belief justified? As Ross and MacKay (2017) point out, there are two problems with this broader exemption.

> First, it is too coarse-grained – it excludes *all* beliefs (irrespective of content) that a sufficiently large number of people share; thus eliminating, by definitional fiat, the possibility of group delusions. Second, this cultural exemption is completely atheoretical – no account is provided of *why* a belief being widely shared counts against it being a delusion.
>
> (p. 317)

Related to the first problem is the question: "how widely held, is widely held *enough*?" There are widely accepted instances of delusions (or delusional sets of beliefs or worldviews) that are shared by two individuals or indeed an entire family (Berrios and Marková 2015). The term for this is *folie a deux*, but the "deux" should not be taken too literally: it can be more than two people. But where do we draw the line? What about a cult of 50 followers? What about a subculture of 2,000 adherents? This first problem, of where to draw the line, is hard to address but note that a solution to the second problem, namely, the one of providing an overarching theoretical framework, could help provide a solution.

It is with this in mind that we turn to the idea that beliefs are in a variety of ways "better" if they are held by a large number of people. Is there, as it were, "safety in numbers" of an epistemic, practical or even biomedical nature?

5.2 Safety in numbers

In Chapter 9, we examine delusion in greater depth, and I don't want to overlap too much with the content of that chapter. Nevertheless, as we will see, in defining delusion, there are three separate kinds of consideration that don't always pull in the same direction. One kind of consideration is epistemic: how plausible, evidentially supported, flexible, etc., is the belief in question? Another kind of consideration is practical: how harmful is the belief to the believer and to those who surround the believer? A third kind of consideration is what we might call a biomedical consideration: is the belief an indication of underlying pathology/

dysfunction (see, e.g. Miyazono 2015)? Let us go through each of these considerations and reflect on their impact for widely held belief *per se*, namely, not specifically religious belief. (Here is one crucial caveat: the mainstream assumption that delusion is, by definition, a medical/pathological term goes unquestioned here. This is not so in the chapter on delusion.)

5.2.1 Epistemic considerations for widely held belief

In traditional epistemology, there is a focus on the sorts of evidence that people have acquired by themselves, as individuals. Thus, we might say that I have evidence for believing that my neighbour is at home because I can see that his car is parked in his drive. I have acquired that evidence by myself, with my own eyes, so to speak. However, a great deal of our reasons for belief, our evidence and indeed our knowledge is received through *testimony*, namely, what people tell us. Although epistemologists working on testimony (e.g. Fricker 1987; Pritchard 2004) acknowledge and theorise about the epistemic vulnerabilities of testimony, they also readily accept that there are good cases where testimony provides justification and even knowledge. In general, testimony can provide at least some form of evidence. The epistemologist is then going to be interested in precisely *how* to think about this evidence (e.g. the conditions under which it *is* high quality or *known* to be high quality etc.).

We're not primarily doing epistemology here, but one feature that in general provides (albeit defeasible) support is the amount of consistent testimony you are receiving. In other words, if everybody is telling you the same thing, that is, in general, good evidence for believing that thing. More to the point, even though there might be elaborate and well-orchestrated group deception, psychologically speaking, you are more likely to believe something the more people tell you that same thing. To the extent that people tend to share, implicitly or explicitly, their beliefs, the more people around you believe something, the more likely it is that you will believe it too, and, all other things being equal, the more likely it is to carry evidential value. The "all other things being equal" clause, coupled with the statement of likelihood rather than certainty, is vitally important since we wouldn't want to rule out mass falsehood or irrationality. Indeed, if we think of distributed, social epistemology (Goldman and Whitcomb 2010), we can also reflect on some less-than-ideal features of certain beliefs that spread widely. We might even theorise about an epidemiology of belief. What kinds of beliefs spread well, which less so and is it always their accuracy that makes them spread well? The answer is: clearly not (see the highly relevant but differently cast "epidemiology of representations" of Dan Sperber (Sperber 1985)). Sometimes, the features that make a belief spread well are the very same features that make it epistemically bad (e.g. it is wacky enough to stick in the mind, or to be used as an in-grouping/out-grouping device).

Moving from the epistemic to the more psychological and cognitive scientific literature, there is a very useful and relevant framework for thinking about these issues in general, and religious belief in particular, namely, that of *epistemic vigilance* (Sperber et al. 2010) (although called "epistemic," it's about the psychology of information-management, rather than epistemic in the traditional philosophical sense, meaning concerning knowledge).

Even if we accept that testimony can yield knowledge (for now, just think of knowledge as high-quality information), there are clearly bad cases where it misleads. In pure cases of testimony, where the testimonial information cannot be independently verified or corroborated, there are two main threats at two extremes of a spectrum: excessive gullibility on the one hand and excessive suspicion on the other. The former will lead you to believing falsehoods, the latter to failing to believe truths. So the trick is to strike the right balance. Striking the right balance, namely, adopting the right degree of suspicion is what Sperber and colleagues call *epistemic vigilance*. A core aspect of their account is the separation of two quite distinct forms of vigilance, namely, *content vigilance* and *source vigilance*.

Content vigilance is about being vigilant about the content of what you've been told. It's about reflecting on the plausibility of what has been said. To do this, you may end up reflecting on whether it's consistent with other things you know about the world, for example. Source vigilance is about being vigilant, not so much to what you have been told, but about where it comes from. The two forms of vigilance can interact, for example if someone very trustworthy tells you something quite implausible, your source vigilance might trump your content vigilance, and you may end up believing them (perhaps adjusting pre-existing beliefs in the process). Conversely, someone you don't trust might tell you something relatively plausible, but you don't believe it because of where is comes from. Considerations that feed into source vigilance include all manner of things. What motives might this person have? What is their track record in truth-telling? Sometimes, it's as simple as: do I respect or trust this person?

There are, however, contexts where the two forms of vigilance don't interact, but rather where the person has no choice but to rely heavily or entirely on source vigilance. One such context is developmental: young children will typically trust their caregivers, and it is beneficial to do so. Note that not only is this relevant for learning about the *world*, but, more basically, it also may have to do with learning *words*. Thus, one hypothesis about how a child might learn a linguistic item is to trust that what is being said is the truth and then work out what the utterance heard must mean given the assumption that it is true in particular contexts. In other words, the utterance is believed *before* it is fully understood.

A very different context where content vigilance might be blocked and the person understandably relies on source vigilance instead is in the context of religion. Articles of religious faith are often unverifiable, or incomprehensible, and hence adherence to them is based on *source vigilance*. Notice, though, that source vigilance is, in itself, a perfectly sensible thing to rely on. Of course, we can reflect on the sorts of social forces at play, emotional pressures of in-grouping, power structures in religious institutions, etc. This leads us onto the next section.

5.2.2 Practical considerations for widely held belief

Talk of "beliefs" is ambiguous between something basic, primitive and action-guiding, on the one hand, and the sorts of things that we express verbally and which we take to be of great personal and cultural significance, on the other. One example of the former is my belief that there is coffee in the mug in front of me. This belief comes into play when, wanting a sip of coffee, I sip from the mug in front of me. One example of the latter is my belief in God or in the intrinsic value of democracy. This belief doesn't so much guide everyday actions about objects in my immediate vicinity, but it may shape more long-term plans and goals and be part of my self-conception.

Both kinds of beliefs, in a sense, involve taking the world to be a certain way, but the latter play a socially binding role for two reasons. First, they are the sorts of things that are verbalised, partly because they are about distal things and partly because they are non-obvious, unlike "I believe this is a mug of coffee." If I started saying that in a context where it is blindingly obvious, people would think I had gone mad (see Wittgentstein's comment "This fellow isn't insane. We are only doing philosophy" (467, *On Certainty* (Wittgenstein 1969))). Second, they are not universally shared. We all believe things like the world won't open up under our feet, and the sun will rise tomorrow. But beliefs in culturally specific things function to in-group and out-group: to cluster human beings together, to forge allegiances. Indeed, as some have argued, this means that there are social pressures that may push our beliefs away from constraints of truth. As Dan Williams puts it:

> Because other agents have reliable access to what we believe and frequently reward us when we hold ungrounded beliefs and punish us when we hold reasonable ones, this creates powerful incentives for otherwise rational individuals to form beliefs in ways that are sensitive to such social rewards and punishments.
>
> (Williams 2021, p. 334)

The practical, social utility of shared belief means that pressures to share belief may lead us to believe things that aren't true, plausible or even coherent. However, one can easily argue that this is adaptive. What is maladaptive for animals that are as social, and socially dependent, as ourselves, is the holding of beliefs that no one else has, beliefs that isolate. Indeed, a belief might be reasonable, plausible and even true, but may be maladaptive compared to a widely held belief that is false.

5.2.3 Biomedical considerations for widely held belief

If we are naturally designed to align our beliefs with others, then, one might argue, someone being isolated in the beliefs they hold – being a "doxastic lone wolf" – might be a sign of underlying pathology. This perspective would make widely held belief, in general, "healthy" and the exact opposite, namely, esoteric beliefs held by one or very few people, "unhealthy."

The strength of this relationship would depend on a number of things. Most obviously, it depends on our view of pathology. If we think that pathology necessarily involves breaking certain kinds of statistical norms, then being the statistical outlier might in and of itself be a sign of pathology. However, this rules out a hypothetical scenario where an individual is made privy to the truth and stands by that truth independently of what all others think and yet is not somehow pathological. It seems that, by stipulation, we are tempted to say that this brave and lonely epistemic pioneer is not pathological.

More plausible perhaps is that such divergence from the crowd, or at least from some peers, is a potential sign of underlying dysfunction. In particular, we can hypothesise that certain general traits, such as grandiosity, paranoia and emotional dysregulation, are the sorts of things that we might call pathological, and these might then give rise to beliefs that do not align with those of others. Here, it is not so much the beliefs themselves that are pathological, but the mental processes and propensities that give rise to them.

One thing worth mentioning at this point is that being alone in your belief, and struggling to get others to agree with you, might be a pretty good (although not certain) indicator that something is "wrong" with you. How this is conceived and whether it is necessarily to do with pathology and how we think of pathology are tricky and multifaceted issues. We will look at this in more detail shortly. Suffice it to say for now that when someone is very much out on a limb in terms of their beliefs, this tends to ring alarm bells with psychiatrists and social workers.

5.2.4 *Taking stock: the intelligence of the masses?*

We have just examined three questions, namely, does aligning your beliefs with the masses tend to (a) make them more likely to be true (an epistemic good), (b) confer practical benefits, such as group cohesion and (c) indicate that you are (psychiatrically/neurologically) "healthy"? The answers to these questions are not clear cut. Roughly, however, we can say that following. With regard to (a), in general, it is epistemically a good idea to align your beliefs with what most people believe. There will be exceptions, for example if you are a scientific pioneer, or if something either false (or incomprehensible) spreads throughout your community, for reasons that may relate to the next point. With regard to (b), beliefs don't only serve a purely informational function, but they serve a social one too: we are drawn to, and bond with, those who believe the things that we do. And in contrast, we are repelled by those who have radically different beliefs, outlooks, worldviews. With regard to (c), and in a way that relates to (a) and (b), although the details will be important, if someone has radically different and extreme views, this is likely to be seen as a potential indicator that something is wrong with them, perhaps specifically medically wrong (e.g. they might have "schizophrenia" or a brain tumour).

5.3 Is religion special?

In the previous section, we examined why, all other things being equal, widely held belief is at least *more likely* to be "good" across three dimensions (namely, epistemic, practical, biomedical) than beliefs held in isolation from others. However, are there grounds for exempting religious belief specifically that are not simply inherited from the fact that they are widely held? This is of course not to say that religious content *always* exempts, but just that it sometimes does, and it might be interesting to reflect on when and why it might. In particular, is there something about the content or nature of religious beliefs *per se* that makes them less pathological? There are two very different angles from which we might approach this and end up answering, broadly: "yes." These two angles depend on whether one is a values-based theorist or an objectivist (see Chapter 2). The former will focus on whether religious beliefs are, roughly, life enhancing for those who have them. The latter will focus on whether they are biologically adaptive. We take these in turn.

First, however, it is worth noting that, if we are to examine religion and whether there is anything special about its relationship to pathology, in the interest of completeness, we shouldn't just limit ourselves to religious belief. Yes, we started by looking at the exemption of religious belief from the status of delusion, but there are in fact two quite different religious phenomena (aspects of religious mental life, as it were) that might be liable to being evaluated on the basis of canonical psychiatric norms, namely, not just religious beliefs, but religious *experiences* too. Note that these would roughly correspond to the two distinct psychotic symptoms of delusions and hallucinations. It's somewhat oversimplified, but the issue of the psychiatric status of religion therefore bifurcates into the issue of (i) whether (or to what extent) religious beliefs are delusional and (ii) whether (or to what extent) religious experiences are hallucinatory (or more broadly psychotic or pathological).

Finally, before moving on, it is important to clarify terminology. In particular, two things that are related but importantly different are the religious and the spiritual. Different theorists take these terms to express subtly different concepts, but I think a useful way to think of them is as follows. Religion is a public and cultural phenomenon. It might be adhered to by a large number of people or it might be a small minority. It is, however, anchored in public things, for example (but not exclusively) rituals, practices, sacred texts, relics, icons, chants, liturgy and doctrine. The spiritual, on the other hand, is a private engagement with the sublime, the profound, the non-material. You can have religious beliefs, but you may have never had a spiritual experience in your life (and also not consider yourself "open" to them, not consider yourself a "spiritual person"). You may notice how the term "spiritual belief" sounds a bit unusual, precisely, perhaps, because beliefs are expressed in the public sphere of language, and the spiritual is private and experiential. Conversely, you might have a spiritual experience that you do not anchor in any particular religion, or in religion at all. However, the two come together where a spiritual experience is interpreted through the lens of religion. In this case, we have religious experience, for example, a vision of the Virgin Mary. All three categories, religious belief, spiritual experience, and religious experience, are of interest to us here.

5.3.1 A values-based approach: religion as life-enhancing

In a now-classic paper, Fulford and Jackson (1997) present three case studies that are intended to illustrate some of the difficulties in delineating psychotic illness and what they call "benign spiritual experience." The case studies themselves are lengthy and detailed, and I won't present them in full here. Suffice it to say that they involve quite florid disconnections from reality in the context of major adverse life events. In a later discussion of that same paper, the authors (Jackson and Fulford 2002) summarise their 1997 paper as follows.

> On the basis of these case studies, we argued that:
>
> 1. spiritual experiences, which are essentially benign, can involve explicitly psychotic phenomenology;
> 2. traditional psychiatric diagnostic methods, which focus on the form and content of beliefs and experiences, are not capable of distinguishing between pathological and nonpathological forms of such psychotic experiences;
> 3. evaluation of such experiences (and claims about "illness" and health more generally) necessarily involves value judgments as well as judgments about matters of fact (specifically, the "facts" of traditional descriptive psychopathology) and
> 4. more specifically, the distinction between pathological and non-pathological phenomena concerns the way in which they are embedded in the structure of the values and beliefs by which the actions of the subjects concerned are defined: "In the case of pathological psychotic phenomena," we argued, "there is a radical failure of action. . . . In the case of spiritual psychotic phenomena, action is radically enhanced" (Jackson and Fulford 1997, 55).
>
> We then proposed a cognitive problem-solving model of psychotic experiences. This model suggests that such experiences can be a normal and adaptive psychological response to existential crises, such as loss of meaning or purpose in life, coming to terms with death or bereavement, and so on (Batson and Ventis 1982; Jackson 1991, 1997, 2001).
>
> (Jackson and Fulford 2002, p. 388)

The central take-home message is that a religious belief arising from a spiritual experience, which may itself be caused by what an objectivist might consider to be an underlying pathology, can be a benign or even beneficial influence on an individual's life. As such, it should not be treated as an illness, or symptom thereof, and indeed, any attempt on the part of a clinician to make it go away would be misguided and potentially negatively impact on the individual's life. By stipulation, what we have here is something that the objectivist would call pathological and the values-based theorist would not.

Before moving on to the objectivist position, we might ask: what is it about religion that seems to give it this life-enhancing role? It's almost a well-worn trope: the criminal, the drug-addict, the alcoholic, who "finds God" and who gets their life back on track. The power of religion for positive personal development is time and again demonstrated. For example Moreida-Almeida, Neto and Koenig (2006) present a large review of 850 studies examining the impact of religion on mental health. They report significant positive impact on psychological well-being, depression, drug abuse and suicide. This should come as no surprise, given that the pursuit of meaning and comfort is a major motivation behind religious engagement. Although it was intended as political critique, Marx famously remarked that religion was the opiate of the people. He may have intended that as a reference to stupefying and demotivating effects of opiates, but, viewed from another more positive angle, they alleviate pain and suffering. In short, there is little doubt that engaging seriously with religious narratives can bring major benefits to psychological well-being.

We will examine this shortly in more detail, but let's for now look at how the objectivist might view religion as special.

5.3.2 *An objectivist approach: the cognitive science of religion*

One thing worth noting at the outset is that the objectivist is, broadly speaking, unlikely to allow exemption from pathology for spiritual *experiences*, many of which would be considered hallucinatory, or quasi-hallucinatory (namely, experiences that aren't fully sensory – more on this in Chapter 8), and hence cast aside at the outset as hardly exemplifying of what human cognition is supposed to do. Religious beliefs, practices and rituals are therefore our focus here.

As we saw in Chapter 2, objectivism about illness and pathology tends to derive its objective norms from the norms of biological proper function, which, roughly speaking, involves thinking about whether an organism is behaving as it should, where this is usually thought of in terms of adaptivity, fitness promotion, etc. So the question arises in this context: is religion natural/adaptive/fitness promoting, etc.?

There is a large body of literature in the cognitive science of religion (CSR), which is directly relevant to this question. CSR (see Barrett 2007 for a review) aims to understand religion in general and certain features of religion (e.g. cross-cultural similarities), in terms of natural (and properly functioning) aspects of human cognition. In other words, CSR asks two related questions. First, given what we know about human cognition, does religion look the way we would expect it to? Second, given what religion looks like, does this tell us anything interesting about human cognition? The answer to both of these questions tends to be a resounding: "yes." Religious phenomena can be fruitfully understood and explained as the expressions of perfectly natural and adaptive aspects of human cognition.

Much of the work in CSR examines specific features of religion and tries to explain them in natural terms. This is interesting but peripheral to (or rather downstream from) our line of discussion. We are interested in whether religion itself (its very existence) is "natural." If it is, then, by the lights of most forms of objectivism, it can't be pathological, at least not *by its very nature*. To reiterate, this is not to say that there can't be instances where religious beliefs or experiences or behaviours are signs of underlying pathology. This simply undermines the *global* claim (held by some, who often take themselves to be the torchbearers for science) that religion is, by its nature, pathological.

Two places where cognitive scientists tend to look to see if a phenomenon is "natural" is first in evolution and second in development. Applied to religion, the evolutionary question becomes: did religion evolve naturally? If it did, it presumably did so due to being adaptive. The developmental question is roughly: are we natural-born theists?

Without going off on too much of a tangent, evolutionary arguments either claim that religious belief and religious ritual are directly adaptive because they facilitate group living and pro-sociality (e.g. Norenzayan et al. 2016) or they claim that religion is a "spandrel" (Gould and Lewontin 1979), namely, a perfectly natural by-product, which is not itself adaptive, but which is the result of processes and mechanisms that are (e.g. hyperactive agency detection (Guthrie et al. 1980)).

In a related manner, developmental psychologists like Deborah Kelemen (2004) and Paul Bloom (2007) claim that we are naturally geared towards religious thinking from a young age. For example children have a strong bias towards seeing the natural world as purposeful in ways that go far beyond what even a very religious adult would think or teach (Kelemen 2004). For example children are inclined to say that rocks are pointy because it prevents them from being sat on, not because of some brute physical process.

Bloom (2007) is particularly interested in the question of the *innateness* of religion, where, roughly speaking, an innate characteristic of an organism is to be contrasted with a characteristic that is a relatively short-term (i.e. at most within the timeframe of that particular organism's life) response to the organism's environment. Starting from the twin observations that religion in general is universal in human cultures, and that there are universals among the diversity of religions, he distinguishes *universality* from innateness. Something might be universal simply because it arises in response to something that all humans face. This would be universal, but not innate, since it would simply reflect a universal environmental influence. To illustrate something similar, Bloom uses the example that every language has a word for hands, but that's because the ability to talk about hands is important, not because hand naming is innate. Religion might arise universally in a similar way, in response to something that all human cultures face, for example making sense of existence, coping with mortality, securing pro-social public morality and so on. This is a plausible, indeed almost obvious claim. But Bloom is making the far stronger (and more interesting, but less plausible) claim that religion doesn't just emerge as an adaptive feature of cultural evolution. It is innate. But what does this mean more precisely?

So far, we have characterised innateness *negatively*, in terms of what it is not, namely, it is not a life-course adaptation in response to the environment. In this sense, "innate" is often opposed to "acquired." But what is innateness more *positively*? This is one of the thorniest issues in philosophy of biology and evolutionary theory, and I cannot do justice to it here. However, it is important to distinguish innateness in biology broadly construed, where we might include brute physiology (that a human being has five fingers on each hand is innate but that those same hands have callouses is not) from innateness in the sciences of behaviour and cognition. I don't want to stray too far afield, but the focus on behaviour and cognition adds a different kind of complexity, which comes from the fact that few if any actual behaviours are innate in the trivial sense that they are all, or almost all, responses to the organism's environment. This is not where the action is in this debate. This can most clearly be seen in the hypothesis about the innateness of language associated with Noam Chomsky (1965). The claim is not that children are born with the ability to speak, but rather that they are born with the capacity to acquire language. But what then becomes interesting about this claim? After all, children are born with the capacity to acquire any number of behaviours, playing tennis for example. But no one claims that tennis is innate. So what is special about language? The point about language is that children have (according to Chomsky) a particular facility with acquiring language and do so far more quickly than you would expect if they were learning it in an all-purpose way. This is analogous to the nature of Bloom's claim. There is something special about the acquisition of religion and religious cognition. We are as human beings especially receptive to the acquisition of religious concepts and religious thinking.

There is much more to be said, but we cannot digress too far. Suffice it to say that anyone who wants to claim that religion is, by its very nature, pathological, where this requires it to be somehow unnatural or maladaptive, is not going to have an easy or obvious task. In short, there is a wealth of literature that supports or indeed successfully trades on the fact that religion is natural. As a result, a blanket insistence that religion is pathological does not seem very defensible, at least not on objectivist, biological grounds.

5.4 The challenge of cultural sensitivity

As already mentioned, when we talk about the exemption of religion from pathological status, we mean that it can enter as a supplementary consideration. The claim is not that any given belief, experience or behaviour that is thematically religious is exempt. It is widely accepted, and indeed entirely unsurprising, that cultural context plays a role in colouring in the thematic content of psychiatric phenomena, and especially psychoses, and so there are bound to be psychiatric phenomena with religious themes. This is partly because the experiential extremes of psychiatric states lend themselves to the thematic extremes of religious framing. Even the most fervent atheist may find themselves turning to God when things are especially bad, and you may feel a special relationship with God when you feel especially good. But this is also partly just a prominent example of a general tendency of human experience to be shaped by cultural context and constraint.

One obvious (almost trivial) constraint is technological. Paranoid delusions in the early twentieth century featured radio waves; they were hardly going to feature Wi-Fi or Bluetooth, as they do today (e.g. Mason, Stevenson and Freedman 2014). Current societal or national concerns can play a role, too: delusions about Russian spies will have been far more common in the United States during the Cold War and so on. It is not just beliefs, but experiences, too, that change with cultural context.

In light of the general claim that cultural context influences how psychotic experiences are experienced and interpreted, as well as the beliefs that come to surround them, it is unsurprising that many cases of psychosis will be experienced and interpreted through a religious lens. This presents a basic practical challenge. As a clinician, how do I know whether someone needs psychiatric help or is simply expressing eccentric religious convictions or otherwise is showing culturally specific experiences or behaviours? How and to what extent do we need to be culturally sensitive? The way in which we address this challenge depends on whether we are broadly objectivist about this or values-based.

5.4.1 An objectivist approach

There are two very different ways that one can be an objectivist about this. One we might call *biological objectivism*, whereas the other we might call *sociocultural objectivism.*

Biological objectivism has a simple, but slightly boring, solution to the problem. Is there some kind of underlying abnormality or dysfunction that has given rise to this person's experiences or beliefs? If so, they need psychiatric help. For example they may have the objective, medical brain condition known as "schizophrenia." Alternatively, they may have a tumour in their temporal lobe. Whether or not these experiences or beliefs have religious or otherwise culturally-specific content is neither here nor there, at least not insofar as it decides pathological status. If these are especially striking, they may serve as a red flag, a warning sign that prompts further investigation, but the ultimate arbiter of pathology would be the outcome of those further investigations. The line between eccentricity and pathology is clearly cut on biological grounds.

Socio-cultural objectivism is the kind of objectivism that we presented before, which is prevalent in the cognitive science of religion and thinks of religion in evolutionary terms, where evolution is, quite rightly, taken to encompass cultural evolution as well as biological evolution. On these views, somebody's religious beliefs might be considered pathological if they fly counter to those very features of religion that make it adaptive. It needn't involve biological pathology but rather cultural pathology. We might (to echo Dawkins) talk about maladaptive memes, as analogues to maladaptive genes (Dawkins 1976).

So, for example, religion has been deemed adaptive in part in virtue of its encouragement of pro-social behaviours. It has on balance been fitness-enhancing. Thus, one might say that religious or otherwise culturally specific phenomena (beliefs, rituals, etc.) that isolate individuals and therefore make them vulnerable to harm (and are certainly not fitness-promoting) will be pathological. To see the relationship between neurological and socio-cultural objectivism, notice that, while this might be an indicator of neurological abnormality or impairment, it doesn't necessarily have to be. There might be objectively toxic and maladaptive, yet distributed, cultural phenomena. To use a crude analogy, this would be a case of pathological sociocultural software running on biologically healthy (intact) hardware.

Of course, this socio-cultural objectivist perspective, though it may appeal to many as a welcome corrective to the short-sightedness of biological objectivism, and therefore may appeal in theory, might be difficult to apply in practice. How, exactly, do we pick out maladaptive cultural phenomena? And how can we really disentangle these judgments from values?

Consider a particularly extreme case: the indoctrination, and subsequent actions, of a suicide bomber. This would, one assumes, come out as maladaptive. It certainly curtails the fitness of all unfortunate enough to be involved. But do we really think of the suicide bomber as someone who is psychiatrically unwell? It seems that no one would want to do this. Those who indoctrinated the bomber in the first place wouldn't want to: they think that what he is doing is good. Those who deeply disapprove of the actions of the bomber don't want to, either, for various possible reasons. They may want to be able to apportion responsibility and blame to him and his mentors. From a more theoretical perspective, it just doesn't seem like violent extremism is something to be pathologised and hence deemed in need of a psychiatric solution. It is, as Scott Atran's (2010) book demonstrates, a problem that is best understood through the lens of the social sciences, and our understanding of it, and solutions to it, will come from there too.

5.4.2 A values-based approach

A values-based approach is going to invite the following question: to what extent does a culturally specific belief or experience negatively impact on the life of the individual and those around her, in a way that encourages us to think of them as being in need of psychiatric help?

Here is where the values-based approach comes apart from the socio-cultural objectivist approach (or where the latter can be deemed to collapse into the former). The suicide bomber is certainly negatively impacting on the lives of others, but not in a way that we deem them to be in need of help. On the contrary, our values encourage us to think of what they are doing as abhorrent, rather than in need of correction and certainly not in need of *psychiatric* correction or treatment.

The values-based approach can be seen as a bit of a difficult balancing act. Central to the whole approach is the idea that we cannot avoid bringing our values to bear on our clinical judgments and practices, and yet it also seems to mandate a degree of sensitivity, as a clinician, to the values of others. How far can we make allowances in light of cultural variation? Values-based theorists seem to think that the answer is: as far as we need to. We may not share their perspective, but as long as we understand where they are coming from and it is allowing them to flourish and make sense of their lives, we should not intervene or deem them psychiatrically unwell.

As Jackson and Fulford (1997) highlight, somebody might have florid and unusual experiences and esoteric beliefs, but still be counted as someone who should not be considered psychiatrically unwell. Indeed, we could even imagine a case where a benign brain tumour produces certain beliefs and experiences that make an individual happy and fulfilled. If, as stipulated, the tumour is benign and does not pose a risk to the person's life, the individual is not ill. Suppose, further, that the removal of this tumour removed these positive beliefs and experiences. The values-based approach would insist that we would be wrong to remove the tumour. Although the scenario is far-fetched, it highlights something important about the conceptual landscape. The aim of medicine is to improve people's lives, and medical concepts (and practice) should reflect that.

5.4.3 *The social kindling hypothesis*

This plea for cultural sensitivity in psychiatry is supported by the values-based approach. However, the case for such a plea would be strengthened still further if it wasn't just argued that there are different standards of what is evaluated as normal in different cultures, but also that there are differences in what is (or can be) experienced in different cultures.

In a landmark cross-cultural interview study, Tanya Luhrmann and colleagues (Luhrmann et al. 2015) compared a sample of 20 voice hearers from the United States (San Mateo), India (Chennai) and Ghana (Accra). There were several striking differences, not only in how the voices were interpreted, but also in how they were experienced.

> In general the American sample experienced voices as bombardment and as symptoms of a brain disease caused by genes or trauma. They used diagnostic labels readily. . . . None of the San Mateo sample reported predominantly positive experiences, although half reported some positive dimensions to the voice-hearing.
>
> (Luhrmann et al. 2015, p. 42)

In contrast:

> More than half of the Chennai sample [11 out of 20] heard voices of kin, such as parents, mother-in-law, sister-in-law or sisters. . . . These voices behaved as relatives do: they gave guidance, but they also scolded. They often gave commands to do domestic tasks. Although people did not always like them, they spoke about them as relationships. . . . Only four of the Chennai sample used the term "schizophrenia."
>
> (p. 42)

And again:

> Although many of the Accra participants understood that hearing audible voices could be a sign of psychiatric illness, their social world accepts that there are human-like non-embodied spirits that can talk. 'Voices [are] spirits,' one man explained. Only two people used diagnostic labels (schizophrenia). . . . When people talked about their voices they were likely to emphasise the positive.
>
> (p. 43)

This study gave rise to what the authors call "the social kindling hypothesis," according to which variations in ways of thinking about minds, persons, spirits, etc., shape the way in which certain things are experienced and the very experiences themselves. If the nature of voice hearing can change so much in different cultures, what must voice hearing have been like in medieval Western Europe, when only the tiniest minority would have cast the slightest doubt on the ever-presence of God? What must Joan of Arc's voices have been like?

This adds further weight to the need for cultural sensitivity. Not only must we be aware that what people from different cultures experience or believe are more likely to be accepted as normal in their own culture: they are far more likely to *happen at all* in that culture as well.

5.4.4 Context of formation versus context of evaluation

One important take-home message, it seems to me, is the following. We need to distinguish between *the context of formation* and *the context of evaluation* when we are reflecting on cultural influences on beliefs and behaviours. Some cultural contexts are more likely to give rise to certain phenomena than others. That's *context of formation*. Some cultural contexts are more likely to negatively (or, indeed, positively) evaluate a given phenomenon than others. That's *context of evaluation*.

These two considerations come apart in the following sense. A context that might be conducive to bringing something about (e.g. stress-induced catatonia) is not necessarily more or less accepting of it. As a result, we need to be careful when we speak of "cross-cultural variation" or "culture-bound syndromes." Are we saying that only in a particular culture is a phenomenon brought about, or that only in a particular culture is a phenomenon (that may well appear elsewhere) given the label or status that it is given in that culture (e.g. of pathology or mental illness)?

Two well-known examples of cross-cultural variation, one historical, the other geographical, highlight the difference.

Drapetomania was the name given by the American physician, Samuel A. Cartwright, to the hypothesised mental illness that was supposed to cause slaves to flee captivity. Needless to say, the desire to flee captivity is perfectly normal. There was nothing about the context that generated a particular pathology. Rather, the socially dominant, slave-owning class found it undesirable that their slaves should run away. This is a particularly extreme and deviant variation in *context of evaluation*. Less extreme, but still highly problematic, is the fact that homosexuality was categorised in the DSM until 1973. Another example (which was touched upon by Luhrmann et al. 2015) that is less clearly about social control, but is no less relativistic, is that voice hearing is deemed almost without exception as pathological in Western context, but can be seen as personally meaningful or spiritual in other non-Western cultures.

In stark contrast, there is the vivid example of Paris Syndrome. Paris Syndrome is characterised by a variety of symptoms, including catatonia, anxiety, delusions of persecution, derealisation, depersonalisation and hallucinations. The condition emerges when Japanese tourists (typically young women) go to Paris and are so disappointed by the fact that it is nothing like what they had expected. Paris syndrome would be unwanted in any culture (there is no culture where catatonia, anxiety, psychosis is desirable), but it is a particularity of the Japanese depiction of Paris and the differences between Japanese and French culture that bring about the highly unwanted catatonic state, an extreme form of culture shock. This is clearly about context of formation, not context of evaluation.

However, once we have made this distinction between context of formation and context of evaluation, we also need to be aware that they frequently interact. One way of understanding this interaction is as follows. The more a given culture causally promotes a phenomenon, the more prevalent it is and the more likely it is to be accepted in that culture. For example shamanic cultures causally promote shamanic experiences and also do not pathologise (or otherwise negatively evaluate) them. The direction goes the other way too. If a given experience or belief is accepted (or at least acceptable) in a culture, people in that culture are more likely to have it, since they won't be socially pressured into abandoning it.

Although there is certainly truth to this, there is another more nuanced and interesting way of understanding this interaction. The individuals themselves who have certain beliefs and experiences are intrinsically sensitive to how they are evaluated in the culture of which they are a part. This can result in stresses associated with stigma and self-stigma, and it means that evaluations become self-fulfilling: initially harmless experiences, in being pathologised, then become harmful and unpleasant. Indeed, we see this in Luhrmann et al. 2015 where the respondents who are embedded in cultures that are less pathologising and who view voice hearing, for example, as imbued in spiritual meaning, find their experiences much less threatening and negative.

In summary, there are two ways in which negative evaluation of phenomena (beliefs, experiences, rituals) can impact upon them. According to the one, it can discourage them from being held and reinforced. According to the other, it can aggravate it and make it more negative through anxiety and self-stigma. There are two important consequences of the latter form of impact. First, this motivates an ameliorativist outlook, which is something that I am going to elaborate on later. Second, it emphasises the very concrete practical importance of cultural sensitivity in clinical practice.

Summary

- Since religion is arguably the quintessential cultural phenomenon, and a significant proportion of psychiatric phenomena have religious themes, religion is a useful lens through which to explore the relationship between culture and pathology.
- The fact that religious belief is exempt from delusional status in the DSM is a high-profile controversy but can be justified on a number of grounds.
- First, religious belief might be exempt simply (as the DSM states) because it is an example of widely held beliefs, and it is healthy to align your beliefs with your peers.
- Second, religion may deserve exemption beyond the "safety in numbers" effect: it might be special in a number of ways.
- Values-based theorists might think of religion (and spirituality more generally) as something that has a special capacity to provide meaning and improve the lives of individuals (especially in times of crisis).
- Objectivists, too, might find that religion, viewed as a naturally evolved phenomenon, has a special status as an adaptive facilitator of group cohesion.
- None of this is to say that religious beliefs and experiences are never pathological (or signs of pathology), and we need to be mindful of when psychosis might take on religious themes.
- More generally, this exploration highlights the crucial need to be sensitive to cultural context but, in particular, to distinguish the context of formation from the context of evaluation.

Questions

- When does a religious belief or experience start to look like it might be pathological?
- Under what circumstances is it appropriate to offer psychiatric help to someone who has certain religious beliefs or experiences?

- Why is it important to align your beliefs with others? Are there exceptions?
- Can you think of examples of variations in the context of formation?
- Can you think of examples of variations in the context of evaluation?

Further reading

Carel, H. & Cooper, R. 2010. Introduction: Culture-Bound Syndromes. *Studies in History and Philosophy of Science Part C: Studies in History and Philosophy of Biological and Biomedical Sciences* 41 (4):307–8.

Fulford, K.W.M. & Jackson, M. 1997. Spiritual Experience and Psychopathology. *Philosophy, Psychiatry, and Psychology* 4 (1):41–65.

Luhrmann, T.M., Padmavati, R., Tharoor, H. & Osei, A. 2015. Hearing Voices in Different Cultures: A Social Kindling Hypothesis. *Topics in Cognitive Science* 7 (4):646–63.

Williams, D. 2021. Socially Adaptive Belief. *Mind and Language* 36 (3):333–54.

6 Scientific explanation in psychiatry

Overview

Psychiatric research is an important area of scientific research. How should we think of scientific explanation in psychiatry? To what extent can the cognitive sciences (neuroscience, neurobiology, cognitive psychology, etc.) help to provide us with explanations of psychiatric phenomena? It is clear that they can make major contributions, but what is the nature of, and the constraints governing, such contributions? Why is psychiatry so challenging for scientific explanation? What is explanation, and what does it mean for an explanation to be good? In order to answer these questions, we need to look at the nature of explanation more generally. An explanatory constraint with a long and distinguished history concerns "levels of explanation," but this means different things depending on your view of explanation. A pragmatic view of explanation will distinguish between *levels* of explanation and *kinds* of explanation. This also sheds some much-needed light on two related issues: the notion of explanatory power and the issue of what reductionism is and whether it is desirable.

DOI: 10.4324/9781315450124-7

6.1 The challenge of scientific psychiatry

We explain things to one another every day. I can explain a variety of things to you. I can explain: how to set the timer for the central heating (you press these buttons, in this order), why I think my neighbours are at home (because their car is in the drive), why I grimaced at you (because you annoyed me) or why I took the ice cream out of the freezer during the main course (because I wanted to give it time to soften).

Scientific explanation, at first glance at least, seems different from everyday interpersonal explanations in that it is precise, regimented and generalised: it doesn't *only* apply on a case-by-case basis. Physics and chemistry are paradigms of exact sciences, hard sciences, and explanations that they provide are both precise and generalisable. You can set up experiments again and again and get the same results and make generalisations and extrapolate predictions about as-yet-untried experiments.

To see the challenge of scientific psychiatry, let's imagine that psychiatry is like these exact sciences. I know that something will conduct electricity because it is metal. I know it will be attracted to a magnet because it is one of iron, steel, nickel or cobalt. Ideally the DSM (or indeed any psychiatric classificatory system, not just diagnostic system) would pick out *kinds* in exactly this way. We would know exactly how someone would respond/react/behave in relevant circumstances, in virtue of them having a particular condition. But, of course, this is not how psychiatry works. Mental illness, disorder, ill health, whatever you want to call it, is deeply complicated, and so exceptions, caveats, multiple factors and kinds of factors need to be taken into account (see Chapter 3).

And then there are questions about whether these factors are correlates, causes, or even constitutive, of the condition. Answering these questions is not easy. Psychiatry is messy. Even the most optimistic biological psychiatrist will admit that it is nothing like physics or chemistry. But this observation is not enough. We need to say more about precisely why it is different.

One option is to say that it's simply a matter of complexity. Some hypothetical beings of vastly superior knowledge and understanding (or perhaps even humans in a however-distant future) could make psychiatry an exact science in the relevant sense. Alternatively, one might say that, no, it is part of the very nature of mental disorder that it simply isn't suited to being dealt with like an exact science, however far knowledge and understanding progresses. If one subscribes to the latter view, we have to explain why that is. We also have to ask ourselves what the inexact – yet optimal – science of psychiatry is supposed to look like. What are we aiming for? What constitutes scientific progress in psychiatry?

Suppose we had some workable kinds (we might even call them natural kinds) in psychiatry. What would that look like? Perhaps with a careful subdivision of existing psychiatric categories, we could zoom in on some law-like generalisations. Imagine a predictively perfect system, where we don't understand the complex underlying nature of what we are trying to predict. This happens daily. As Murphy (2010, p. 606) points out,

> we can predict that if Rex is in the kitchen then leaving the roast beef there unattended is a bad idea, as Rex is a dog and dog-meat interactions take a predictable form . . . and we can safely make this prediction despite total ignorance of the evolution, metabolism or olfactory biology of dogs.

Similarly we might say: Sam suffers from an anxiety disorder. If I invite him to a big drinks party full of strangers, he's not going to have a nice time. Is this what we want from scientific explanation in general, and from explanation in psychiatry in particular?

I have raised far more questions than I have answered, but that was the point. I want to show how the very notion of a proper scientific psychiatry is challenging. Things quickly become dizzyingly complex, and we need to make sure that we don't run before we can walk. So let's start with the basics.

6.2 The nature of explanation

A standard distinction in philosophy of science (including philosophy of science as applied to psychiatry) is between *nomological explanations*, namely, explanations that work by pointing out laws or law-like regularities, and *mechanistic explanations*, namely, explanations that work by pointing to mechanisms. However, in my view, we need to start with something even more fundamental. We need to distinguish not only what we take to be doing the explaining (namely, laws versus mechanisms), but also what explanation more fundamentally *is*. What I mean by this will become clear as we progress.

Different accounts of what explanation is fundamentally, the *nature of explanation*, allow different answers to the question of what is doing the explaining. Some prioritise laws and others, mechanisms. Still others allow for explanations that go beyond both laws and mechanisms. As we have seen, it is not simply that psychiatry seems to lack graspable laws: well-understood mechanisms seem elusive too. The extent to which this is bad news, however, depends on one's views about the nature of explanation, so we examine this now.

So let's take a big step back. What is explanation? As Achinstein (1983), among others, has noted, "explanation" (like so many other words that end in "-tion," e.g. "construction") is subject to a "process-product" ambiguity, which is to say that it can refer either to an act of explaining or to the product of such an act. What philosophers of science tend to be interested in is the product. However, as we will see, there are different views about the relationship between the act of explaining and the explanatory product, and the importance of the former when considering the latter.

Different views about the fundamental nature of explanation (as a product) can be illustrated in terms of different answers to the following two questions:

1) What kinds of things are the *relata* in explanations? (viz., when we say that x explains y, what kinds of things are the values for x and y?)
2) What does it take for x to successfully explain y?

There are surely other ways of carving up the territory, but I find it useful following Faye (2007) to distinguish between three kinds of views of explanation, namely: formal-logical, ontological and pragmatic views of explanation. I will now present these in turn. My aim is not to criticise or evaluate them but rather to show that each makes key issues about scientific explanation in psychiatry look rather different.

6.2.1 The formal-logical view

On the formal-logical view, perhaps most famously put forward by Hempel and Oppenheim (1948), an explanation is an abstract entity; in particular, it is a logically valid argument. A valid argument in Hempel and Oppenheim's view is a series of propositions, structured bearers of truth and falsity that stand in logical, and inferential, relationships to one another. Without going into too much detail into the philosophy of logic and language, propositions are (roughly) what declarative sentences (roughly, well-formed statements that can be true or false) express. Crucially, propositions are not sentences themselves, since different sentences can express the same proposition and vice versa. Obvious examples include sentences as uttered by different people or in different contexts. Thus, if I say, "I have brown hair" and you say to me, "You have brown hair," we are expressing the same proposition (the same state of affairs makes our assertions true, namely, the state of affairs where SW has brown hair). Conversely, if we both say "I have brown hair," we are expressing different propositions (mine concerns me, yours concerns you).

An *explanandum* or the thing to be explained, according to Hempel and Oppenheim, is a proposition that follows *deductively* from an *explanans* (the thing that does the explaining). To use a simple example, suppose you want to explain the proposition expressed by: "Rex ate the steak." You might do so like this:

P1. Rex is a dog
P2. All dogs, when left with an unattended steak, eat the steak
P3. Rex was left with an unattended steak
Therefore
C. Rex ate the steak

Here, the premises entail and therefore explain the conclusion. There are many further developments of this overall picture. Recall, for example, that we distinguished what an explanation *is fundamentally* with what actually does the explaining within such a view. Within the formal-logical view, it has often been noted that a specific feature of *scientific* explanation (as opposed to everyday explanation) is something that generalises and that offers a secure foundation for knowledge, understanding and the progression of these.

This is often put in terms of laws, or law-like generalisations (the difference being one of metaphysical robustness rather than simply observed regularity), which above is played by P2. P2 is a rather silly (and inaccurate) law, as far as scientific laws go, but it has the relevant logical properties to allow one to draw general conclusions about dogs and steaks. As we briefly touched upon, explanations that rely on laws (or law-like generalisations) are called *nomological explanations*. They feature prominently within the formal-logical view but are not constrained to it, as we are about to see.

Let's sum up by answering the two aforementioned questions.

1) What kinds of things are the *relata* in explanations? Answer: propositions or sets of propositions.[1]
2) What does it take for x to successfully explain y? Answer: x needs to follow deductively from y, where y is the *explanandum* (a proposition) and x is the *explanans* (a set of propositions.)

A few things should be noted about this overall view.

i) Scientific and ordinary (everyday) explanations are profoundly different in nature. The things we call "explanations" in daily life never, or at best rarely, pick out logically related propositions and certainly rarely appeal to laws or law-like generalisations. (Some of these may still be explanations, and even good ones in relevant respects, but just not scientific ones – i.e. some may be too specific, while others may not even be good explanations, i.e. aren't even deductively related in the right way.)
ii) This way of thinking about explanation is prescriptive rather than descriptive. In other words, it is not interested in capturing how we use the word "explain" nor even in capturing what scientists are actually engaged in doing when they explain things. It aims to tell us what something *ought* to be if it is to count as an explanation in this refined, ideal, sense.
iii) Explanations are objectively "out there" to be discovered.

What are the consequences of this for our thinking about explanation in psychiatry? The first thing worth noting is that, although one could accept that explanations in psychiatry (as elsewhere) are good to the extent that they approximate this ideal, psychiatry will rarely, if ever, get close to it. Physics might, but psychiatry is a long way off. In other words, if you subscribe to a formal-logical view, you will likely be pessimistic about psychiatry's explanatory prospects or at least its good, solid, scientific explanatory prospects.

More problematic, however, is that the formal-logical view offers little in the way of guidance or constraint when we come to look for explanations in psychiatry, and that's arguably what we are after here. In other words, psychiatry is so different from physics that, if we attempt to hold the former to the standards of the latter, we will give up immediately and make no progress at all.

With that in mind, let's look at two other ways of thinking about explanation, which hold more practical promise for thinking about explanation in psychiatry in general and trying to produce good explanations in particular.

6.2.2 The ontological view

On the ontological view (sometimes called the "ontic view"), explanations aren't made up of abstract, logically related propositions. They are made of concrete entities like, for example, objects, states of affairs or events. For example you might think that events explain other events. In particular, it is common within this approach to think of causes explaining their effects. An instance of fire explains an instance of smoke.

So, in answer to aforementioned questions 1 and 2, we get:

Events (or states of affairs) explain other events (or states of affairs), and they do so by standing in predicable law-like causal relations.

Like the formal-logical view, a number of things should be noted about this view:

i) Again, scientific and ordinary (everyday) explanations are different in nature.
ii) Again, explanations are out there to be unearthed. You discover them. You find a particular event, and you unearth the explanation of that event, namely, its cause or causes.

One recent theorist, who buys into this account in philosophy of science generally, is James Woodward (2003). Another, who applies a related view specifically to psychological explanation, is Donald Davidson. To simplify somewhat, Davidson (1970) takes events to cause other events. He also takes explanation to require the picking out of a cause (which is an event) to explain an effect (which is also an event). However, events are only explanatory "under a certain description." He is sensitive to the fact that picking out events that are causally related is not sufficient to be explanatory: you have the pick them out in a *causally relevant* way. For example to explain why the scales go down when weighing some plums, you appeal to the weight of the plums, not their colour, even though those are two aspects of one and the same event (namely, the putting of plums on the scales). This has some affinities with the pragmatic view that we are about to examine.

Within the ontological framework, explanation is fundamentally causal (Psillos 2002). However, there are different kinds of causal explanations. Things get complicated very quickly since the details depend on one's views surrounding the metaphysics of causation (heavy-duty issues we don't want to get into here), but, roughly, there are, first, laws (or law-like regularities) of cause and effect. This may look like the formal-logical view, due to its appeal to laws, but note that it isn't abstract propositions doing the explaining but actual concrete things in the world. Appeal to laws, or law-like regularities, allows for predictability and, as such, allows us to construct theories that are "empirically adequate," which is just a fancy way of saying that they fit with what we observe (where "what we observe" includes an indefinite variety of possible experimental manipulations).

One may wonder whether predictability or empirical adequacy is all we really want from our scientific explanations, or whether we should be more demanding. In any case, there is, within the ontological framework, a different kind of explanation altogether that is thought to confer not simply predictability but a higher degree of *understanding* (we will touch upon the notion of scientific understanding later). Such explanations are the *mechanistic explanations* we mentioned earlier. To illustrate, we may know that CFCs cause holes in the ozone layer, or that smoking causes cancer, and that would rely on observed regularities (perhaps carefully controlled for confounding variables to establish causation rather than mere correlation). This would, we can stipulate, provide nomological explanation and empirical adequacy. However, we might not know the mechanisms whereby these things happen (e.g. the precise way that tar in cigarette smoke is carcinogenic, disrupting cellular DNA). Once we know this – the "cogs and pulleys" – we can provide a *mechanistic explanation*. And this confers more understanding, rather than mere prediction. Also, it has a different logic. It doesn't just say, if you take away this phenomenon, you will no longer get this phenomenon (or, if you *add* this, you'll get that). This very binary structure, though viewed ontologically (as involving concrete things in the world), can be translated into abstract propositions. This is why both formal-logical theorists and ontological theorists can provide nomological explanations. Mechanisms are different. They involve a dynamic sketch (Craver 2006) of how things work: how the teeth of the cogs fit together in a clock (or how voltage-gated ion channels let ions in and out of neuron). These are things that we use diagrams, and imagination, to understand: not structured, abstract propositions.

What are the consequences of this for our thinking about explanation in psychiatry? In stark contrast to the formal-logical view, the ontological view forms the overarching backdrop for a great deal of theorising about explanation in psychiatry. I use the word "backdrop" deliberately, since it is rarely, if ever, explicitly stated. I have yet to encounter philosophers of psychiatry explicitly say "I'm adopting an ontological or ontic framework," and there are good reasons why. What you do see, however, is that researchers in cognitive science and psychiatry look for, and prize highly, mechanistic explanations (and when they sketch the mechanism, it is the mechanism itself and not the sketch that is doing the explaining).

This is no bad thing in itself, but it is worth noting that this is the case and that a different backdrop makes central issues about explanation in psychiatry look rather different.

6.2.3 The pragmatic view

According to a pragmatic view of explanation, an explanation is a good answer to an explanation-demanding question (van Fraassen 1980; Achinstein 1983). Note that not all questions are explanation-demanding. Some are requests for facts ("What's your name?") or favours ("Can you please pass the salt?") or whatever. Explanation-demanding questions (E-questions, for short) tend to start with "Why" or "How." As we shall see, a variety of factors, both psychological and environmental, may contribute to the "goodness" of an answer to such questions.

So, an explanation is the product of a communicative act, and the relata of explanations are not propositions, nor events: they are utterances, or speech acts, that are heavily dependent on context. To clarify, whereas an "utterance" is a more generic category, a "speech act" is a term of art within speech act theory, which, starting with Austin (1962), seeks to elucidate a taxonomy of speech acts and the conditions, both social and psychological, which are necessary for something to count as one speech act as opposed to another. Examples of speech acts include, asserting, requesting, questioning, baptising and, perhaps, *explaining*.

Contextual factors that are often deemed relevant in explanations include, among other things, conversational context, the epistemic state of the demander of the explanation, etc. However, the most important in the current context are the explanatory concerns of the demander of the explanation (which I will henceforth call "the consumer"). An important constraint, which is notably absent in the two previous frameworks, is that an explanation has to address the explanatory concerns of the consumer and has to be considered satisfying by the consumer if it is to count as a good explanation. This subjective satisfaction is a necessary but not a sufficient condition. The reason for this is clear: there are many objectively bad explanations that we may wrongly consider satisfying (e.g. "just-so stories"). So they have to be satisfying in a non-illusory way. However, and here's where the pragmatic view has a valuable point of difference, an explanation that is objectively good by the standards of either the ontological or formal-logical view, but which leaves the consumer unsatisfied, is not considered a good explanation.

So, to reiterate, explanations are relative to a particular instance of a question being asked and have to cater to the consumer's epistemic state. The consumer, it must be noted, isn't necessarily an individual (and in scientific explanations typically won't be) but could be a collective. And the "question" doesn't have to be explicitly asked but could be asked implicitly by the scientific community as a whole (or a subset of that community). Or perhaps a more accurate way of putting it is that we talk in terms of questions as a way of expressing explanatory concerns, but people can have such concerns without explicitly formulating questions in a natural language. Sometimes a good explanation comes along, which is satisfying, and it feels like a good answer to a question you never knew you had.

Another point of clarification is worth making here. The pragmatic view can easily account for the following apparent intuitive support for the ontological view. We often use the word "explain" to talk about events "explaining" other events. We say, for example, that the presence of fire explains the presence of smoke. But for the pragmatic view, this is just a manner of speaking. In actual fact, a *person* explains the presence of smoke by *telling you* about the presence of fire. Or, alternatively, the fire explains the presence of smoke, if you see it for yourself, by "telling" you of its own presence. The latter is not so closely tied to linguistic expression, but it is nonetheless about information and about how consumers of explanations can package and make sense of information. Implicit in this being explanatory for you is that you have an understanding of how fire relates to smoke.

There are a number of variations on the pragmatic view. The *locus classicus* is the chapter in van Fraassen (1980) entitled "The Pragmatic Theory of Explanation." Another important variant has been proposed by Achinstein (1983) relying on the tenets of speech act theory. In our current context, where we are thinking about explanation in psychiatry, we are not interested in the finer details but more in what all versions of pragmatic view in general have in common, in particular, in contrast to the formal-logical and the ontological views characterised before. Then we can see how this casts issues about explanation in psychiatry in a different light.

i) Scientific and ordinary explanations are essentially the same. The former simply has a more regimented context (viz. the explanatory concerns are regimented and shared across a community, namely the scientific community).
ii) Explanations, being the products of communicative acts, aren't unearthed, fully formed, but carefully expressed. They are certainly answerable to how things stand in the world, but they need to be selective and comprehensible to the consumer of the explanation. Explanations, *in the relevant sense*, simply do not exist in a possible world devoid of inquiring beings that produce and consume explanations. In such a world, there is just stuff happening. Only with an inquiring agent, who finds some of this stuff interesting and mysterious (due to gaps in knowledge or understanding), does explanation appear on the scene. Furthermore, these explanations are consumed within a wider pragmatic context, whether it is everyday life, the court of law, the lab or the clinic.

Although, as I've just said, explanations emerge as a result of inquiring agents who are ignorant (an omniscient being surely has no use for explanations), explanations cannot be requested from a background state of total ignorance either. The epistemic or informational state of the demander of the explanation will in part determine her explanatory concerns, and her explanatory concerns will dictate the kind of explanation that would be satisfying.

Now we can clearly see a consequence of this that I feel is far-reaching and of great importance. It is that, according to the pragmatic view, there is no such thing as a de-contextualised "Explaining X" or "X Explained" or (perhaps worst of all) "*The* Explanation of X" (and yet you can find – some truly excellent – books with titles that make such claims, for example, Craver's "Explaining the Brain," Dennett's "Consciousness Explained" or Bentall's "Madness Explained"). At best, within a pragmatic framework, X is a phenomenon of interest about which we are relatively unified in our explanatory concerns. A theorist purporting to "Explain X" may well appear to be doing so by addressing such concerns.

On a pragmatic view, one event can arouse different explanatory concerns, each of which demands different explanations. Suppose that there is a plane crash. If one is conducting a legal investigation, one will, for example, ask for an explanation of the plane crash in terms of poor decision-making. If one is an aeronautical engineer one may ask for an explanation in terms of the weather conditions, and how they impacted on the flight of the aircraft. Depending on certain facts about the crash, either of these explanations may be unavailable. However, this does not involve forcing somebody who is asking the former question to accept an answer that is suitable to the latter question. Rather, it involves getting the person to see that they need to ask a *different question* because their particular question, in the circumstances, has no answer. For example suppose the weather conditions were so severe and unpredictable that nobody is to be held accountable for the plane crash. Somebody looking to hold someone accountable for the plane crash is, given the situation, asking the wrong question. Things like this happen in countless fields of enquiry. We will see that this will turn out to be extremely relevant to psychiatry.

Another very important feature of pragmatic approaches is that, since explanations are good answers to E-questions, we need to think a bit about how these questions are individuated (i.e. what make two E-questions the same or different). One way of thinking about how E-questions are individuated is by explanatory concerns, and these are contrastive. Cutting out the intermediary of explanatory concerns, we can say that E-questions are *contrastive*. What does this mean, and why is it important?

Consider the question:

Q. "Why do birds fly south?"

This may look like one question, but it's actually ambiguous between several questions, three of which are most pertinent. This is to do with different ways in which these may be contrastive, as expressed by emphasis. So compare:

Q1. "Why do *birds* fly south?"
Q2. "Why do birds *fly* south?"
Q3. "Why do bird fly *south?*"

The emphasis draws attention to the relevant point of contrast. Q1 is asking: "Why do birds, as opposed to other animals, fly south?" Q2 is asking: "Why do they fly, as opposed to, say, walk or swim, south?" Q3 is asking: "Why do birds fly south, as opposed to north, east, or west?." These are different questions. And indeed, there might even be more fine-grained ways of breaking them down.

We'll look at this notion of contrastive explanation later on, but it's a crucial contribution to the point about "Explaining X." One problem with "Explaining X" (or "X Explained") is not only that it doesn't target anyone's explanatory concerns, but also that, in a related manner, it doesn't *rule out relevant contrast classes*. You can tell me what is going on in the brain when something (let's say, experiential) happens. But an explanation arises when you explain to me why we get one thing *rather than another*. So, for example you can say that someone is having an auditory experience because they have activation in auditory cortex. But how explanatory is that? Well, for the pragmatist, it's not an explanation at all. What would count as a good explanation would depend on the consumer's explanatory concerns. "Why is an auditory experience taking place, rather than a visual experience or no experience at all?" It might go some way towards answering that (albeit with some circularity – why auditory cortex is there rather than somewhere else requires yet another story). "Why is it an auditory experience of a voice, saying something specific, rather than a sound, or indeed a voice saying something else?" The explanatory concern expressed there asks for more than simply to point to activation in auditory cortex.

Before we turn to psychiatry more specifically, it is worth noting that a pragmatic view can encapsulate lots of the good explanations constructed within the two other frameworks. The only difference is that there is the constraint that they have to be intelligible to the consumer and cater to her explanatory concerns. So, if the consumer understands the propositions within a formal-logical explanation, and they cater to her explanatory concerns (we will unpack the notion of explanatory concern in more detail later), then this would be a good explanation by the pragmatist's lights too. Similar can be said about causes and mechanisms. Indeed, a consequence of pragmatism is pluralism, since there is a plurality of explanatory concerns and ways they can be catered for. I may just want to know whether smoking might give me cancer, and whether I should stop or not. Alternatively, I might want to know exactly *how* smoking causes cancer and how the particles of tar from tobacco smoke have a carcinogenic effect. Each explanation will be good depending on the consumer's explanatory concerns. The everyday smoker and the research oncologist have different epistemic starting points and different informational (as well as practical) concerns. Pointing out causes and pointing out mechanisms aren't competing theories of explanation: they are just different explanations that are good explanations for different consumers, with different explanatory concerns.

What does this mean for explanation in psychiatry? We'll answer this in more detail later, but, ultimately, it both frees up and constrains psychiatric research. We can explore correlations, causes and mechanisms and embrace the contributions of myriad disciplines. Furthermore, there is no reductive imperative nor a hierarchy that places the "exact" or "hard" sciences above the "inexact" or "human" or "social" sciences. That's the liberating side of the pragmatic approach. However, it is also highly demanding: as well as having to be accurate, we need to be careful that our explanations are intelligible and cater to the consumer's (often our own) explanatory concerns. We, in particular, need to be wary of illusory explanation, cases where we feel that our understanding has been furthered when in fact it hasn't, leading to complacency and a premature end to enquiry. We'll see more on this later.

6.2.4 *Taking stock*

So far, I've presented three different frameworks for thinking about what explanations are, fundamentally. This feeds into our thinking about what makes explanations good. It also tells us about what form explanations can take, namely, in terms of what is doing the explanatory work. Let's just recap and clarify the main points.

Within a formal-logical framework, since you need a deductive, inferential, relationship between propositions, what you ideally want is a law (or "law-like generalization"). Here, explanations take the form of the sorts of arguments that we might teach in a logic class. Why did Socrates die? Answer:

P1. All men die
P2. Socrates is a man
C. Therefore Socrates died.

Although laws (or generalisations) might well come in handy in psychiatry, the formal-logical framework is really too austere and stringent to be of use to psychiatry. This is not a criticism of the framework as a framework in general philosophy of science: it points to an ideal, not a code of scientific conduct.

Within the ontological (or ontic) framework, things in the world explain other things. So, for example an event explains another. What does the explanatory work is our understanding of those things, either in terms of the laws that govern them or in terms of how they work, namely, the *mechanisms* whereby they operate. So what we end up with are causal explanations and mechanistic explanations. Causal explanations are law-like, in that they exploit regularities. Fire explains smoke, because fire causes smoke and because (to simplify) "All fire produces smoke." The ontological framework is in the background of most work on explanation in psychiatry and in the cognitive sciences. In particular, the popularity of mechanistic explanations, and also of various statistical approaches, often seems built on the idea that concrete things in the world explain other concrete things in the world in terms of what they contribute causally, either by being a cause (or *the* cause) or by contributing to a complex mechanism.

This all sounds very plausible, but it is also correct – and accommodated – within a pragmatic framework. Within a pragmatic framework, explaining is an exchange of information, constrained by relevance and intelligibility. A common way that one agent will explain something to another is by appealing to causes. Or, if the consumer's explanatory concerns request it, they could appeal to mechanisms. But there is (potentially) more besides. For example there are appeals to reasons, or narratives, that don't necessarily fall into either causal or mechanistic explanations, and that might be not only subjectively satisfying, but also objectively good. We will see more on this later, especially when we come to explore explanatory virtues that extend beyond mere accuracy.

6.3 Levels and kinds of explanation

One big question in philosophy of explanation, as applied to psychiatry, is: "what are the constraints that we can reasonably place on explanation in psychiatry?"

In other words: "what principles can we use to rule out candidate explanations as poor or inadmissible?" A standard answer to this is that "levels of explanation" provide some sort of constraint. I want to present what has traditionally been meant by this and then show how the pragmatic framework casts it in a different light.

6.3.1 *Levels of explanation: an ambiguity*

It is very common in both psychology and psychiatry to admit that there is a constraint concerning levels of explanation. It is a constraint insofar as you shouldn't mix up these levels on pain of conceptual confusion. But, of course, what you think "levels of explanation" are will depend heavily on what you think explanation is in general. What I aim to show now is how these "levels of explanation" can be interpreted ontologically or pragmatically (i.e. epistemically). I will illustrate, by means of a particular example, how this can lead to some people talking past each other.

The classic presentation of "levels of explanation" is not within psychiatry, but in cognitive science, and is to be found in David Marr's classic text *Vision* (Marr 1982). In this he specifies three levels. The highest level is the *computational level*: it tells us what the system is trying to achieve. The middle level, the *algorithmic level*, tells us which algorithms and representations carry out those goals. The lowest level, the *implementational level*, tells us how all of this is (presumably physically) implemented. These levels rest on an idea that is central to functionalism, namely, multiple realisability. The higher levels can be multiply realised in the levels below. Once what is going on at the computational level has been fixed, how this is *achieved* at the algorithmic level is an open question. Different algorithms and representations could achieve the same computational goals. Furthermore, some algorithms and representations could be realised by different physical states and processes.[2]

What is the relevance of this very abstract picture for psychiatry? In psychiatry, we can think of phenomena at different levels, and it is hugely important to not mix these levels unthinkingly. But what does mixing these levels of explanations look like? For the proponent of the ontological view, this is about appealing to events (real things in the world) at one level to explain events at another. For the proponents of the pragmatic view, it is about answering questions that are demanding answers at one level, with answers that talk about things at another level. When theorists don't make it clear what their commitments are with regard to ontological or pragmatic views of explanations, we can get people talking past each other, namely, seeing disagreements where in fact there are none.

Here is an excellent illustration of this. Campbell (2008) criticises a claim that Frith (1992) makes about constraints on explanations in psychiatry. Frith's claim (1992, pp. 26–7) is as follows. Suppose we were to find a really strong and reliable relationship between two phenomena, one being an underlying biological phenomenon, for example, the amount of dopamine in the brain, and the other phenomenon, a psychiatric phenomenon of interest, for example, thought insertion. Suppose, too, that we found that we could reliably measure the amount of dopamine in the relevant part of someone's brain and accurately predict the presence and severity of their thought insertion. Then we might be tempted to say that the former (amount of dopamine) explains the latter (presence and severity of thought insertion). However, Frith continues, the phenomenon of dopamine levels is "simply not admissible" (p. 26) as an explanation of thought insertion. Campbell's gloss on this is that Frith wants the relationship between dopamine and thought insertion to be transparent, not opaque, to reason. Campbell finds this far too demanding, and scientific progress would have been hugely hampered had this constraint been heeded. Like in other areas of medical research (e.g. Campbell refers to John Snow's early work on cholera), Campbell's suggestion is that causation, and causal explanation, in psychiatry ought to be conceived through the isolation of "control variables." Control variables are things that you can manipulate, and they cause an entirely predictable change in the phenomenon that you are trying to explain.

What we see here is an implicit commitment from Campbell, not only to an interventionist theory of causation, but also to an ontological view of explanation. The former is relatively unproblematic, but the latter leads to a misunderstanding of what I take to be Frith's important and valid point. Frith would probably concede that the notion of a control variable is interesting and important. However, it seems to me that Frith's point about dopamine and thought insertion can be usefully made within a pragmatic view of explanation. Within this view, the utility of the control variables can be accommodated (and further elucidated), while also making it clear that the interventionist approach does not exhaust explanation in psychiatry.

From the pragmatic perspective, the control variables are promising to the extent that your concerns are about intervening on the phenomenon in question (more simply, for example, lessening or eliminating it). Now, of course, in a cholera epidemic, you are strongly motivated to eliminate cholera. Knowing what you have to do to achieve that is extremely valuable, and time is of the essence, so the fact that you don't fully understand cholera is something you are willing to put aside. Having said this, later, a greater understanding may be requested (why is it that water carries cholera and not air?). With thought insertion, we don't only want to know how to make it go away: we also want to understand various things about the phenomenon.

As Frith rightly points out, this so-called explanation "does not say anything about the role of dopamine neurones within the physiological domain" (p. 26). I would also add that it fails to answer the question that comes most naturally to us when presented with thought insertion: on what possible evidential grounds could someone deny ownership of their thoughts? If I want to answer this, it is not dopamine, or any other neurobiological phenomenon, that I could turn to. None of these things provides evidence for the individual with psychosis. They needn't know the first thing about dopamine! But it is not irrelevant *either*. The work on dopamine and other neurobiological discoveries contribute a potentially vital piece of an overarching puzzle. We have numerous questions that need answering. It will help, in different ways, to answer each of them. For example understanding what dopamine does in the brain might help to explain what the experience is like and why, which, in turn, may provide understandable grounds for disowning one's thoughts.

6.3.2 Levels and kinds of explanation: two separate constraints

If we think of levels of explanation pragmatically then it's not just levels of explanation that we need to be careful about, but also *kinds* of explanation. Let me explain exactly what I mean by this.

In psychiatry, there are so many different questions to be asked about the same subject matter (the same "chunk of reality"). Questions can differ for different reasons. It could be because we are looking at the phenomena with a different fineness of grain. Or we can ask different *kinds* of questions because we have different *kinds* of explanatory concerns; for example, one that we will see crops up, in particular, when we are dealing with persons or agents and we want to know why someone has acted the way they have or believes the things they do.

Since on a pragmatic view, an explanation is a good answer to an E-question (it is satisfying in a non-illusory manner to the demander of the explanation), this opens up at least as many explanations of a given phenomenon as there are explanatory concerns, and as many *kinds* of explanations as there are *kinds* of concerns. Some concerns in psychiatry are about understanding a causal system. Other concerns are about understanding people; why they do things, why they believe things, why they feel the way that they do, etc. Different sorts of answers will address these sorts of concerns. You can't force someone who wants an explanation of a plane crash in terms of poor decisions to accept an explanation in terms of the weather. What you can do, however, is try to redress their concerns. This will be especially important when the type of explanation being asked for is simply not available.

Within the context of this chapter, what a pragmatic view of explanation makes room for is the claim that "levels" of explanation in cognitive science are about *causal* explanatory concerns. There are many different such concerns, but they are, at heart, the same kind of concern. In addition to this, however, there are important explanatory concerns that aren't causal in the same sense. Addressing these concerns produces not another level of explanation but a different *kind* of explanation (Wilkinson 2014).

Within explanations of a certain kind, which have causal and predictive explanatory concerns, one can distinguish between causal explanation and mechanistic explanation. Roughly, in asking for a causal explanation one is asking for a cause, without inquiring into the mechanism whereby that causes functions. This provides some degree of understanding, and a sufficient degree of understanding for some situations, for example, if one wants to avoid a particular effect. Thus, we might establish that smoking causes cancer. We might not know exactly how it does so, but knowing that, at least, is enough to suggest that (*ceteris paribus*) if we don't want cancer, we shouldn't smoke. We can think of causal explanations as answering a certain kind of question, namely, a "How come?" question. "How come he got cancer?" This is often expressed with a causal use of "Why," as in "Why did he get cancer?" This causal use of "why" is very different from a justificatory use that we are about to encounter.

A mechanistic explanation answers instead a "How?" question. It provides not just the cause, but also the mechanism whereby a certain causal process operates. Using the cancer example, it isn't enough to know that smoking causes cancer: what is required is a description of the mechanism, for example, in terms of carcinogenic disruption of genetic material through radiation emitted by particles present in tobacco smoke. Mechanistic explanation provides a greater degree of understanding than merely citing causes, and it is of central importance in various areas of science, for example in biology, cognitive psychology and neuroscience. Mechanistic explanations are a sophisticated brand of causal explanations, since the kinds of concerns addressed are of the same kind, namely, of understanding how a brute causal system will behave in relevant counterfactual circumstances.

There is a third kind of explanation, which is variously called rational explanation (Davidson 1963), personal explanation (Dennett 1969) (although one could argue that the latter is a slightly broader category). They involve answering a *justificatory* and opposed to *causal* "Why" question (and answering them with justificatory rather than causal uses of "because"). So, within the pragmatic view, you end up with three kinds of explanations. We can illustrate these in terms of different kinds of E-questions:

1. "How come?" (a.k.a. "causal Why?") – causal explanation/nomological explanation.
2. "How?" – mechanistic explanation.
3. "Why?" (a.k.a. "justificatory Why?") – rational/personal explanation.

From within the pragmatic view, these kinds of explanations operate as a further and crucial constraint in addition to the levels constraint. So, there are two constraints, the "levels constraint" and the "kinds constraint," and both are extremely important for explanation in psychiatry.

The levels constraint can be expressed as follows: "you should not address an explanatory concern in a way that is unintelligible to the demander of the explanation." For example my concerns might be mechanistic, but if you tell me about dopamine, and I don't know the relevance of dopamine for what I want to explain, then (as Frith 1992 very rightly points out) that is an infringement of that constraint.

The kinds constraint can be expressed as follows: "you should not answer one kind of E-question with an answer that is appropriate for another kind of E-question." Even if your answer is intelligible to me, that is not the question that I was asking, and so those are not my explanatory concerns. An example of this is if I ask: "why (namely on what grounds) might someone deny ownership of their thoughts?," and you answer: "they have too much dopamine in their brain." Even if this statement proved to be accurate in a free-standing manner (namely, removing that excess dopamine made the thought insertion go away) it wouldn't be an answer to my question, to *that particular* "Why?" question. On the other hand, an account of how their thoughts felt alien might.

6.3.3 A comment on personal explanation

Although the "personal/subpersonal" terminology was introduced by Dennett (1969), as Dennett himself acknowledges, it has analogues in Ryle and Wittgenstein.[3] Roughly, whereas subpersonal explanation is mechanistic or causal, answering "How" and "How come" questions (respectively), personal explanation answers a "Why" question, where "why" is understood in a certain way. We do sometimes use "why," when our explanatory concerns are causal or mechanistic. For example when we ask "Why did he get cancer?" we mean, "By what cause or process did he get cancer?" We know he did not get cancer for a *reason*. However, when we ask, "Why is there a STOP sign at the end of that road?," we are (usually) asking for a reason, a justification, a *rationale*, for its being there. Along with the distinction between justificatory and causal uses of "why" in the *question* ("Why is there a STOP sign there?/Why did you raise your hand?" versus "Why did he get cancer?"), we have the distinction between justificatory and causal uses of "because" in the *answer* ("Because there tends to be fast-moving traffic in the main road/Because I wanted to ask a question" versus "Because he smoked heavily"). Answering such a "why" question involves citing a person's reasons or grounds for believing certain things and acting in certain ways (or the general, publicly agreed, reasons, not attributable to a specific person, as in the case with the STOP sign). If I ask "Why is there a STOP sign there?" and you answer "Because someone put it there," I would understandably think you were being facetious!

With beliefs and actions, we often ask questions of one another: "Why do you believe that?" and "Why did you do that?." In doing this, we are asking a very particular kind of question, and one that requires a very particular kind of answer. This answer is commonly called a *rational* explanation (Davidson 1963; McDowell 1985). The way "rational" is used here is categorical, not evaluative: you can have "rational explanations" of *irrational* phenomena. It is in part because of this and in part because the word "rational" is used in so many different ways (and in part because I see my view as sharing much with Dennett's early view) that my terminological decision is to use "personal" rather than "rational."

If you ask me "Why did you raise your hand?" and I answer, "Because I wanted to ask a question," that's normally a satisfying explanation. If I tell you a full physiological story about what happened up until the point when my hand went up, that may be interesting, but it's not an answer to *that* question. For starters, it doesn't even describe me, the person, performing the *action* of raising their hand; it describes in very accurate detail a bunch of physiological movements. Not only this, but also a description of a causally related sequence of events is not what you asked of me. You were after a *reason*. The same applies when you ask the question "Why do you believe this?" You are after *reasons* for my belief, not any mechanistic story.

Now, why is this not a mechanistic explanation? It is not a mechanistic explanation because when I give you my reasons for raising my hand, it is not thanks to understanding a mechanism that you come to understand my action. It is in virtue of "meaning not mechanism" that the explanation I provide you with is satisfying. Now, you may object that we understand how meaning causally operates on people, so, in a sense, we do understand the mechanisms of meaning, the mechanisms of reasons and rationality. Two things can be said in response to this. First, this may be correct, but it is an unusual use of the word "mechanism." What the objection describes more closely resembles causal explanation. We don't know the mechanism by which certain reasons operate, but we know the sorts of things that they tend to give rise to. We know that certain beliefs and desires in certain contexts will give rise to certain actions. We certainly don't know by what *mechanism* they do so. Second, I would say that, although it looks superficially similar, personal explanation is not causal explanation either, since the explanatory concerns of the consumer are subtly different. Many theorists will disagree with me here, but I will attempt to quickly explain this now.

It is worth noting that there are those who are fully accepting of the explanatory autonomy of reason-giving explanation, who claim that reasons are causes. For example unlike, say, Elizabeth Anscombe (1957) (and the early Dennett), Davidson sees reasons as causes. Now, although I side with Anscombe on this, I am willing to accept for the sake of argument that reasons are causes, especially if one accepts a counterfactual theory of causation. If I hadn't wanted to ask a question, I would not have raised my hand. In that very simple sense, my *desire* to ask a question was a cause of my *action*. However, it does not follow from something being a cause that something is explanatory *in virtue of being a cause*. I think that since our explanatory concerns when we ask that special variety of why-question are not causal concerns, the explanation given in terms of reasons is not accurately characterised as a *causal explanation*, even though reasons may well be causes.

Why do I think that our concerns aren't causal? First, consider that causal explanatory concerns are closely tied to prediction. A causal understanding of a system allows us to predict how that system will behave in relevant possible circumstances. Now consider a very common, retrospective, case of personal/rational explanation. We want to understand why somebody did something (e.g. killed someone), so that we can hold him or her accountable. Finding out about this does not confer any future predictive power. (In fact, we already have the predictive power that we need: we already know how beliefs and desires give rise to and sometimes justify actions.) However, it enables us to understand the individual. It renders them intelligible. Alternatively, consider a forward-looking case, but where the subject behaves in a deeply surprising way. When this happens, are we just bemoaning a failure to predict, to causally understand, the subject? I would suggest that there is more to it than this: we are perplexed by this *person as an agent* by the fact that we find them unintelligible.

To take an example, suppose someone holds up a poisonous mushroom and announces: "this mushroom is deadly poisonous and I have no intention of killing myself" but then pops the mushroom in his mouth. This behaviour is *surprising*, but it is not just surprise that you feel, but perplexity and confusion. In particular, you don't understand why this person ate the mushroom. You can hypothesise that they were lying, either about the poisonous nature of the mushroom or about their intentions to stay alive. Or they were demonstrating an antidote. However, at face value, this action is perplexing.

6.3.4 Some examples

We can use explanations of different kinds about the same phenomenon of interest. To see how this works, it is worth running through some examples. These are all examples that will have entire chapters dedicated to them later on. However, the aim here is to sketch how the overall explanatory structure works.

Delusion

Take, for example, the Capgras delusion, often described as the delusion that a loved one has been replaced by an identical-looking impostor. We can ask about a patient with the Capgras delusion the subpersonal, causal question, "How has their brain damage disrupted normal cognitive functioning?" A really good answer to this will make it altogether unmysterious why (how come) this particular damage has disrupted functioning in this particular way and not in any other way. This will require causal and mechanistic understandings at different levels. However one can also ask, "Why (on what grounds) do they believe what they do?" In answering this, you can't use the same vocabulary as when answering the first, subpersonal, question. Dopamine dysregulation, modular damage, etc. – none of these is even the right kind of thing to provide grounds for the subject. One hypothesis which *does* speak to this question is: "because although this person looks just like my mother, she feels really uncanny, and I'm getting unsettling emotional responses to her."

Now consider thought insertion. When I ask the person "Why do you believe that someone else is putting their thoughts into your head?," as we've mentioned, I am not after a story in terms of dopamine. I want to know on what possible *grounds* someone could deny ownership of their thoughts! They might say something like, the thoughts aren't the sorts of thoughts that I might have: they are incongruous with my understanding of who I am. Or they might say, they are incongruous with the context I am in. Or they might say that, although they are congruent both with their personality and the context in which they find themselves, the phenomenology of the thinking feels odd: it feels loud, heavy, alien, uncontrolled. Then, once that question has been answered, other questions present themselves: "How come this person's thought processes feel the strange way that they do?" or "How come they are out of character, or out of context?" And here, the answer might appeal to dopamine, but the answer needs to be given carefully. We need to know about the contribution that dopamine makes, in the relevant part of the brain, to the process of thinking. For that, of course, we need a good account of what thinking even is!

Although it is correct that subpersonal vocabulary shouldn't feature in personal explanation, this is not to say that the cognitive and brain sciences can't make very important contributions to personal explanations. They certainly can. The nature of the contribution is that it can give us an idea of the nature of the grounds that a subject might have (e.g. what experiences or emotions they might be undergoing) and how it is that they have them, or rather, why (how come) they have these experiences "and not others." For example as we will see in Chapter 9, subpersonal psychology can suggest that the Capgras patient is experiencing a feeling of unfamiliarity towards the misidentified individual. It can also suggest (e.g. by understanding the function of the damaged region) why it is that the Capgras patient experiences lack of familiarity to faces and not, say, to voices, or places or objects. Once we understand what the subject may be experiencing, there is scope for their claims to be rendered *intelligible*, namely, to be subject to personal explanation.

In other cases, the contribution of subpersonal psychology may be very different. It may be able to warn us when personal explanation is *unavailable* (this would be the lesson of the top-down accounts of Chapter 9). That is to say, it may warn us when any attempts at understanding the subject in terms of subjective grounds would be a waste of time. We saw with our plane crash analogy that an understanding of the situation may lead us to conclude that no blame is to be attributed. Similarly, an understanding of some cases may lead us to realise that there are no reasons for certain behaviour or perhaps even beliefs: they are simply caused.

Depression

Moods, such as sadness, happiness and elation, can also be given personal explanations. "Why are you feeling down?" someone might ask, using a justificatory rather than merely causal sense of "Why." "Because I've just lost my job" is a pretty good answer to this question. "Because my serotonin levels are unusually low," although it may accurately describe what's going on in the person's brain, is not an appropriate answer to that question. Part of what makes the first answer a good answer is not just that it's the right kind of answer, providing a personal explanation, but also because we want moods to be proportionate to the events that give rise to them. Both of these constraints are central to our understanding of depression, as we'll see in Chapter 10.

Addiction

For addictive behaviours, which are cases that *look* like goal-directed actions that warrant personal explanation, some theorists insist that that would involve the wrong kind of explanation. Don Ross (2010) seems to take something like this view with regard to addictive gambling: "[we may] wonder what it is that keeps attracting [the subject] to gamble more. Is it thrills or money or relief from boredom or what? This is the wrong question to ask if we want to understand gambling *addiction*" (Ross 2010, pp. 140–1). What Ross is advocating here is a move from one kind of explanation to another, because one kind of explanation is unavailable. As we'll see in Chapter 11, other theorists, on the other hand, may wish to provide a personal explanation in terms of strong urges or even the biased treatment of evidence via an attentional abnormality. These might in turn require a subpersonal explanation: How come these urges and biases are there?

Hearing voices

You can't ask a justificatory "why" question about the mere presence of an experience. It's not a belief that is grounded in evidence, nor is it an action that is done for reasons. However, that doesn't shift our explanations and explanatory concerns immediately to causal ones. For example hearing voices may sometimes relate to certain moods and emotions that may be partly explicable in terms of reasons. Traumatic emotions of guilt, shame or fear aren't simply caused brutely: they are responses to certain events and in virtue of the meaning or significance of the events. Furthermore, these sorts of explanatory concerns may involve fine-grained contrastive explanation. We might be interested in knowing not just why someone hears a voice (as opposed to no sound at all or as opposed to a sound that's not a voice), but more specifically why someone hears the voice of a particular agent (e.g. an abusive step father) or a voice saying particular things (e.g. one-word aggressive expletives as opposed to a running, emotionally neutral commentary).

6.4 Related issues

To finish off this chapter, I'd like to mention a couple of important related issues. The first is about distinguishing accuracy from explanatory power. Ideally you need both, but too often preoccupation with the former eclipses the latter. The second is to do with the concept of reduction and reductionism. "Reductionism" is often seen as pejorative: we sometimes call people reductionists as a criticism. I think it's important that we get clear on exactly what this can mean. On some interpretations, it's not such a bad thing.

6.4.1 Accuracy and explanatory power

Quite independently of one's views on explanation, philosophers of science routinely distinguish between the two virtues of *accuracy* and *explanatory power* (see Rapoport 1972; Ylikoski and Kuorikoski 2010). This distinction is especially relevant in domains where knowledge and understanding are relatively limited, like psychiatry, and all the more important since it is often overlooked in psychiatric scientific theorising. I won't have space to do justice to debates about the precise nature of explanatory power, but even a rough presentation of the notion will prove useful.

The basic idea is that scientific theorising can be good in different ways and along different dimensions. In particular, it is important to distinguish between explanatory virtues and evidential virtues. These virtues are, in principle, independent of each other. So, as a first attempt, explanatory power is a virtue of scientific theories, viz. something that makes them good, independently of their accuracy. In other words, all other things being equal, a theory that has explanatory power is better than one that lacks it. At a first approximation, you are getting at explanatory power when you bracket (put aside) the question of accuracy and ask: "how explanatory *would* this theory be *if* it were true?"

The main point for our purposes is that theories can score highly on accuracy, but not so highly on explanatory power, and vice versa. It is the former permutation that is of greatest concern, since accuracy is often over-weighted *at the expense of explanatory power*. In other words, too much theorising is accurate but lacking in explanatory power, and yet it is praised for its accuracy without anyone (seemingly) noticing the explanatory deficiencies.

So, what is explanatory power in more detail? First, explanatory power is often appealed to in the context of competing theories, where we might ask: "which of two theories has more explanatory power?" This ultimately means: "which theory generates better explanations?" But "better" is, of course, much too vague. Three important explanatory virtues (which make explanations "good") are precision, relevance and integration. All other things being equal, an explanation that is more precise, more relevant, more integrated is better than one that is less of those things. Let's look at these in turn.

What is explanatory precision? One intuitive way of thinking about it is in terms of *modal proximity*, namely, closeness in the space of relevant possibilities. In particular, an explanation has more precision, the "closer" in the space of possibilities, where the possibilities ruled out are to the actual world. To borrow an example from Ylikoski and Kuorikoski (2010), suppose someone buys a navy blue car. We can compare different E-questions of varying precision.

Q1. "Why did this person choose blue (as opposed to yellow, red, etc.)?
Q2. "Why did this person choose navy blue (as opposed to any other shade of blue)?

Q2 is clearly more precise. All other things being equal, answers to Q2 are more explanatorily powerful than answers to question Q1. However, there is a further complication, which stems from the word "relevant" in "relevant possibilities," and this brings another virtue into play (that can, but doesn't always, compete with precision). This is to do with how the demander's explanatory concerns are catered for. Answering Q2 when the demander is still in the dark about Q1, and wants an answer to Q1, would be an explanatory failing. It would be an irrelevant degree of precision.

Nevertheless, scientific inquiry is not simply about asking questions and answering them. It is about dynamically moving through an interrogative space of questions in a sensible manner that enhances our understanding of a domain of interest. In light of this, the virtue of precision is not entirely negated here. It is not only incumbent of the provider of the explanation to cater to the explanatory concerns of the demander, but also, in scientific contexts at least (since in everyday contexts one is not normatively bound in the same way to care about, or find relevant, anything in particular) it is incumbent on the demander to progress towards greater precision and understanding, to fully present why we end up with X and not Y, where X and Y are relatively close. This imperative towards greater understanding is related to the third virtue, namely, integration. Explanations that are well integrated into a theory or framework stimulate new and unforeseen E-questions and can feed back into the theory and improve it.

Let me now use an example from an explanation of a psychiatric phenomenon to illustrate how accuracy and explanatory power can come apart. Consider the theory that auditory verbal hallucinations are the result of spontaneous activation in auditory cortex. Let us, for the sake of argument, grant it total accuracy, as far as it goes. It would still lack explanatory power. It doesn't carve up the space of possibilities in a satisfying way. More specifically, it lacks both precision and relevance.

It lacks *precision*, since, while it tells us why the individual is having an auditory experience rather than no experience at all, it doesn't tell us why somebody hears a voice rather than any other sound, let alone why someone might hear a voice in any particular qualitative features, as opposed to any other qualitative features (see Chapter 8 for more details on the many dimensions of variety that might need explaining).

It lacks *relevance* since it doesn't cater to (what one can safely assume) are our explanatory concerns. It fails to plug a gap in our ignorance. We know that activation in auditory cortex correlates with auditory experiences. Indeed, that's almost a triviality, since that's why we call it auditory cortex in the first place. (On a side note, we also need to unpack the notion of "spontaneous activation": what does that mean, and where does this spontaneity come from?)

6.4.2 Two kinds of reductionism

Finally, I'd like to address the question of reductionism. It is important to distinguish two kinds of reductionism: "metaphysical reductionism" and "explanatory reductionism." The latter then subdivides, perhaps unsurprisingly, depending on one's views of explanation.

In the most general sense, "reduction" is about reducing something to something else. By extension, being a reduction*ist* is to subscribe to the view that this can and should be done. But what is being reduced to what? And what does "reduction" mean anyway?

A metaphysical reductionist is talking about a particular chunk of reality and is claiming that something rather complicated (e.g. mind, mental illness, the stock market, a pandemic) *is* nothing but something more basic (e.g. atoms moving at varying speeds). This kind of metaphysical reductionism is what many people think of when they think of reductionism, especially when they use it pejoratively, and yet it is a view that actually few philosophers hold, whether in philosophy of mind, psychiatry, economics or medicine. This kind of reductionism is not to be confused with physicalism (the view that there is nothing other than the physical). More reputable physicalist positions that eschew reductionism make use of concepts like realisation, grounding, supervenience and emergence. For example few, if any, philosophers will just flatly state that the mind just is the brain, although many will accept that is grounded in, supervenes on or is even realised in the brain.

An explanatory reductionist will often be openly so, and it is a far more plausible position that its metaphysical counterpart. Indeed, a reductive explanation is often thought to be a very good thing. It involves explaining one complex thing you understand relatively poorly in terms of something that is simpler and you understand rather better. Indeed, far from being something bad or even controversial, this is an integral part of much of scientific explanation. For example we may explain the phases of matter, and phase transitions, in terms of bonding between molecules.

In some respects, then, reductive explanation is simply one kind of explanation that many theorists will accept to be useful in some situations. But that doesn't make them explanatory reductionists. An explanatory reductionist is rather someone who prioritises reductive explanation or perhaps who denies that other forms of explanation have as much, or indeed any, validity.

The appeal or plausibility of this view will depend on what we think explanations are, since it is a claim about explaining one thing in terms of another. This is more appealing within an ontological framework. Granted, then it starts to look like metaphysical reductionism but that would be a mistake. It is not to say that, for example, the mind is the brain but that you explain certain mind events in terms of brain events, which are more comprehensible (in a way very much analogous to how we might explain phases of matter in terms of molecular bonding).

Things look rather different within a pragmatic view. My explanatory concerns could be of a reductive nature, and that might make extremely good sense within my overarching scientific enterprise. This would be mechanistic in most cases. But my concerns could be causal or justificatory. Someone died because they were poisoned with arsenic. That could satisfy me, if I were a detective, without knowing exactly how arsenic kills (causal rather than mechanistic). Someone raised their hand because they wanted to ask a question (justificatory). There is nothing reductive about those explanations, and they are none the worse for it. Ultimately, a pragmatic view of explanation makes explanatory reductionism unappealing although reductive explanations clearly play an important role within the explanatory pluralism that the pragmatic view encourages.

Summary

- Scientific psychiatry presents various challenges for thinking about how our scientific understanding of psychiatric phenomena (for both illness and treatment) should progress.
- The challenges look rather different depending on one's view of explanation more generally.
- There are three very broad families of explanation, namely, the formal-logical, ontological and pragmatic views.
- A common constraint surrounding explanation, not only in psychiatry, but also in the cognitive sciences, concerns levels of explanation.
- From a pragmatic view, there aren't only constraints on levels of explanation, but also on different kinds of explanation.
- We ended by looking at two related issues: the notion of explanatory power and the different senses in which someone can count as a reductionist.

Questions

- In everyday contexts, what counts as a good explanation? What sorts of things do you expect from an explanation? Does this seem different from explanation in scientific contexts?
- Suppose you were having unusual experiences (e.g. hearing voices), to what extent would an account in terms of neurotransmitters in your brain count as a satisfying explanation?
- Does an explanation in terms of reasons (either epistemic, "I believe this, because of that evidence," or practical, "I did that because I wanted this") work as a good explanation, or does it need further fleshing out?
- Can you think of explanations that have good explanatory power (they have both precision and relevance) but that are inaccurate portrayals of the actual world?
- Can you think of explanations that are accurate portrayals of the actual world but that lack explanatory power (because they lack precision or relevance)?

Notes

1. Although note that a set of propositions, for example a set of premises, can simply be conjoined into one long proposition that would have the same logical properties – so, logically speaking, a set of propositions just is a proposition: a rather long one.
2. Marr is seen as the classic source of levels' explanation, and this is, often implicitly, interpreted within an ontological view: state or events at one level of existence explain other events at that level of existence, unless you can explain how higher

levels are realized in lower levels. But Marr's actual presentation of levels is in terms of *questions:*

> "*Computational theory*: What is the goal of the computation, why is it appropriate, and what is the logic of the strategy by which it can be carried out?
> *Representation and algorithm*: How can this computational theory be implemented? In particular, what is the representation for the input and output, and what is the algorithm for the transformation?
> *Hardware implementation*: How can the representation and algorithm be realized physically?" (Marr 1982, p. 25). One wonders: was Marr actually a pragmatist?

3. As Dennett writes: "the lesson to be learned from Ryle's attacks on 'para-mechanical hypotheses' and Wittgenstein's often startling insistence that explanations come to an end rather earlier than we had thought is that the personal and sub-personal levels must not be confused" (1969, p. 95).

Further reading

Bolton, D. & Hill, J. 2005. *Mind, Meaning and Mental Disorder: The Nature of Causal Explanation in Psychiatry*, second edition, New York: Oxford University Press.
Cooper, R. 2007. *Psychiatry and Philosophy of Science*, Stocksfield: Acumen Publishing.
Murphy, D. 2006. *Psychiatry in the Scientific Image*, Cambridge, MA: MIT Press.
Sober, E. 1999. The Multiple Realizability Argument Against Reductionism. *Philosophy of Science* 66:542–64.

PART 2

Philosophy and Psychopathology

7 Schizophrenia

Overview

Schizophrenia is perhaps the quintessential disorder of modern psychiatry (Szasz famously called it the "sacred symbol of psychiatry" (Szasz 1976)). But what is schizophrenia? This question is ambiguous, with at least two possible readings. On one reading, this is a question about the *concept* or *category* of schizophrenia. How do we define and delineate "schizophrenia"? In other words, when do things count as instances of schizophrenia? On the other reading, this is a question about what the phenomenon in the world that we have already agreed to call schizophrenia fundamentally is: what is its nature and aetiology? We will address both of these in turn.

First of all, "schizophrenia" is a word, a word that entered into circulation at a particular time and got used in a particular way for a particular purpose. As a result, we will begin with a brief history of the word in the hope of getting a sense of the concept that the word expresses. As we will see, the concept of schizophrenia is closely associated with psychosis, and so we will examine that concept too and the relationship between psychosis and schizophrenia.

We will then move on to examine the phenomenon itself. What are the symptoms? How prevalent is it? Who tends to get it and why? Then we

DOI: 10.4324/9781315450124-9

look at hypotheses about the underlying mechanisms at play, which we present and assess. We then examine phenomenological approaches that focus, not so much on the underlying mechanisms as on the lived experience of psychosis. Since the significance of schizophrenia is as a disruption to lived experience, these accounts need to be taken seriously at the very least as crucial to setting the explanatory targets: what is it that needs explaining? We end by presenting and assessing critiques of the concept of schizophrenia. For a variety of reasons, these critics would prefer it if the term "schizophrenia" dropped out of circulation altogether.

7.1 The concept of schizophrenia

In this section, I present a brief history of the word "schizophrenia" in the hope of shedding some light on the concept it expresses. Schizophrenia is closely related to psychosis, and I will clarify what that is too, but schizophrenia and psychosis are not only *not* the same thing, but they are also not even the same *kind* of thing. And that is crucial to a proper understanding of schizophrenia. To make the claim plainly now: schizophrenia is (purportedly) a disease, whereas psychosis is a state of mind, that may or may not be caused by that, or indeed any, disease.

7.1.1 A brief history of "schizophrenia"

Etymologically speaking, the term "schizophrenia" comes from the Greek words "schizein" (to split) and "phren" (mind). This has led some to think that it is the same phenomenon as "split personality," but this is far from accurate. Sometimes, people use the term figuratively to mean that they are in some way divided, torn or in two minds, which does little to help curb the perpetuation of this inaccuracy.

Eugene Bleuler, who first introduced the word in 1910, intended the term to refer, not so much to a splitting of the personality (a Jekyll and Hyde phenomenon where someone would be one person at one moment and someone else the next (which actually sounds more like "dissociative identity disorder" (DID)), but rather to a splitting *up* of mental capacities, a disintegration, rather than a division in two or more distinct personalities or minds.

Although Bleuler introduced the term "schizophrenia," he attached his new label to a pre-existing category introduced by Kraepelin in 1887, under the Latin term "*dementia praecox*" – (roughly, "early life insanity"). Note that the Latin word "dementia" just means madness or folly (literally "out of your mind," *de-mentia*) and lacks the narrower meaning that it has in English, namely, a deterioration of mental capacities most commonly associated with senility and sometimes with a specific disease, as in the case of Alzheimer's.

Although Kraepelin is credited with isolating and identifying the condition and Bleuler with giving it the long-lasting label it still bears to this day, the kinds of things that today (and in Kraepelin's and Bleuler's days) would be called schizophrenia have, unsurprisingly, existed and been documented throughout history. How these conditions were conceived of in the cultures and societies of the past is clearly extremely varied, not least as a reflection of overall conceptual frameworks and world-views within which mental illness has had to find a place. The sorts of things that we might call "schizophrenia" today have been seen, variously, as punishment from the gods, demonic possession, imbalances of humours and so on and so forth.

How and why was this category first isolated in the nineteenth century by Kraepelin and his contemporaries? To understand it fully, it is crucial to understand what it was supposed to be differentiated *from* (one of Kraepelin's main aims, after all, was differential diagnosis; see Chapter 3). When someone has schizophrenia (or *dementia praecox*), they have it *as opposed to what*? In early theorising on mental illness, there was a very broad distinction drawn between psychoses on the one hand, and neuroses on the other. Neuroses are conditions of psychological distress that do not involve a disconnection from reality. Psychoses, in stark contrast, quite simply denote some kind of disconnection from reality. So schizophrenia (or *dementia praecox*, as Kraepelin called it) was introduced as a *subspecies of psychotic condition*. In particular, it was to be distinguished from " affective psychoses" that occur as a result of mood disorders (e.g. major depression or bipolar disorder, both of which can, but needn't, involve disconnections from reality).

7.1.2 "Psychosis"

It is testament to how fluid the concepts expressed by these words are that, if we look at the history of the word, "psychosis" was not always taken to be in *opposition* to neurosis (as it was for Kraepelin). In fact, in its first usage (Canstatt 1841), it was taken to be a subspecies of neurosis. Indeed, unlike schizophrenia or *dementia praecox*, which were derived directly from Greek and Latin, respectively, "psychosis" is simply a contraction of "psychic neurosis." In this context, however, "neurosis" simply refers to any pathology of the nervous system and so also carries a different and broader meaning than the sense of neurosis that was conceived in opposition to psychosis.

But enough of history and etymology! One take-home message from this is that words get introduced, and the conceptual landscape associated with these words is constantly shifting (and often in ways that the very users of those words fail to appreciate). However, we can get some purchase on the concepts of schizophrenia and psychosis by reflecting on their relationship to one another. In particular, this comparison will give us an initial sense of the type of concepts that we are dealing with here.

"Psychosis" means a "disconnection from reality." But there are different ways of being disconnected from reality. Indeed, one might start by asking: what is it like to be *connected* to reality? We can examine what is meant by "reality" (e.g. whether reality is objective, inter-subjective or merely subjective), but this involves getting into heavy-duty metaphysics, and a more pressing focus is on this notion of being "connected." We are primarily connected to reality via our senses, via our perceptual experience. Sometimes our experiences are accurate; at other times they are misleading. However, we are also connected to reality in virtue of our beliefs, beliefs about things that we aren't currently perceiving. Of course, we also make perceptual judgements: when we take the testimony of our senses at face value, we come to believe things about the world as immediately perceived (e.g. that the coffee cup is in front of me on the table thus and so). All three of these count as the relevant kind of "connection" to reality and hence can also be involved in the relevant kind of "disconnection." Certain kinds of false belief (i.e. delusions) count as psychosis. Certain forms of misleading experiences (i.e. hallucinations, in whatever modality, but most prominently visual and auditory) also count as psychosis. Finally, false beliefs that are based on hallucinations also count as psychosis.

One interesting observation at this point is that two individuals might both count as "psychotic" even though one might be hallucinating, but be aware that they are hallucinating and staying "connected" to reality at the level of belief and judgement. Clearly, there is a sense in which they are less disconnected, indeed to the extent that they even are at all, than someone who takes their hallucinations at face-value. This is sometimes put in terms of "insight." The hallucinating subjects who know that they are hallucinating and dismiss the hallucinations (e.g. "as merely the products of their brain") have "insight," namely, an awareness of their condition. They should, in my opinion, not just be considered *less* psychotic than a hallucinating individual who lacks that insight: they should not be considered psychotic at all. Unfortunately, however, hallucinations, however they are appraised, are widely considered to be components of psychosis.

Another interesting observation concerns catatonia or unresponsiveness. When this occurs in the context of depression, it is just seen as extreme despondency, but in the context of schizophrenia, where it is seen as a negative symptom (although some have noted that antipsychotics can seem to cause catatonia (see, e.g. Choi, Kim and Koo 2011)), it is considered psychotic. But to what extent is a total lack of connection to *any* reality comparable to a disconnection from consensus reality via an adherence to a different reality? In other words, surely it is wrong to characterise the mental state of someone who fails to take themselves to inhabit any possible world to be in any way like the mental state of someone who takes themselves to inhabit a rich and florid world that is far removed from the actual world? In this sense, one could suggest that catatonia in fact *undermines* psychosis.

7.1.3 The relationship between schizophrenia and psychosis

As we will see towards the end of this chapter, the very concept of schizophrenia has come under attack. This idea that a concept can come under attack might strike you as bizarre, but it will become clear what I mean by this. In any case, as a result of the dislike of schizophrenia, many clinical practitioners prefer to use the more neutral term "psychosis" to refer to cases that might be called "schizophrenia" by mainstream psychiatry. However, this interchangeability can be misleading. First of all, there are a number of psychotic disorders (e.g. schizoaffective disorder, schizophreniform disorder, bipolar disorder with psychotic features) that are recognised to not be schizophrenia. But second, this demonstrates something extremely important: schizophrenia and psychosis are *two different kinds of concept*. Whereas "schizophrenia" is supposed to refer to an underlying *disease* ("a chronic, severe and disabling brain disease," according to the US National Institute for Mental Health's 2003 online public information page about schizophrenia), "psychosis" refers to a particular psychological state that could have any number of underlying causes and explanations, of which one could (in principle) be schizophrenia.[1] In other words, psychosis is (potentially) a symptom of a number of diseases, one of which might be schizophrenia. Thus, if you encounter "the symptoms *of* psychosis," it might seem pedantic to say so, but it is a category mistake. You can talk about "psychotic symptoms" (symptoms we might characterise as a disconnection from reality, as per the previous section) or "the symptoms of schizophrenia" but not "the symptoms of psychosis," since psychosis isn't a disease, it doesn't have symptoms, but rather is an experience, which might be a symptom (to the extent that it might be caused by, and taken to be indicative of, an underlying disease).

To sum up, then, psychosis refers to the experiential state of being disconnected from reality. Thus, if there is such a thing, "schizophrenia" is a disease – a biological, developmental disorder – whose purported properties we will examine shortly. Now we can get a sense of what being a "critic of schizophrenia" might mean. Psychosis, certainly, exists; we witness it. But the underlying disease entity "schizophrenia," the criticism goes on, does not (or at least not in anything like the way in which it has been presumed to exist). More on this topic will follow later in the chapter, in section 7.5.

7.2 The phenomenon itself

According to psychiatric orthodoxy (e.g. as seen in the DSM, ICD and public discourse), there's a phenomenon in the world that we can encounter, and that phenomenon gets called "schizophrenia." What is that phenomenon supposed to look like? First of all, I'd like to present how someone might attract a schizophrenia diagnosis in the present day. This involves looking at the latest version of the DSM (DSM-5). Then, I want to look at the epidemiology of schizophrenia. What is the statistical landscape? Who is it that is particularly at risk of getting this purported disease?

7.2.1 Diagnosing schizophrenia

The diagnostic criteria in DSM-5 are to be found in Box 7.1.

Box 7.1

A. Two (or more) of the following, each present for a significant portion of time during a 1-month period (or less if successfully treated). At least one of these must be (1), (2), or (3):

1. Delusions.
2. Hallucinations.
3. Disorganised speech (e.g. frequent derailment or incoherence).
4. Grossly disorganised or catatonic behaviour.
5. Negative symptoms (i.e. diminished emotional expression or avolition).

B. For a significant portion of the time since the onset of the disturbance, level of functioning in one or more major areas, such as work, interpersonal relations or self-care, is markedly below the level achieved prior to the onset (or when the onset is in childhood or adolescence, there is failure to achieve expected level of interpersonal, academic or occupational functioning).

C. Continuous signs of the disturbance persist for at least 6 months. This 6-month period must include at least 1 month of symptoms (or less if successfully treated) that meet Criterion A (i.e., active-phase symptoms) and may include periods of prodromal or residual symptoms. During these prodromal or residual periods, the signs of the disturbance may be manifested by only negative symptoms or by two or more symptoms listed in Criterion A present in an attenuated form (e.g. odd beliefs, unusual perceptual experiences).

D. Schizoaffective disorder and depressive or bipolar disorder with psychotic features have been ruled out because either (1) no major depressive or manic episodes have occurred concurrently with the active-phase symptoms, or (2) if mood episodes have occurred during active-phase symptoms, they have been present for a minority of the total duration of the active and residual periods of the illness.

E. The disturbance is not attributable to the physiological effects of a substance (e.g. a drug of abuse, a medication) or another medical condition.

F. If there is a history of autism spectrum disorder or a communication disorder of childhood onset, the additional diagnosis of schizophrenia is made only if prominent delusions or hallucinations, in addition to the other required symptoms of schizophrenia, are also present for at least 1 month (or less if successfully treated).

A number of observations can be made at this point. First, like all diagnostic criteria, it's like a sorting algorithm. You take any individual, and this sorts them into the category of having schizophrenia or not having schizophrenia. This is unsurprising, given the nature of diagnosis. However, a follow-up observation is that this sorting algorithm is *disjunctive*. By this, I mean that it's a bit like taking a room full of people and dividing them into two groups. In one group, you have people who have brown hair or wear glasses (the logical disjunct is an "inclusive or," its extension here being those who have brown hair, glasses and both brown hair and glasses) and in the other group, everyone else. There's nothing wrong *in principle* with this sort of disjunctive sorting. For example this is rather common when we know about the underlying cause or pathogenic process. Thus, a respiratory virus may cause a sore throat, a dry cough or both. This means that two patients might rightly get the same diagnosis but not have overlapping symptoms (one has a sore throat, the other a cough). However, if you reflect on the absence of this level of understanding in the case of schizophrenia, the fact that two individuals might present with no overlapping symptoms at all, and yet still get the same diagnosis (and hence the same prognosis, treatment plan etc.), looks more worrying. Since we'll dedicate an entire part of this chapter to the case against the diagnostic category of schizophrenia, we'll leave this observation here.

Another observation worth making is that, in spite of what we have just said, there is quite a strong focus on differential diagnosis (see sections D, E and F in Box 7.1), and, hence, there appears to be a significant concern that someone shouldn't be misdiagnosed. One might then ask: Why is it so important that the schizophrenia diagnosis is handed out with such care? Why, for example, is it so important that someone with schizoaffective disorder shouldn't get a schizophrenia diagnosis? This speaks perhaps of the perceived severity, and seriousness, of the schizophrenia diagnosis. As we will see in more detail shortly, it comes with huge amounts of stigma (including self-stigma), and it is also deemed to be a very serious condition, with a poor prognosis, and far-reaching implications with various institutional entities. It could affect your career, your life insurance, car insurance (if you've been allowed to drive, that is) and so on.

7.2.2 Epidemiology of schizophrenia

Who gets schizophrenia? There are a number of dimensions along which interesting statistical patterns have been observed and reported.

For example along dimensions of sex/gender, Saha et al. (2005) report that schizophrenia is diagnosed 1.4 times more frequently in men than in women. (Note that I say "sex/gender," since it is unclear whether this effect is due to biological sex or the social gender roles or a combination of the two.) Age of onset is also typically earlier in men (the peak of onset being 20–28 years for men and 26–32 years for women), and the outcomes (and likelihood of full recovery) are worse. Although in general, onset tends to be in early adulthood, some "very early onset schizophrenia" before the age of 13 has been reported (Da Fonseca and Fourneret 2018). There are also categories of "late onset" (40–60) and "very late onset" (after 60). The very late onset cases are more prevalent in women and associated with the menopause (Gupta et al. 2012).

Other interesting patterns include what's called "urbanicity." In other words, those born in cities are at greater risks: the bigger the city, the greater the risk, and the longer you have lived there, the greater the risk. There also appears to be an increased risk among immigrant populations.[2]

Other risk factors, which are often seen as conflicting explanatory approaches, are trauma and genetics. The studies vary, but, for example Kendler, Masterson and Davis (1985) report that having a first-degree relative (parent or sibling) with schizophrenia increases one's risk of also developing schizophrenia by at least 18-fold (up from 1 per cent in the general population, so up to 18 per cent). Ten per cent is the figure more widely reported. As for trauma, Morgan and Gayer-Anderson (2016) report that childhood adversity increases risks of developing psychosis by two to four times.

Although these are often seen as pulling in opposite directions (and as being promoted by different "camps," with trauma-based psychologists talking about psychosis, and genetics-based psychiatrists talking about schizophrenia), they are integrated in the well-known diathesis-stress model (Meehl 1962). According to this model, schizophrenia arises from the interplay of genetically disposed vulnerabilities that get activated by certain "stressors."

At this point, it is useful to make some observations regarding the epidemiology. Cross-cultural variations and different patterns of onset point in a similar direction, namely, that it could seem that different phenomena are being grouped under one category. One obvious fault line appears to be between internalistic and externalistic factors. But perhaps these aren't competing explanatory models of the same thing: perhaps they are good attempts at making sense of different phenomena. For example there might be the purely "biological cases": clear and sudden age of onset, first-degree relatives and no discernible childhood trauma. Then there might be trauma-based cases, where there might be no visible genetic factors but clear and extreme traumatic personal history. These differences might not just relate to different ways of causing the same psychotic condition that might be called schizophrenia – it might also align with differences in presentation. For example the trauma-based cases might have psychotic symptoms with themes and content that are un-mysteriously linked to the precise trauma in question. A purely biological case, in not having a content-laden, meaningful trigger is less likely to have such a link.

Having said that, perhaps a greater understanding is achieved once we start theorising about underlying causes and mechanism. Furthermore, this may tie seemingly disparate presentations together as different symptoms of the same underlying disease or condition. We turn our attention to such mechanistic explanations now.

7.3 Mechanistic accounts

Before presenting attempts to explain the mechanisms underlying schizophrenia, there are three points that I'd like to make.

First, we need to be mindful of whether the accounts in question are intended as accounts of *psychosis* or accounts of *schizophrenia.* In other words, are they accounts of a mental state that is disconnected (experientially and doxastically) from reality? Or are they accounts of a particular underlying disease, which is one of many potential causes of being so disconnected? Sometimes, it is not entirely clear. What makes the difference between an account being specifically about schizophrenia or more generally about psychosis is the extent to which i) other routes to psychosis are countenanced (if all psychosis is seen as variants of this one account, then the account begins to look like an account of psychosis rather than schizophrenia *per se*), and also ii) a commitment is made to some kind of underlying condition. As we look at different accounts, it will be clear how well they work as (candidate) accounts of psychosis or schizophrenia. Second, not all of these accounts are necessarily mutually exclusive. They may be compatible with or even feed into each other. Third, when these are accounts of schizophrenia, they are often driven by a unifying objective. In particular, they want to show how the different symptoms and symptomatic presentations can be explained in terms of one underlying deficit. This doesn't only carry explanatory power (see Chapter 6), it also goes some way towards justifying why so much variety gets grouped together under the label "schizophrenia." However, as many critics have pointed out, this arguably puts the cart before the horse. You don't *first* group disparate phenomena together and *then* look for what they could possibly have in common. To the person with a hammer, everything looks like a nail, and if you look for long enough, or at the right level of abstraction, you can find similarities between almost anything.

7.3.1 Source-monitoring accounts

One desirable feature of an explanatory account is that it can account for a variety of symptoms in terms of one overarching factor (e.g. deficit or abnormality). One very popular candidate for such a factor is the inability to attribute events to their correct source, in particular, to recognise them as coming from oneself. Although we will go on to talk about other "symptoms," let's start with arguably the most canonical symptom, namely, auditory verbal hallucinations (a.k.a. hearing voices, which we will focus on exclusively in the next chapter). The question arises: where do these voices come from? And the answer is: they come from the hallucinating individuals themselves, but are not recognised as such. Then a follow-up question emerges: why are they not recognised as self-produced? Note that we could ask this with both justificatory and causal uses of "why" (see Chapter 6; What is the cause?" or "What are the grounds?")

This is all by way of introduction, and I will present some answers to these questions shortly, and also in much more depth in the next chapter. However, I'd like to start by clarifying two different accounts, from different methods and traditions, that all-too-often get conflated. They are certainly similar and, indeed, *compatible*, but for the sake of clarity, it is worth distinguishing them. They are: *source-monitoring* and *self-monitoring accounts.*

Source monitoring is most clearly understood in the context of a memory task. As opposed to recollecting the *content* of a memory (what it is about), you can recollect the *source* of the memory. What counts as a "source" varies enormously. For example Vinogradov et al. (1997) write, "the source of information is the spatial, temporal, and contextual characteristics of an event as well as the sensory modalities through which it was perceived" (p. 1530). In a classic review of source monitoring (in general, not in the context of schizophrenia or psychosis), Johnson, Hashtroudi and Lindsay (1993) suggest that the term "source monitoring" subsumes at least three distinct abilities.

1. *Internal source monitoring* – distinguishing one's real actions (verbal and bodily) from merely imagined ones.
2. *External source monitoring* – distinguishing between outer sources (e.g. one third party from another).
3. *Reality monitoring* – distinguishing between self-generated and outer events.

Since at least Bentall and Slade (1985), there has been the hypothesis that this notion of source monitoring could help to explain schizophrenia/psychosis, especially, the third category of reality monitoring.[3] In other words, the hypothesis at the centre of source-monitoring accounts is that people with schizophrenia/psychosis are bad at distinguishing between self-generated and outer events. Not only that, but they also have a bias in a particular, externalising, direction: they have a *general propensity* to mistake self-produced events for external events. For example a self-produced piece of imagery, either in the form of inner speech or episodic memory, is liable to be misattributed to an outside source. A number of psychological studies have investigated the performance of different populations on various tasks that are deemed to be source-monitoring tasks (see Brookwell, Bentall and Varese (2013) for an excellent review).

For example Johns and McGuire (1999) got participants (clinical participants and controls) to read out words into a microphone and instantaneously played their voices back to them via headphones, with varying distortions in pitch. Sometimes, what the participant hears is indeed their voice, distorted, whereas in other conditions it is not. Subjects with a diagnosis of schizophrenia were significantly less good than control participants at correctly identifying their own speech. Interestingly, within the schizophrenia group, voice hearers fared less well on the task than non-voice hearers.

One concern with these sorts of studies is that, although the data are fascinating, they are open to interpretation. Source monitoring is a task, not a process in itself, and failures in source monitoring could be caused by more fundamental deficits that are central to schizophrenia, rather than being explanatory of schizophrenia itself. For example what if deficits in working memory, or attentional control, could explain why these individuals are bad at source monitoring, but it is actually these deficits that are more central to the phenomenon of schizophrenia? What you would then have is a basic underlying deficit that causes source-monitoring deficits as well as other symptoms of schizophrenia.

This hypothesis would make source-monitoring deficits more of something that *needs* explaining, rather than something that *does* the explaining.

Before moving on to what I am calling self-monitoring accounts, I would like to make a caveat of a terminological nature. Some of the literature (e.g. Johns and McGuire 1999) refers to behavioural source-monitoring tasks as "self-monitoring." That is fine, but it is not what I am calling self-monitoring, since that is a neurocognitive mechanism rather than a behavioural task. The reason these theorists refer to "self-monitoring" is that these tasks are the sorts of source-monitoring tasks that involve judgements about self or non-self (me versus not me), that is what I prefer to call (following Johnson et al. 1993) "reality monitoring."

Now we look at the importantly different self-monitoring neurocognitive mechanism, problems with which may or may not be responsible for reality monitoring deficits.

7.3.2 *Self-monitoring accounts*

Unlike source monitoring, which has its roots in personal-level behavioural experimental tasks, self-monitoring has its roots in the subpersonal, namely, in physiology and cognitive neuroscience. In other words, whereas source monitoring is about a (personal-level) *task*, self-monitoring is about a (subpersonal) mechanism. There may be compatibility in the sense that the task may be underpinned by the mechanism, or, more accurately (especially when considering psychosis/schizophrenia), errors in a source-monitoring task may be caused by underlying issues with the self-monitoring mechanism. Nevertheless, this is far from being a trivial claim, and there isn't a straightforward translation between the two frameworks.

The postulated mechanism of self-monitoring can be traced back at least as far as Helmholtz (1866). His explanatory concern was not with schizophrenia, nor psychosis, nor indeed with memory, but with the following foundational problem presented by vision. When an image moves across the retina, for example, as your eyes pan across a room, how does our brain know whether it is the world moving across our eyes or our eyes moving across the world? Helmholtz's suggested solution is as follows. Our brain can tell the difference because when our eyes move, there is a motor command. More specifically, information about the motor command, which Sperry (1950) later dubbed the "corollary discharge", is used by the brain to predict the sensory consequences that would be produced by the eye movement. If the predicted and actual sensory consequences match, then the brain infers that the change was self-generated, and the conscious percept is adjusted accordingly.

We can experience for ourselves what happens when there is no such motor command, and hence no such adjustment, when we press on our eye with our finger (I'd recommend for reasons of hygiene that you do this gently through your eyelid, not touching your eyeball directly!). When we do this, since the movement is not produced by the usual endogenous ocular processes, there is no motor command and no correction of the percept and hence the world itself seems to move (even though it is of course only our eye moving).

It took over a hundred years for Helmholtz's ideas to be applied to psychosis (Feinberg 1978). Feinberg's initial paper was on "thought" (which he took to involve "motor mechanisms") and thought insertion. However, the symptoms of schizophrenia where the self-monitoring account is most clearly introduced are delusions of control, since it is clear that, if anything involves motor commands, bodily actions do (Feinberg's idea that thoughts are motoric is, by comparison, somewhat controversial, although see Jones and Fernyhough 2007).

In delusions of control, a subject may perform actions that are in keeping with her plans and intentions (e.g. she might brush her hair), but she claims that somebody else is controlling her (and, of course, because that's deemed a preposterous notion, the person is deemed delusional). Frith and Done (1989), in a now-classic paper, took this to be a problem with self-monitoring. In particular, there is a mismatch between the predicted and actual sensory consequences of the bodily movement, and so (as with Helmholtz's ocular example) the movement is attributed (N.B. by the nervous system, not the person) to an external source. This, in principle, could be a problem with the generation of the prediction itself (e.g. the corollary discharge) or with the mechanism that compares expected and actual sensory (including proprioceptive) input, what became known as "the comparator". Later manifestations of the self-monitoring account (Frith, Blakemore and Wolpert 2000) saw comparator-based self-monitoring as the human body's way of meeting a computational challenge, in particular, involving skilled reaching and online correction (see Wolpert 1997 for a review of these computational approaches to motor control). This connection, in the late nineties, between the computational neuroscience of healthy cognition and the neuropsychology of schizophrenia undoubtedly contributed to the appeal and credibility of self-monitoring accounts.

Whereas in Helmholtz's example, the recognition by the nervous system that a certain stimulus is self-produced causes a correction of the conscious percept (in line with "I'm moving, not the world"), in more typical bodily motor control, it results in *sensory attenuation*. In other words, sensory consequences that are deemed self-produced are felt less strongly. The evolutionary benefit of this is clear enough: your nervous system needs to pay attention to stimuli that come from the outside, not the endogenous stimuli that (in a well-functioning system) will be harmless and irrelevant. Various data suggest that something goes wrong with this monitoring and subsequent attenuation (Blakemore et al. 2000). The most striking is the reported finding that subjects with diagnoses of schizophrenia can tickle themselves. The postulated explanation for this is that there is a mismatch between expected and actual sensory consequences, and the sensory consequences are not attenuated: the tickling sensation is as intense as being tickled by somebody else. "Typical" subjects can't tickle themselves because their nervous systems accurately monitor, and successfully attenuate, the sensory (in this case tactile) consequences of the tickling movements (Blakemore, Wolpert & Frith 1999).

To summarise then, it looks at first glance as though source monitoring and self-monitoring are synonymous. One coherent position is that source-monitoring failures are explained by failures in self-monitoring, and you'd be easily forgiven for thinking that by looking at how the research within the two paradigms gets put together. Closer to the truth is the claim that self-monitoring and reality monitoring (subset 3 of source monitoring) are the same, but even that is not quite correct. Self-monitoring refers to a *mechanism* whereby your *nervous system* uses information from its own motor commands to generate predictions about the consequences of action and then compares the predicted and actual sensory consequences as the action takes place. Reality monitoring refers to the *task*, performed successfully or unsuccessfully by *you as a conscious experimental participant* of determining whether something was generated by you or by the outside world (and hence providing genuine information about "reality").

One area where these two come apart is the milder, merely experiential (not doxastic, i.e. doesn't lead to belief) version of delusions of control, namely, *experiences of passivity*. In these cases (Balzan et al. 2013), patients claim that it is as if their actions are being controlled by others, or they feel odd, more passive than usual. And yet they don't go on to form the delusional belief that their actions aren't their own or are being controlled externally: they recognise that, in spite of this strangeness, the actions are their own. This points towards the fact that the low-level self-monitoring associated with Helmholtz and Frith generates primarily experiential distortions, and when these are intense enough, perhaps, or a wider interpretative system takes shape (see Fierro, Hernández and Malcolm 2018), these give rise to beliefs (it is worth noting that patients will frequently transition from passivity experiences to delusions of control (e.g. Frith, Blakemore and Wolpert 2000)).

This, in turn, nicely articulates why I find it problematic that the Johns and McGuire (1999) study (the vocal distortion replay study mentioned before) is presented in terms of "self-monitoring." On the contrary, in this study, the experiential distortion is *digitally achieved*, and, yes, it seems like participants with schizophrenia diagnoses (and especially voice hearers among them) are more likely to misidentify their voices on that basis. But the experiential distortion is actually there in the auditory stimulus that the participant is being presented with. It is not generated by unusual processing in the participant's nervous system. For Frith, in contrast, self-monitoring deficits are what yield the experiential distortion in the first place.

In terms of explanatory power (namely the explanatory virtues that a hypothesis has *regardless of accuracy*), the self-monitoring hypothesis is more powerful, since it postulates a mechanism. Nevertheless, suppose we grant that it is correct: people with schizophrenia (or rather their nervous systems) have self-monitoring deficits. The question then arises: *why do they*? (clearly this is a *causal* rather than *justificatory* use of "Why?," equivalent to "How come?" see Chapter 6).

What sort of answer would address this? Potentially, an appeal to differences in anatomy or brain chemistry might. A hypothesis of the latter type is what we explore now.

7.3.3 The dopamine hypothesis

Working within a framework where the focus is on neurotransmission (rather than, say, neuroanatomy), there have been several versions of a hypothesis involving dopamine, as well as some other hypotheses involving other neurotransmitters, too, most notably glutamate. For ease, we will focus on the simplest version of the dopamine hypothesis. This is simply the hypothesis that hyperdopaminergia (excessive amounts of dopamine in the brain) is involved in the aetiology of schizophrenia. This hypothesis emerged from the discovery of anti-psychotics in the 1950s (Delay 1952) and the later observation in the 1960s that these drugs (whose mechanism of operation was, at the time of their discovery, poorly understood) increased the metabolisation of dopamine (i.e. the rate at which it dissipates) in mice. Then, in the 1980s, it was shown that amphetamines, dopamine agonists (i.e. substances that elevate dopamine levels) can aggravate or even induce psychotic symptoms (Kokkinidis and Anisman 1981). This led Laruelle and Abi-Dargham 1999 to memorably hypothesise that dopamine is "the wind of the psychotic fire." At this point, the dopamine hypothesis was in its infancy but gave rise to decades of intensive research.

Although it would be extremely interesting to review this important research, it would take us too far afield (interested readers might like to look at Howes and Kapur (2009) for an accessible review detailing three successive developments of the dopamine hypothesis). We must not lose sight of what is philosophically interesting here, and that is the very notion that, regardless of the precise details, a neurotransmitter might hold the key to understanding schizophrenia. How, even if we suppose the hypothesis to be true, do we get from dopamine to schizophrenia and, in particular, the experience of psychosis? Suppose psychosis can indeed be caused by too much dopamine in the brain (or excessive activation of D2 receptors or whatever your favoured version), how do we get, in an explanatorily satisfying way, from one thing to the other? To make matters more complicated, dopamine has also been appealed to (as we will see in more detail later in Chapter 11) in understanding addiction. How can this one magical substance hold the key to both psychosis and addiction, problems which, on the face of it, seem very different, with one being about informational issues, and the other about motivational ones.

A first vital step involves understanding what dopamine does in general. Dopamine levels fluctuate in our brains all of the time in ways that don't induce psychosis (or addiction) and indeed underpin healthy neural functioning. So what does dopamine really, fundamentally do? In the public imagination, dopamine is associated with pleasure. This might partly be because it's a conceptually easy equation, and also the sorts of dopamine-depleted states after heavy use of dopamine agonists (e.g. a cocaine come-down) are deeply unpleasurable, indeed dysphoric.

More recent scientific opinion ties it not to pleasure but to motivational salience. Although there must be some truth to this, this helps to account for addiction, but sheds little light on psychosis (not to mention other conditions associated with dopamine depletion, like Parkinson's). Dopamine appears to be an all-purpose molecule that is used for different things depending on how and where in the brain it is used. In other words, dopamine plays a generic role in something utterly fundamental to cognition. This is why in certain contexts, it might play a role in accounting for the presence of bizarre experiences and beliefs (psychosis); in other contexts, it might play a role in accounting for a lack of agentive control (as in the case of addiction); in others, it may cause tremor and motor impairment (as in Parkinson's). But how can we think of cognition in such basic all-purpose terms, so that dopamine can play this generic role, with these specific applications?

One promising way of answering this involves adopting the Predictive Processing Framework (PPF) (Friston 2005; Clark 2013; Hohwy 2013). To my mind, the kind of shift in thinking that the PPF prompts (whether rightly or wrongly) is a bit like discovering the basic principles of elementary charge in chemistry. It doesn't seem to us in daily life like the matter in the universe organises itself according to these principles, but the discovery that it does is extremely powerful. It helps us not only explain chemical reactions that seem utterly miraculous, but also unite a large variety of disparate phenomena under unifying principles and mechanisms. Similarly, it doesn't seem to us like cognition works in the way that the PPF describes, but the realisation that it does, if indeed it does, is a powerful one.

I will state the position quickly here, and then flesh it out in the next section. According to this view, at some level of description, *all the brain ever does is minimise prediction error*. But the brain is a complex and hierarchically arranged prediction-error-minimising machine. Not only that, but it is embodied as well: it is the control-centre of a mobile organism. Thus, as we will see in far more detail, prediction error is sometimes minimised by updating your model of the world (yielding changes to percepts and beliefs), other times minimised by changing the world through action. Within this framework, dopamine is hypothesised to boost or turn the gain up on prediction error. No wonder, then, that it might force the adoption of strange models in the brain (psychosis) or force actions that the agent might more rationally not fully endorse (addiction). Let me now present the PPF and PP accounts of psychosis, in more detail.

7.3.4 Introducing the predictive processing framework

Since predictive processing is just a way of thinking about cognition in general, let's put psychosis to one side for the time being. We will bring it back in once we've seen what role dopamine is supposed to play in this framework. The predictive processing framework (PPF) rethinks how cognition fundamentally works, and so to best appreciate its importance, it pays to start with the "standard framework" in cognitive science against which it is pitched.

According to the standard framework, the brain's main task is to process information from the outside world. This information comes in the form of impacts upon our sensory receptors. When such information comes in, it gets passed along a processing chain, which, little by little, hones the information. One illustration of this way of thinking is in the diagram from Marr's (1982) book *Vision* as shown in Figure 7.1.

Clearly, there is a huge amount of variation among the theories and models that fall within this standard framework.[4] This standard framework is as intuitive as it is broad. We know, for example, that when we switch off the lights, we can no longer see. It is therefore natural to assume that our visual experiences are constructed out of information we get from light. The same can be said for sound and soundwaves. However, this reasonable intuition, as we are about to see, is importantly mistaken according to proponents of the PPF.

There are, in any case, good reasons to be dissatisfied with the standard framework. First of all, it requires the brain to wait for events in the world to happen and for information from those events to reach it. One might expect the brain to have evolved to be more proactive, more anticipatory. Second, according to the standard framework, the information that comes in early on in processing is relatively sparse and noisy, and then, as it gets passed on, it gets more and more complex and fine-grained. Given that on this model, the information at each stage is represented in neural activity, this means that, as we go up the cortical hierarchy, things get progressively more and more complicated and more and more costly in terms of the using of neural metabolic resources. One might expect the brain to be more efficient.

The PPF emerges as an alternative framework that presents the brain as both proactive and cost-effective. Instead of thinking of the brain as responding

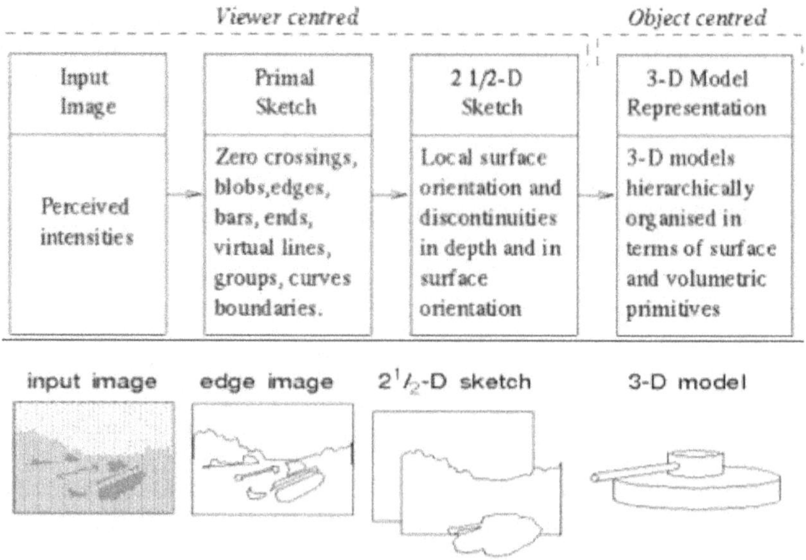

Figure 7.1

to – and constructing experience out of – impacts on the nervous system, we can instead think of the brain as predicting (or attempting to predict) rather than responding to sensory information. Conscious experience, on this view, is not constructed out of sensory impacts after the relevant event but is constituted by your brain's best predictions and, thereby, stays one step ahead. This of course is not to say that the environment plays no role: it plays a crucial, but somewhat different one. It triggers our predictions and then keeps them in check; it keeps them accurate. Thus experience, though massively flexible (and more flexible than in the standard view), isn't an "anything goes" situation: it is still, at least usually (hallucinations and delusions aside), answerable to what is going on in the world.

To understand the predictive processing framework in a bit more detail, it is worth reflecting on two things: *ambiguity* and *efficiency* – in particular, the ambiguity present in (even the cleanest) sensory signals and the fact that the brain has to operate as efficiently as possible.

The brain's main task is to settle on a hypothesis about what is going on, based on *ambiguous input*. This ambiguity means that a number of different hypotheses are compatible with the input (and, indeed, the same hypothesis is compatible with different inputs). Given this, how does the brain settle on one hypothesis rather than another? It needs to take into account: not just the fit of the hypothesis with the input, but also how statistically likely that hypothesis is, independently of that particular input. This is called the "prior probability" of the hypothesis and will often, but not always (it is arguably sometimes a function of innate biases), be assigned as a result of past experience (i.e. statistical learning). As a result, a hypothesis could fit the input extremely well, but its prior probability could be so low that it isn't even considered. Conversely, a hypothesis could have such a high prior probability that, even though it doesn't fit the input well, it is settled upon. As you may have noticed, what I have just described is Bayesian inference, and the predictive processing framework can be helpfully thought of as a neural implementation of this Bayesian strategy. (I note in passing that not all Bayesian accounts of cognition are predictive processing accounts. These agree broadly with the strategy if not necessarily the implementation (see Griffiths, Kemp and Tenebaum 2008 for a review).)

One nice illustration of a case where a hypothesis has such a high probability, but doesn't fit the input very well, is the case of the Hollow Mask Illusion. When you are presented with a rotating mask that is slowly turned to present you with the concave back of the mask, your brain "corrects" the concave stimulus into a convex stimulus. You experience the concave back of the mask as convex. This is due to the "convex face hypothesis" having such a high prior probability that it overrides the incoming signal. This is because faces are extremely important stimuli, and your brain has the expectation that the faces you will encounter will always be convex (a fair expectation!). The prior probability of the "convex face" hypothesis is so high that, even though the "concave face" hypothesis would better match the input, it is never selected.

It is fair to assume that, as a result of selection pressures, the brain will implement this disambiguation strategy as *efficiently* as possible. "Efficiently," in this instance, means performing the inference as quickly, as accurately and with as little energy expenditure as possible. This is where the notion of prediction comes in. Predictive strategies are used to maximise efficiency in informatics. In fact, it forms the basic principle of data compression. You save on bandwidth, when passing a message, by only passing on what is newsworthy. What counts as "newsworthy" is simply what the receiver of the message hasn't already predicted, namely, prediction error. Of course, what count as the "sender" and "receiver" of the message in the case of the brain are predominantly *within it*, namely, they are different parts of the cortical hierarchy.

This counts as an implementation of a Bayesian strategy insofar as the Bayes-optimal selection of a hypothesis determines a set of predictions about subsequent inputs, namely, inputs that are compatible with the hypothesis. If the hypothesis does a good job of predicting inputs, it will be kept. If it does a bad job, it will be tweaked or abandoned altogether in favour of another hypothesis. In other words, one hypothesis is selected rather than another if it better *minimises prediction error*.

There are now two important ways in which the PPF can be developed further. One concerns "precision weighting," and the other, the "hierarchical arrangement of hypotheses." We take these in turn. The first is particularly important for understanding the role of dopamine within the PPF.

Although incoming signals are ambiguous, in different contexts, the degree of ambiguity will differ. To maximise its predictions, the brain needs to accurately estimate how much ambiguity (uncertainty) there will be. In other words, it needs to make second-order predictions, namely, predictions about how much it should rely on its predictions (which amounts to how much it should pay heed to the prediction-error).

In contexts where low ambiguity is expected (high signal-to-noise ratio), higher precision will be demanded, and the prediction error will be taken more seriously. Conversely, when there is high ambiguity (or, through some top-down influence, there is no interest in the stimulus), low precision is demanded, and the prediction error will be taken less seriously. This is called "precision weighting" (Hohwy 2012), and it amounts to turning up (or down) "the gain" on prediction error. It is this precision-weighting that dopamine is fundamentally involved in (Corlett et al. 2010). This is central to all forms of brain function, since "all the brain ever does is minimize prediction error" and weighting precision is central to this.

I haven't the space to go into technicalities, but, so far, we have only mentioned how the brain generates hypotheses that fit the world. This is known as "perceptual inference." However, within the PPF, the brain also generates non-actual hypotheses that it then actualises through movement. For example it generates proprioceptive predictions about my arm going up, which then get fulfilled by my arm actually going up. This is called "active inference," and these action-ensuing predictions play the role of motor commands within the PPF. Again, without going too far afield, generating the non-actual hypotheses in a decoupled way (i.e. that doesn't have to align with reality) requires the turning down of precision. Executing the action requires in turn a rapid boost of precision, forcing the prediction to be fulfilled by bodily movement. If, as the PPF hypothesises, dopamine is involved in precision, we can see how a deficit in it might be used to account for Parkinson's. We will soon see how, in the context of perceptual inference, it can be used to account for psychosis.

The second important further development of the PPF is that hypotheses are hierarchically organised, with the hypotheses of one level providing the inputs for the next. "Higher" parts of the hierarchy are, roughly, those parts that are further away from the sensory stimulus. These tend to be at longer timescales and a higher level of abstraction. "Lower" parts of the hierarchy are closer to the sensory stimulus. These tend to be at shorter timescales and at low levels of abstraction. These, for example, correspond to early stages of visual processing: your brain's early statistically driven attempts to make sense of noisy inputs. Of course, in order to express these neurally encoded predictions, we need to use rough-and-ready descriptions in natural language (in this case English), but there is nothing linguistic about the priors/hypotheses ("light comes from above"/"This is a face") themselves.

Let's take an example (adapted from Pezzulo (2014)) to illustrate the predictive hierarchy. Suppose that, on the basis of a noise, which you take to be a squeaking window, two hypotheses present themselves about what is going on: either the wind blew the window, or a thief is clambering into your house. At the stage where those two hypotheses are competing, a great deal of ambiguity has already been resolved, in a Bayesian fashion, at lower levels of the hierarchy. For example in early stages of auditory processing, the qualities of the sound will have been settled upon, giving rise to the conscious experience being a certain way, qualitatively speaking. Higher up the hierarchy, that sound gets interpreted as a creaking window, as opposed to something else. The direction of causation is from the (events represented in the) lower regions of the hierarchy to the (events represented in the) higher regions of the hierarchy. However, the direction of the Bayesian inference is from the effects to the causes. What we get can be represented in the diagram given in Figure 7.2.

7.3.5 Predictive processing accounts of psychosis

With the lengthy interlude explaining the PPF in general terms out of the way, what has this all got to do with schizophrenia/psychosis? In a now-classic paper by Fletcher and Frith (2009), the PPF was taken to account for both hallucinations and delusions. A central tool in making this claim was this aforementioned hierarchy. Hallucinations are aberrant hypotheses *low* in the hierarchy; and delusions are aberrant hypotheses *high* in the hierarchy.

Since then, many papers of both theoretical and experimental varieties (see Griffin and Fletcher 2017 for an excellent overview) have explored psychosis through the lens of the PPF. I cannot do justice to the wealth of this literature

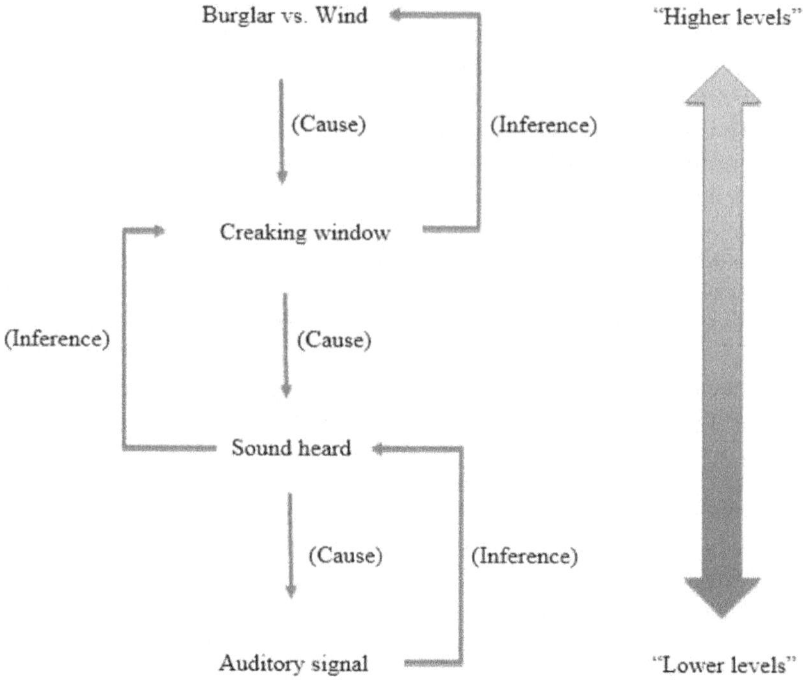

Figure 7.2

Source: Taken from Wilkinson, Dodgson and Meares (2017)

here, but I would like to make some very general observations, of a philosophical nature, about the powerful implications that the PPF has not only for understanding psychosis in particular, but also for understanding psychiatric illness more generally (it is no coincidence that the PPF will reappear in several subsequent chapters).

First, as mentioned, it sheds light on the role of dopamine in setting the "precision weighting" or "gain" on prediction error. It's a hugely important insight that dopamine is "wind of the psychotic fire," but *why* is hyperdopaminergia associated with psychosis? The answer (which is consistent with Kapur's "aberrant salience" dopamine hypothesis (2003)) is: because dopamine sets high precision weighting, which amounts to turning up the gain on prediction error. Dopamine fluctuates all the time and remains *appropriate* depending on context. Crucially, there is no such thing as having too much dopamine *full stop*, only having an inappropriate amount *for a particular context*. Precision (underpinned by dopaminergic activity) plays the functional role of second-order prediction: predictions about your predictions, namely, how seriously you should take your predictions (and hence generate more prediction errors), given the context. Thus, with too much dopamine, you take your predictions *too* seriously for the context. It is the equivalent of being in a poor informational context (e.g.

a darkened room), while your nervous system "thinks" that there is lots of high-quality information to be extracted. You then get hypothesis selection that is freewheeling, unconstrained and desperately trying to minimise all of this extra prediction error. *That* is why too much dopamine results in psychosis, according to the PPF. (Later, in Chapter 11, we'll see why in different parts of the brain, precision and dopamine play a role in action and hence addiction.)

Second, according to the PPF, conscious experience is much less reliant on input. Indeed, conscious experience is simply a function of your brain's best hypothesis, so, in principle, doesn't need any input at all. The role of input, as we have seen, is only to keep the brain's hypotheses on track. This means that when we come to explain experiences that correspond badly to what is going on in the world, like hallucinations, we don't so much need to ask ourselves, where did all of this input (or the illusion of this input) come from? Instead we ask: why did the brain adopt this erroneous hypothesis? And because our brain's hypothesis building is largely self-generated and only kept accurate by a very delicate process of correction, this doesn't only make the explanatory task more palatable, it also has a welcome normalising effect. The normally perceiving brain and the hallucinating brain are much more similar to one another within the PPF than within the standard framework. Indeed, according to the PPF, perception *just is* controlled hallucination. All of this, which applies to a better understanding of hallucination (see Wilkinson 2014), also applies to flashbacks in psychological trauma (Wilkinson, Dodgson and Meares 2017). Indeed, as we are about to see, flashbacks can be understood as your brain generating a very specific hypothesis in contexts where it is inappropriate to do so.

A third important consequence of the PPF is that it presents processing and learning (e.g. perception and perceptual learning) as two sides of the same coin. There is no sharp division between a programming phase and a deployment phase. Our brains "learn on the job," so to speak. On many standard models, especially classical computationalist ones, the brain is like a computer: it takes inputs and generates outputs, and the input/output function, at whatever stage in processing, is determined either by something innate or learnt. The processing and the learning are kept very separate. This is particularly relevant when we think about psychological trauma. On these accounts, trauma can only be understood as an event that has somehow broken the system. Traumatic experience and non-traumatic experience are radically different in kind: while the latter uses the system, the former breaks it. The PPF, on the other hand, views all encounters with the world as involving learning to some degree. It's just that a traumatic experience involves a great deal of learning for a very specific context that is maladaptive in the vast majority of contexts that the person finds herself in. But the fundamental process of a traumatic and non-traumatic experience is fundamentally the same. This has potential consequences for thinking about how trauma and certain predispositions (perhaps genetically acquired) may interact in the genesis of psychosis.

A closely related fourth benefit of the PPF is that it can (in principle) account for psychosis (and other symptomatology) that is the result of psychological trauma (or adversity) and psychosis that is not. This is because dopamine can function as a result of learning (namely, undergoing certain experiences), but there may also be more directly biological, perhaps genetic, factors in dopaminergic production and metabolisation. Indeed, this may help to explain why two very different etiological pathways (namely adversity and genetics) can end up with similar end results. Indeed, there may also be scope to explain why in different phases of the life-course, the nervous system is more susceptible to trauma. As we just said: processing and learning are two sides of the same coin: our brains learn on the job. But, they tend to learn more, and more rapidly, the earlier on in the job they are! Therefore, the younger someone is when they are subjected to trauma, the more profound and generalised the maladaptive learning will be.

7.4 Phenomenological approaches

Some theorists approach psychosis from a radically different perspective. Whereas mechanistic accounts try to hypothesise about what is going in the nervous system so that the presence of the experienced symptoms can be explained, other theorists take a close examination of the experienced symptoms as their starting point. What is it like to have schizophrenia? What is it like to experience psychosis? These, roughly speaking, are phenomenological approaches.

7.4.1 Big P versus Little p phenomenology

What exactly are phenomenological approaches, and what do they offer? First of all, I think it is important to distinguish two senses of phenomenology that I like to call "Big P" and "Little p" phenomenology (see McCarthy-Jones et al. 2013).

Little p phenomenology just means paying attention to lived experience. What is it like to undergo a particular experience? What are the qualitative features of a particular experience? The results of this are modest and targeted. Indeed, Nayani and David (1996), McCarthy-Jones et al. (2014) and Woods et al. (2015), all refer to themselves as "phenomenological," and what they do is ask people about their experiences (in survey or interview form).

Big P Phenomenology is an entire method of philosophy, incorporating various branches (or "schools"), whose inception is usually associated with Edmund Husserl. We haven't the space for a digression on the various schools of phenomenology, although Heidegger (Husserl's one-time student) focused on being and existence, giving rise to what might be called "existential phenomenology." Sartre and Merleau-Ponty were in turn heavily influenced by Heidegger, with the former giving rise to existentialism (which uses but goes beyond phenomenology in its methods), and the latter becoming distinctive, and very fashionable to this day, for being the phenomenologist of *embodiment*.

Ultimately, whatever branch of Phenomenology you look at (and they vary significantly in method, focus and ontology), the results are of a broad and very general nature. They are not (in contrast to little p phenomenology) typically about this or that experience. Indeed, Phenomenology (especially in Husserlian traditions) often presents itself as uncovering the fundamental structure of experience.

Although the theoretical distinction between Big P and little p phenomenology is clear enough, this is not to say that research and theorising always neatly falls into one camp or the other. For example phenomenological approaches to psychosis (or "phenomenological psychopathology" more generally (see Stanghellini et al. 2019)) can make use of insights from the Big P phenomenologists but use them to make targeted claims about specific experiences of psychosis. At other times, insights from psychosis, or other experiential exotica, are taken to shed light on fundamental but normally unnoticed aspects of experience.

7.4.2 *Ipseity accounts of schizophrenia*

Sass and Parnas (2003), like Frith and many others, seek to unify the "disparate assortment of signs and symptoms" associated with schizophrenia under one explanatory abnormality. However, unlike mechanistic accounts, the abnormality in question is not to do with cognitive processing, or neurobiology, but in conscious experience (i.e. phenomenology). The precise nature of this abnormality is in the experience of *self*. To state their hypothesis in their own words, schizophrenia is fundamentally an "ipseity disturbance" (from the Latin *ipse/ipsa/ipsum* meaning himself/herself/itself). What does *this* more concretely amount to?

To simplify somewhat, they claim that there are two features that are central to schizophrenia: "hyper-reflexivity" and "diminished self-affection." These two features "are best conceptualised not as separate processes but as mutually implicative aspects or facets of the intentional activity of awareness" (Sass and Parnas 2003, p. 430). Roughly, hyper-reflexivity involves an excessive and disruptive awareness of self which interferes with the interactions and mutual construction of self and world. Usually in our experience of the world, the self is tacit, something *through* which we experience the world. This very same process (of hyper-reflexivity) can be reported as diminished self-affection because that which was tacit is now no longer being inhabited smoothly as a medium of taken-for-granted selfhood. This diminishes the "self-feeling of self" (p. 430).

They then isolate "three syndromes of schizophrenia" (i.e. symptom clusters) and proceed to show how hyper-reflexivity and diminished self-affection can shed light on all three. Those are positive symptoms, negative symptoms and disorganisation symptoms. Let us quickly take these in turn.

For Sass and Parnas, positive symptoms are

> defined . . . by a loss of the sense of inhabiting one's own actions, thoughts, feelings, impulses, bodily sensations, or perceptions, often to the point of feeling that these are actually in the possession or under the control of some alien being or force.
>
> (p. 431)

Negative symptoms (which include lack of motivation, lethargy, inability to derive enjoyment from life, asociality) "are in fact not seen as straightforward deficits but involved the presence of positive aberrations of various kinds" (p. 433). In contrast to patients suffering from depression who "report a quantitative decline in energy, mental intensity and the ability to think efficiently, schizophrenia patients typically report a qualitative alteration of thought and perception that is far more difficult to describe" (Cutting and Dunne 1989) (p. 433).

Through a variety of highly suggestive phenomenal logical descriptions and theorising (drawing, for example, on Blankenburg's (1971) account of schizophrenia as "the loss of natural self-evidence," namely, obviousness, taken-for-grantedness) they support the view that, contrary to orthodox approaches, negative symptoms are not a "lack" of something but the positive presence of deep alterations to self-experience (ipseity) which can from the outside look like a lack. It is impossible to summarise the rich tapestry of phenomenological anecdote, but there is a very striking description from Antonin Artaud (1965), exemplifying the opacity and object-like nature of what is typically transparent and taken for granted: "the limbo of a nightmare of bone and muscle, with the sensation of stomach functions snapping like a flag in the phosphorescence of the storm" (Artaud 1965, pp. 28–9).

Third, the disorganisation symptoms, namely, "abnormalities in the organisation of thought, speech and attention," are also explained as surface features of underlying disturbances in ipseity: "these disruptions of the sense of existing in a coherent perceptual or cognitive universe are especially likely to involve hyper-reflexive distortion of the normal forms of awareness" (Sass and Parnas 2003, p. 436).

7.4.3 *Psychosis as disruption to the anticipatory structure of experience*

Other phenomenological accounts aren't claims about the fundamental nature of *schizophrenia*, and indeed it might strike one as noteworthy that Sass and Parnas seem to be making such strong essentialist claims about a purported disease on phenomenological grounds. Other accounts seek to understand psychotic experiences and experiential disturbances in phenomenologically sophisticated ways, but without any commitment to diagnostic entities or disease entities.

One leading figure in this kind of phenomenological investigation is Matthew Ratcliffe. Although there is some focus on the self, Ratcliffe (2017b) has denied that schizophrenia fundamentally involves disruptions to the minimal self (ipseity) and has gone as far as to deny the tenability of the distinction between minimal and interpersonal self. He suggests that schizophrenia should be understood relationally rather than individualistically. In any case, Ratcliffe's focus is on psychosis and on experiential disruptions, more generally, rather than specific diagnostic categories.

Ratcliffe's central insight in his excellent book *Real Hallucinations* (2017a) is that different intentional states (states that have "intentionality" or "aboutness") differ in virtue of their intentional structure or profile. Typically, we recognise our experiences as perceptual, imaginative, recollective, etc., because of this structure and how they unfold. In psychosis, there is a disruption to this structure, and there is a blurring of the boundaries between these kinds of states. A central claim of this book, as reflected in the title, is that the *real* hallucinations that we actually encounter in psychiatry, aren't (as philosophers of perception have explicitly stated in thought experiments, but philosophers of psychiatry have too readily assumed) straightforwardly sensory-perceptual states, potentially qualitatively (phenomenologically) indistinguishable from veridical perception, but simply where the worldly stimulus or object is missing. Instead, they involve a blurring of boundaries, a perceptualisation of the typically non-perceptual. Hallucinations (especially verbal ones) are the topic of our next chapter, but in his book (and in several papers), Ratcliffe advances crucial insights about psychosis more generally, especially in his Chapters 5 and 6.

To give a brief flavour, Ratcliffe shows how the structure of intentionality depends on a "global style of anticipation" (p. 136). This global anticipatory style (drawing judiciously on Husserl's notion of "horizons" and Wittgenstein's *On Certainty*) forms an affectively laden "bedrock" of "habitual expectations" (p. 160). When this anticipatory structure gets disrupted, either neurobiologically or through trauma (there is an excellent discussion of trauma, and torture, in the book and how it can shatter this trust, this bedrock), you get the kinds of experiential exotica, the disconnections from reality associated with psychosis and its prodrome (its gradual onset, prior to full-blown, a.k.a. "frank" psychosis). (You may have noticed an affinity with the predictive processing accounts mentioned earlier in this Chapter, and Ratcliffe points to this explicitly (p. 185).)

7.4.4 The explanatory contribution of phenomenological approaches

What is the role of phenomenological approaches? There are two kinds of contributions, the first of which is utterly uncontroversial, and they are as follows.

Contribution 1. Phenomenology gives us a more fine-grained appreciation of what it is we are trying to explain. This might result in subtyping, or in correcting misconceptions about what it is we are trying to explain.

Contribution 2. Phenomenology doesn't just give us descriptive tools to better understand and categorise the experiences that are *to be explained* by other means. It can, at its best, do a fair amount of the explanatory legwork itself. Depending on your views of explanation (see Chapter 6), it can render certain things less mysterious. It may also play an important unifying role in showing how seemingly disparate phenomena are actually different manifestations of one and the same phenomenological disturbance or feature (e.g. ipseity or anticipatory style). We may even allow ourselves to refer to phenomenological mechanisms. At its best this could give us vital clues for treatment since it could suggest a unified underlying target for intervention rather than a dizzying array of surface phenomena.

As I mentioned, *Contribution 1* is uncontroversial, and, although it is more modest that *Contribution 2*, it is a hugely important part of interdisciplinary psychosis research. Many will find *Contribution 2* obscure: ultimately, they will think, to really understand what is going on, we will need to do some neuroscience. I point the reader to Chapter 6 for reasons to be suspicious of this way of thinking about explanation. According to an appealing view, we do not "explain phenomena," we ask explanation-demanding questions that pertain to certain phenomena and do our best to answer them. Viewed in this way, there is no reason why phenomenology cannot be deeply explanatory.

7.5 The case against schizophrenia

There are a number of theorists who think that the label or category or "schizophrenia" should be dispensed with. There are several different kinds of such critique, motivated by different kinds of considerations. Jeffrey Poland (2006) helpfully distinguishes socio-political critiques and scientific critiques, and I will do the same. However, I think there is a third category that can all-too-easily be conflated with the socio-political category, and that is what I am calling pragmatic critiques. As we will see, both socio-political and scientific critiques want to invalidate the category of schizophrenia, but from very different perspectives, and with very different motivations. In contrast, a pragmatic critique doesn't focus on validity at all. The focus is on what a particular diagnosis *does*, and if what it does is damaging, then it shouldn't be used. The key point here is that a diagnosis can be "valid" and yet deeply damaging.

Briefly, before looking at these three kinds of critiques, let's remind ourselves of what the concept of schizophrenia is. Unlike psychosis, which is a state of mind, schizophrenia is an underlying condition, and, at a minimum, two separate diagnoses of schizophrenia need to pick out two underlying conditions that are either the same or sufficiently similar to warrant diagnostic practice (Chapter 3). What do we mean by the claim that schizophrenia is an "underlying condition"? Without wanting to repeat too much of Chapter 3, the strongest version of this claim would be to repeat the statement from the NIMH website we quoted earlier (and is quoted disapprovingly in Read, Mosher and Bentall 2004, p. 3) that schizophrenia "is a chronic, severe and disabling *brain disease*" (my emphasis). This is a particularly strong claim, and it is relatively easy to disagree with (see Chapters 2 and 3 for good conceptual reasons to not equate mental illnesses with brain diseases). However, we must note that disagreeing with this does not entail jettisoning schizophrenia altogether. According to a more moderate position (which many practicing psychiatrists would be sympathetic towards), there may well be a mental disorder (not a brain disease) that corresponds to schizophrenia. Furthermore (again, as we saw in Chapter 3), we mustn't take the "underlying" in "underlying condition" too geographically literally (i.e. as being in the brain). It is underlying in a causal/explanatory sense, and, if externalists are to be taken seriously (and they should be), the underlying pathology can extend into the environment.

7.5.1 Socio-political critiques

As Poland aptly puts it (2006, p. 167), socio-political critiques "have been concerned to un-mask the non-scientific and non-clinical interests, forces, ideas and processes that lead to the introduction of mental illness concepts like schizophrenia and to the maintenance of social practices organised in terms of them." As he goes on to explain, this serves to undermine the credibility and validity of the concepts. They purport to be robust, scientific concepts, but they are nothing of the sort: they are the products of "social and historical contingency" (ibid.). A valid and robust scientific concept would remain stable in nearby possible worlds: it would be the same if history had turned out differently and would not admit of cross-cultural variation. Not so concepts like schizophrenia: it is far too modally fragile (i.e. variable across the space of possibilities) to count. There are a great many critics who employ this method. I will briefly summarise two: R. D. Laing and Michel Foucault.

Laing's focus is on showing how, in spite of suggestions to the contrary, "schizophrenia" is not something "in" a patient. His central claim is that it can only be understood in its social context and, in particular, the context of the family. Though consistent with much of Laing's work, this is (in my opinion) most memorably presented in Laing and Esterton's (1964/2016) book *Sanity, Madness and the Family.*

This book reports and comments on lengthy interviews with a series of women who had been diagnosed with schizophrenia, along with their families. The central result, or rather claim, that emerges is that the experiences and behaviours that led to diagnosis only failed to make sense when taken out of context. The extent to which this is plausible will depend on the case in question, but, to give an example (one that Cooper 2017 also picks out), Maya seems to get her diagnosis largely on the basis of her belief that she can read people's thoughts. When the interviewers speak with her parents, it turns out that *they* seem to think that she can read people's thoughts too, so there is a clear reason for her beliefs (see Chapter 5 on the rationality and, all other things being equal, healthiness of aligning your beliefs with those around you).

The more wide-ranging applicability of the "family-context" theory is questionable, but the central thrust of the critique, that "schizophrenia" is not a disease, disorder or illness "with" the patient, but is only attributed as a lazy way of making sense of that which has been seen out of context, has significant appeal.

Foucault's focus, in contrast, is historical. In "*Folie et deraison: Histoire de la folie a l'age classique*" (Foucault 1961) (most commonly read in English under the title "Madness and Civilization"), he conducts a genealogy (a story about the emergence over time) of the modern concept of mental illness, including "schizophrenia," with a view to debunking orthodox narratives.

To summarise, and inevitably simplify, whereas mainstream historical narratives presented the nineteenth-century medicalisation of "madness" as a great progressive and scientific achievement, and saw a humane reversal of what Foucault calls "The Great Confinement" of the seventeenth century where behavioural undesirables were simply locked up to protect society, Foucault thought instead that this was the height of moral hypocrisy. Far from being a scientific *discovery* that madness was medical and needed treatment, it was a political *choice*. In other words, it was simply another more morally disguised way of controlling social undesirables.

Although different in the details, Laing and Foucault (and others like Szasz, Scheff, Sedgwick) share the overarching aim of presenting psychiatric categories in general, including schizophrenia, as being not what they seem, namely, not as robust and scientific. This appearance is a stunt. They are instead contingent, fragile, the product of social context and dubious political choice.

7.5.2 Scientific critiques

In stark contrast, scientific critiques don't suggest that there are any nefarious socio-political forces at play, whether accidental or deliberate. On the contrary, schizophrenia is indeed a scientific concept (rather than a socio-political concept that is disguised as one), it's just a rather *bad* one.

There are many examples of scientific critiques (see in particular contributions to Read, Mosher and Bentall 2004, and Chung et al. 2007). An especially explicit and self-conscious, scientific critique is Mary Boyle's *Schizophrenia: A Scientific Delusion?* (1990, 2002). Her entire first chapter is dedicated to a general elucidation of what it means for schizophrenia to be a scientific concept and of the sorts of standards, given that it is treated as such, it might be subject to. In other words, the question is: given that "schizophrenia" has the status of a scientific concept or construct, in terms of both the seriousness and objectivity with which it is taken in clinical contexts and the way in which it structures scientific research (e.g. functions as an inclusion criterion, is treated as a variable in clustering participants and analysing data, etc.), does it have validity commensurate with that status?

To simplify Boyle's argument somewhat, the claim that schizophrenia is a scientific concept implies that it was derived from the observation of a pattern of regularities (Boyle 2002, p. 8). The question then arises as to what kind of scientific concept we're talking about; what is being postulated? Interestingly, Boyle writes in her second edition (2002) of the now-universal rejection of schizophrenia as a disease, claiming that the term "schizophrenia" is now said to refer to a syndrome (and not a disease) which is usually described as a "clustering of symptoms and signs" (Morris 1978).[5] I am not sure how unanimous this is in the psychiatric community. Consider, for example, in addition to the NIMH quotation from earlier, Nancy Andreasen's claim: "schizophrenia is a brain/mind disease" (Andreasen 2001). Having said that, the NIMH appears to have dropped mention of brain disease, opting instead for the term "severe mental illness."

Nevertheless, whether "schizophrenia" as a concept (henceforth SZ) is a disease or syndrome is not of primary importance. The main point is that it is a *scientific construct*, and, regardless of its ontological status as disease, syndrome etc. (see Chapters 2 and 3 for discussion of some of these issues), it at a minimum requires "construct validity." Boyle emphatically denies that SZ has this kind of validity. She presents the criteria of validity in two parts, one of which is upstream, the other downstream. They are:

a) There exists a pattern of intercorrelated observable features from which the existence of an associated construct have been inferred, and
b) The construct has predictive power.

Against a), Boyle argues that at no point did those who introduced the concept of schizophrenia have the kind of evidence of patterns of observable features that would have justified the inference to a construct. Indeed, it seems that the cart was put before the horse. The construct was hypothesised first as a shot in the dark, which then acted as a distorting lens that reinterpreted clinical observation, which then became interpreted as different manifestations of what had been stipulated to be the same disease (whether its status as a disease is dropped is largely irrelevant since it still builds on an illegitimate genealogy). In a related manner, this and similar considerations are problematic for b). Boyle bolsters this with an extended critique of the genetic data. Other researchers also criticise both the unity and predictive utility on various grounds, from hypothesised patterns in differences to cognitive capacities, and task performance, to differences in brain structure (see contributors to Read, Mosher and Bentall (2004)).

In short, there is a great deal of heterogeneity in "symptomatic" presentation, treatment response and general outcomes. To paraphrase Bentall (2003), once you've explained the so-called symptoms, there's nothing left to explain. And if this is the case, we aren't really dealing with symptoms, since they are not the symptoms of anything. Bentall's suggested alternative is that we should talk of "complaints."

7.5.3 Pragmatic critiques

Both socio-political and scientific critiques target the validity of the concept of schizophrenia, the former by showing that it is the modally fragile product of non-scientific socio-political forces. It is worth noting an interesting assumption here: that purely scientific forces and motives must be a good thing (since their being polluted by non-scientific factors is *bad*). Scientific critiques accept that

SZ is the product of scientific forces at work, but it is just science done *badly*. Of course, the divide between the two kinds of critique is not so clear-cut, since, as Boyle (1990) and Poland (2006) rightly point out, the bad science can be explained in terms of the science being derailed by these non-scientific forces.

In stark contrast, pragmatic critiques aren't focuses on the validity of schizophrenia as a scientific construct. What they care about is what, whether scientifically valid or not, the schizophrenia label *does*.[6] Does it do more harm than good? And how does it interact with other aspects of society and institutions? As a result, are there societal improvements that can be made by changing the pragmatics of the discourse surrounding what is now called schizophrenia? What are the impacts of schizophrenia narratives on individuals with the diagnosis and their families? Across what other dimensions can we talk about its impact?

Notice that these pragmatic considerations can be marshalled in the service of defending the label as well as attacking it. Presented thus, a vast array of different views will fall under this pragmatic category, and I couldn't possibly present these comprehensively. However, I'll briefly present what I think is a particularly interesting avenue within this area. Following Tekin (2011), there is a hugely important question about how the schizophrenia label and psychiatric labels, more generally, impact on self-narratives. This is particularly interesting in the context of the social stigma of schizophrenia. The way in which getting a diagnosis may range from cataclysmic (see, e.g., Eleanor Longden's TED talk for a particularly powerful expression of this), to the providing of relieved sense-making ("I *knew* something was wrong with me"). Indeed, some have argued that a *simple renaming* of schizophrenia would have a significant effect on stigma. For example George and Klijn (2013) argue that renaming schizophrenia "Psychosis Susceptibility Syndrome" (PSS) would reduce stigma, especially self-stigma, and improve outcomes. Whether this is actually the case would remain to be seen, but it is certainly an interesting proposal, especially when you think about the tarnished and arbitrary history of the word "schizophrenia." (Although see Gaebel and Kerst (2019) for a sceptical perspective on the benefits of renaming schizophrenia.)

7.5.4 Where does this leave us?

First of all, let's distinguish a number of different claims.

(i) There is no such thing as schizophrenia, because that is a disease concept, and none (not even a subset) of those who are now, or have historically been, deemed to have "schizophrenia" has anything that resembles a disease.

This could be motivated by a broad antipathy to psychiatry, including a general dislike of the idea of underlying diseases or disorders. To paraphrase Richard Bentall (again), once you've accounted for the so-called symptoms, there's no underlying disease left to explain. This is a total paradigm shift that leaves no room for schizophrenia at all.

There could, however, be more moderate adherents to (i), who would agree that schizophrenia is not a *disease*, but it is a disorder or a syndrome. That may be more plausible, but if you make that move it is hard to avoid the historical fact that the word was introduced and used for the best part of a century to express a disease, an underlying pathology giving rise to various symptoms (and varying ones in different individuals). And we still talk of the symptoms *of* schizophrenia. And symptoms are always symptoms *of* something. Nevertheless, even if we grant these moderates that schizophrenia is not a disease, but is rather a syndrome or disorder, they are still susceptible to the following criticism.

(ii) There is no such thing as schizophrenia, because the category is hopelessly broad and vague. The set of people, today and in the past, who get the diagnosis are extremely heterogeneous.

This could be held by someone who is in principle on board with underlying mental disorders but simply bemoans the woolly heterogeneity of "schizophrenia." Note that this criticism of heterogeneity could be more or less extensive. At one end of the spectrum, one may claim there is no potential for clusterings at all: each patient is unique. Alternatively, there may be valid sub-clusterings that would form a valid disease category or, more tentatively, a syndrome category. In short, this approach bemoans that schizophrenia is a confused category, but there is nothing wrong in principle with psychiatric diagnoses in general, when they are clear. This leads us on to the next, weaker, criticism.

(iii) There is no such thing as schizophrenia, because the observable "presentations" of these patients are far too heterogeneous to hypothesise or discern a unified construct. However, if we embrace the heterogeneity, we can embrace a different label that is as broad, perhaps even broader, that is a spectrum syndrome.

This is precisely what George and Klijn (2013) point towards with their "Psychosis Susceptibility Syndrome" and also van Os (2016) with his "Psychosis Spectrum Syndrome." This is a "rebrand and relax" strategy.

(iv) Schizophrenia does exist, it is a disease, but diagnostic categories have acted in ignorance and cast their net too wide. We might for example think of there being a relatively narrow category of "schizophrenia proper," with which other forms of psychosis are too often conflated.

According to such a view, perhaps those "pure biological" cases that aren't trauma, adversity or substance-abuse related, and that have a clear link with family history (see Jablensky 2010, on the "familial type"), unified symptoms (thought disorder, prominently), a narrow window of age of onset, etc., are good candidates to occupy this status. Suppose, further, these cases had the homogeneity and underlying essence to constitute a disease. This constitutes less of a criticism of schizophrenia and more of a "tightening up." This raises two questions. First why should this (or any particular) subset retain the label, and not some other clustering or even the disjunctive remainder? One answer to this is that schizophrenia is a purported disease category and should track the discovered diseases. Second, why should we retain the word schizophrenia at all, given its baggage of stigma and its mangled conceptual history? Couldn't we divide and rename on all sides of the divide? Murray, for example (Murray 2006) has suggested "dopamine dysregulation disorder" as the label of a subset of what is currently called "schizophrenia."

To sum up, then, there are lots of different ways of being dissatisfied with the *status quo* when it comes to schizophrenia. At one mild end of the spectrum, our diagnostic category of schizophrenia needs tightening up, and perhaps a rebranding, or perhaps a careful subdivision. At the other end, schizophrenia, the label and the framework that gives rise to it as a concept, should be jettisoned entirely, probably along with many other psychiatric diagnostic categories.

Summary

- The word "schizophrenia" (which was more or less synonymous with what Kraepelin called "*dementia praecox*") was introduced by Bleuler to express a disease concept, in particular, one of a number of psychotic diseases/disorders.
- Schizophrenia is diagnosed, like any disease, on the basis of clear, but controversially disjunctive, diagnostic criteria.
- There are a number of mechanistic accounts of schizophrenia, some of which appeal to cognitive mechanisms of self-monitoring, others to neurotransmitter deficiencies and others to general processing within the brain.
- Phenomenological approaches to schizophrenia are important and insightful, although there are different opinions about their explanatory role.
- Finally, several theorists find that the concept of schizophrenia should be abandoned and that the word "schizophrenia" should be avoided. Some of these critics are motivated by broadly speaking sociological considerations, while others by theoretical ones.
- Two things are certain, though. First, the term "schizophrenia" carries some very undesirable historical and socio-political baggage. Second, the set of things that get called "schizophrenia" is too broad. You can find two individuals who arguably have the same condition, but you can also find two individuals who both have diagnoses of schizophrenia who clearly do not have the same condition.
- In light of this, it might be advisable to more carefully subdivide the current category of "schizophrenia" and/or to jettison the word "schizophrenia," not only for socio-political reasons, but also because it is not clear which of the new subdivisions should inherit the label.

Questions

- What does the word "schizophrenia" make you think of? (Do you think of voices, delusions or disordered speech? Or none of these?)
- How would you feel if you received a diagnosis of schizophrenia? How would you feel about people thinking of you as a "schizophrenic"?
- Why do you think we have held on to a word that has come into circulation in such an unregimented way?
- Does the dopamine hypothesis of schizophrenia seem plausible to you?
- How many different ways can you imagine being "disconnected from reality" (having "psychosis")? Can you try to be as varied and creative as possible? (What sorts of beliefs, emotions, experiences could be involved?)

Notes

1. It is worth noting that not all psychiatrists are comfortable thinking schizophrenia as a disease, but it certainly picks out an underlying condition, which produces symptoms.
2. Data suggests that, in the UK, African-Caribbean and Black African groups are at greater risk of psychosis (Fearon et al. 2006).
3. Although note that Richard Bentall has been one of schizophrenia's most vociferous critics.
4. Many models, which would still fall within the standard framework, allow for various top-down influences. It is therefore inaccurate to say that the standard framework is characterized by a uniquely bottom-up direction of processing. What does, however, characterize the standard framework concerns what the relevant causal antecedents of a given conscious experience are. The standard framework will say that the relevant causal antecedents are the environmental information. Models or theories that allow top-down influence simply say that the top-down influences impact on how that incoming information is received, processed and passed on. There is no question, top-down influences notwithstanding, that any given chunk of conscious experience is preceded by the relevant environmental event (which the experience is often thought to somehow represent) and constructed out of information (causally construed) from that event.
5. The observant reader will have noticed that the question then arises: "a symptom of what (if not a disease)?"
6. At this point, we might distinguish *labels* from *concepts*. The "schizophrenia label" is the public use of that word as it attaches to individuals and triggers a particular societal response.

Further reading

Frith, C., Blakemore, S.J. & Wolpert, D. 2000. Explaining the Symptoms of Schizophrenia: Abnormalities in the Awareness of Action. *Brain Research Reviews* 31 (2):357–63.

Griffin, J.D. & Fletcher, P.C. 2017. Predictive Processing, Source Monitoring, and Psychosis. *Annual Review of Clinical Psychology* 13:265–89.

Poland, J. 2006. How to Move Beyond the Concept of Schizophrenia. In M.C. Chung, B. Fulford & G. Graham (eds.), *Reconceiving Schizophrenia*, Oxford: Oxford University Press.

Read, J., Dillon, J., Read, D.J., Bentall, P.R. & Mosher, L. (eds.). 2013. *Models of Madness: Psychological, Social and Biological Approaches to Psychosis*, second edition, Abingdon, UK: Routledge.

8 Hearing voices

Overview

The phenomenon of hearing voices, sometimes (although, as we shall see, not optimally) called "auditory verbal hallucination," is a high-profile psychiatric symptom that both benefits from philosophical scrutiny and functions as a useful lens through which to think about the human mind. Voice hearing is far more common than many people realise and is subject to a number of other misconceptions. A related misconception is that, if you hear voices, you have a serious mental illness and probably schizophrenia. There are two further common misconceptions. First, hearing voices is a relatively unified phenomenon, both in terms of the phenomenon as it is in itself and in terms of its cause and aetiology. Second, "hearing a voice" is just like hearing a voice in the everyday case, except that there isn't anyone talking, namely, there isn't a speech sound out there in the world. In other words, it resembles a canonical hallucination: an experience that is very similar (even phenomenologically identical) to the sort of experience you might get in the presence of an environmental stimulus, except that there is no such environmental stimulus.

In this chapter, we will correct these misconceptions and then assess different attempts to understand the phenomenon of voice hearing. There is an account of voice hearing that probably deserves to be called the orthodox or mainstream account. This account is attractive but faces

DOI:10.4324/9781315450124-10

some challenges. We will then examine alternative accounts and see how they fare. Perhaps inevitably, there remain some explanatory challenges, but, ultimately, voice hearing is a fascinating lens through which to think about the nature of the human mind more generally.

8.1 The varieties of voice hearing

The phenomenon of hearing voices, often called "auditory verbal hallucination" (AVH for short), is a high-profile psychiatric symptom. Its most common association (unhelpfully, as we shall see) is with schizophrenia, but it occurs in other diagnostic contexts and, most importantly and overlooked, in contexts where the voice hearer is not in need of care and carries no diagnosis. Voice hearing occurs with an almost dizzying array of variety, and making some progress in presenting the variety and trying to extract some useful order from it are the purposes of this first section.

Voice hearing varies across a number of dimensions. In this respect, one might say that there are *varieties of variety*. There are, for example:

i) Varieties in the context in which they occur
ii) Varieties in the apparent nature of the phenomenon itself
iii) Varieties in the underlying nature of the phenomenon.

Let's address these in turn.

8.1.1 Varieties in context

When we talk about varieties in context, we could mean two quite different things. One operates at rather long timescales and has to do with the overall state or history of the voice hearer. For example this could include the *autobiographical context*. We could look at various forms of trauma and adversity, or alternatively, family history. On the other hand, we might mean the *diagnostic context* in which the voices occur. Voices are commonly associated with schizophrenia diagnoses, but this is unhelpful. Although voices will often incur a schizophrenia diagnosis, voices occur in many other diagnostic contexts, including, but not restricted to, schizoaffective disorder, bipolar disorder, major depressive disorder, post-traumatic stress disorder (PTSD), obsessive compulsive disorder (OCD), borderline personality disorder (BPD), dissociative identity disorder (DID) and a variety of anxiety disorders. Most importantly, however, voice hearing can occur in individuals without any clinical diagnosis and who are not in need of care (see, e.g. Larøi et al. 2012). This includes not only one-off occurrences under conditions of extreme stress or exhaustion, but also repeated voice-hearing experiences over extended periods of time. There is in fact a growing and supportive community of "non-clinical voice hearers," individuals who regularly, sometimes constantly, hear voices but who do not seek help and do not consider themselves psychiatrically ill. Some individuals have turned to these supportive networks (the most well-known being the Hearing Voices Network) after having gone through the ordeal of being "processed by" psychiatric services. Many report being saved by the realisation that there is nothing medically wrong with them.

The other thing that we might mean by "context" operates at short timescales. "Context" in this sense means what is going on immediately preceding and during a particular episode of voice hearing. Although correlation doesn't always imply causation, part of what we might be talking about here are "triggers." While the two clearly interact, we might here distinguish "psychological triggers" from "environmental triggers." Exhaustion, stress and anxiety (the first of these could also count as somatic) all count as psychological triggers. There are many potential environmental triggers, but one that is especially relevant is whether an environment is noisy or quiet. Some voices are only present in, or aggravated by, silence, while others are only present in, or aggravated by, noise (Dodgson and Gordon 2009). This is certainly something that needs explaining. Other environmental triggers include, for example, places that are reminiscent of past trauma. However, all of these environmental triggers require (of course) the subject to be in a particular psychological state, and so, somewhat trivially, the proximal trigger is always psychological.

8.1.2 Varieties in phenomenology

As we've seen in the previous chapter, phenomenology can have a very technical meaning ("Big P Phenomenology"), which is associated with the philosophical practice of "phenomenology," founded by Edmund Husserl. There are different schools of phenomenology, and it looks slightly different depending on which tradition you work within. But the word "phenomenology" has also acquired a far less technical meaning, according to which it simply means paying attention to the qualitative aspects of experience or sometimes what people refer to as "what it's like" to undergo an experience or "what the experience is like" (although see Stoljar 2016 on a different approach to the meaning of "what it's like" parlance). This is the sense of phenomenology that especially interests us here. Note, somewhat confusingly, that it can refer to paying attention to the qualitative aspects of experience (as in "doing phenomenology") and the qualitative aspects of the experiences themselves (as in "the phenomenology of voice hearing").

Experiences of hearing voices vary across a number of phenomenological dimensions. The classic paper on the phenomenology of voice hearing is by Nayani and David (1996), but a highly influential and more recent study is by Woods et al. (2015). These studies show that the voice-hearing experience exhibits a large array of variety. Here, I present a (non-exhaustive) list of dimensions of variety.

Loudness. Voices can be loud, quiet or even soundless. (For an attempt to understand what "Hearing Soundless Voices" might actually involve, see Wilkinson 2019).

Location. Voices can be experienced as coming from somewhere in the room, or from inside your head, or even from your leg or from all around you.

Complexity. Voices can be extremely simple, one-word experiences (e.g. expletives (Dodgson and Gordon 2009)), or they can be long discursive commentaries (Woods et al. 2015).

Type of speech act. Voices are not usually meaningless experiences of speech sound: they often take the form of speech acts (i.e. "doing things with words," Austin 1962). So, for example, they can assert, question, insult and command.

Mode of address/pronoun use. Voices can vary in their pronoun use (first, second, third person pronouns). However, delving beyond the superficial use of pronouns is the mode of address. To what extent, for example, is the hearer just "listening in" to something being said, or to what extent are they the voice's intended addressee? Related to this is the experienced capacity to interact with the voice: to what extent can the voice hearer feel like they can talk back – and to what extent might that be relevant or efficacious?

Agency and personification. Voices aren't usually disembodied speech sounds, they are the expression of underlying agency (what Deamer and Wilkinson (2015) call the "speaker behind the voice").[1] The richness of this represented agent can vary. Wilkinson and Bell (2016) have presented a taxonomy of how this can vary, as follows.

> 1) *Absent agency* (the hearer doesn't experience the voice as coming from an agent).
> 2) *Agency without individualisation* (the hearer experiences the voice as coming from an agent but has no way of individuating the agent or of re-identifying them over time).
> 3) *Internally individualised identity* (these are re-identifiable but either anonymous "incognito voices" or those given purely internally generated names/identities).
> 4) *Externally individualised identity* (these are also re-identifiable over time, but their perceived identity, whether fictional (e.g. Sherlock Holmes) or real (e.g. the Queen of England) is pre-existing and "borrowed").
>
> (Wilkinson and Bell 2016, p. 107)

Linguistic versus non-linguistic. One very important dimension of variation is whether the experience is even experienced as linguistic. So the experience can be of shouting, screaming, etc., namely, a voice, and potentially agentive, but not linguistic. Experiences that are often clustered with voices, construed as "auditory verbal hallucinations," are simply auditory hallucinations that are neither linguistic, nor agentive, nor even voice-like. Here people might experience the sound of a klaxon, a revving car engine, the bang of an explosion, etc.

How do we cluster these phenomena together in the face of all of this variety? In a theoretical landscape where there are: (i) auditory but non-verbal hallucinations (call these AH); (ii) auditory *and* verbal hallucinations (the classic AVH) and (iii) non-auditory verbal "hallucinations" (soundless voices) (which we might call VH, for verbal hallucinations), there is a genuine question about whether (i) and (ii) are more similar to one another, or whether (ii) and (iii) are. This may seem like a trivial matter, but it's really not: it profoundly shapes our explanatory concerns and questioning. The former position, taking (i) and (ii) to be more similar, and (iii) as more of an outlier, would take these experiences to be primarily auditory experiences that are sometimes (perhaps even canonically) verbal. The phenomena of primary interest would be in the shaded area of Figure 8.1.

The latter position, taking (ii) and (iii) to be more similar, and (i) as the outlier, would take these experiences to be primarily verbal (or as Deamer and Wilkinson (2015) say, communicative, or as Wilkinson and Bell (2016) say, agentive) experiences that are sometimes (reported as) auditory. The phenomena of primary interest occupy the shaded area in Figure 8.2.

We will later see which of these two positions is more plausible, but at this point it is worth noting that the first way of looking at things (as represented in Figure 8.1) encourages an approach where the central E-question (see Chapter 6) is "Why does this person experience an auditory experience in the absence of a sound?" In turn, this encourages an enquiry into sensory and individualistic aspects of cognition (out of which more social and agentive things might be constructed

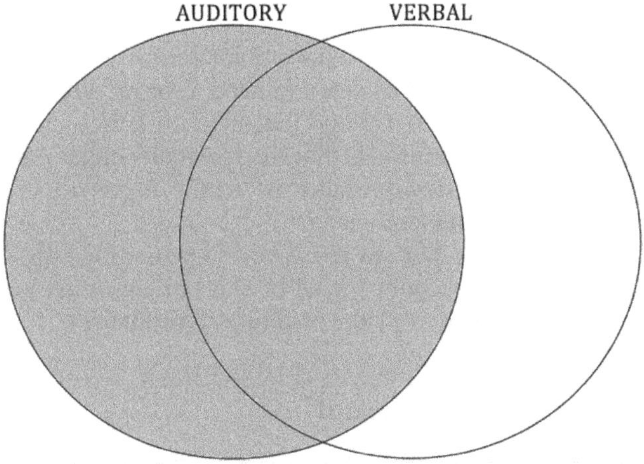

Figure 8.1

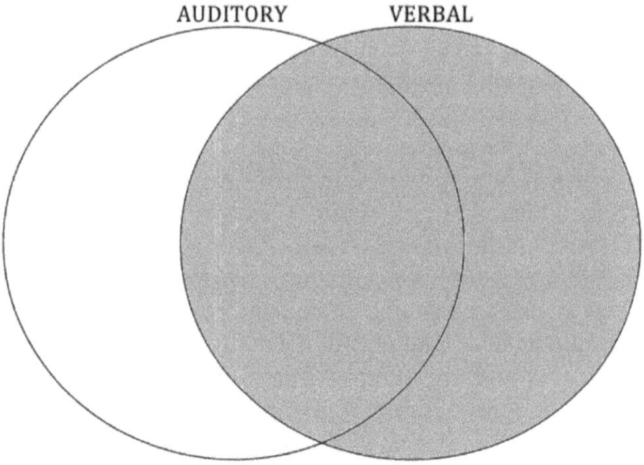

Figure 8.2

or inferred). The second way of thinking about these phenomena elicits the E-question: "why does this person experience illusory communicative agents?," and this in turn encourages a social, relational, agentive approach from the outset.

8.1.3 Varieties of underlying mechanism

Given the varieties, in terms of context and phenomenology, we might well expect varieties in the underlying nature of the phenomenon. In particular, we might think that a useful approach to voice hearing is to explore the possibility that it involves a "normal" everyday process, that we (sort of) understand, going wrong in some way. So it could involve as its "raw material" (namely, the base process that has gone wrong), memory, imagination, inner speech, etc. As McCarthy-Jones et al. (2012) importantly point out, it is likely that different hypotheses about different mechanisms will need to be recruited to explain varieties in context (in both short-term and longer-term) senses and also phenomenology. Another way of putting it is that the phenomenon, as currently picked out, is so varied that, deep down, what it is "really" is going to be a heterogeneous collection of different phenomena.

We will return to this later, but one thing worth mentioning is that any account is either going to need to account for all of this variety or accept that it only serves as an account of a subset of the phenomena of interest.

8.2 The self-monitoring account

Perhaps the most influential account of voice hearing is what we might call a self-monitoring account. There are a number of variants, but I will sketch it broadly to begin with.

8.2.1 Introducing self-monitoring

This term "self-monitoring" may strike you as familiar, and that is because you will recall from Chapter 7 that one mechanistic account of schizophrenia is in terms of self-monitoring deficits. These traditionally function as powerful unifying accounts of *schizophrenia* (Frith 1992). Their ambition is to explain the various symptoms of schizophrenia in terms of one underlying deficit, a self-monitoring deficit. According to these views, the symptoms of schizophrenia differ insofar as what is failing to be properly monitored differs. So, what is failing to be monitored correctly in the case of voice hearing? The most popular answer is: *inner speech*. However, note that, as an account of schizophrenia *per se*, this, by stipulation, only works as an account of voice hearing *in the context of schizophrenia*, or, more precisely, as a *symptom* of schizophrenia. And yet, we have seen that voice hearing does occur to a very significant degree outside of schizophrenia diagnoses. Presumably these non-schizophrenia-related voice-hearing experiences will not fall under the remit of these self-monitoring accounts of schizophrenia. Furthermore, as we have seen (from Chapter 7), there is a more global concern with the validity of schizophrenia as a diagnostic category.

This is precisely why it is so important to clarify that self-monitoring accounts of voice hearing (of AVH) are not typically intended as accounts of an underlying condition called schizophrenia but of voice hearing in general (see Jones and Fernyhough 2007). It is a "symptom-based" approach. The core suggestion (which we will unpack shortly) is that voices are misattributed episodes of inner speech, and that this misattribution is the result of self-monitoring abnormalities, but there is no commitment, however, to these being the result of (still less definitive of/essential to) a clear-cut disease, or even syndrome, called "schizophrenia."

8.2.2 Inner speech-based self-monitoring accounts

At first glance, this might seem like a promising suggestion. Bracketing the truth or accuracy of the hypothesis, it certainly has some *explanatory power* (see Chapter 6). To see this, compare the hypothesis with another one that some people have proposed, namely, that AVHs are to be understood in terms of spontaneous activation in auditory cortex (Cho and Wu 2013). Although this hypothesis is perfectly meaningful (at least once we've clarified what we mean by "spontaneous") and could even be *true*, it isn't by itself very explanatory. That is because, as we've seen (Chapter 6), to explain something is not simply to describe what is going on when that thing is happening. It has to cater to our explanatory concerns, and explanation is *contrastive*. You don't simply explain a phenomenon *full stop*. You explain *why* you get a given phenomenon as *opposed to another relevantly similar phenomenon*. What counts as "relevantly similar" will depend on (or indeed reflect) your explanatory concerns. But whatever we take the relevant contrast to be, the spontaneous activation account doesn't rule out any plausible candidates. It tells us (given what we know about auditory cortex) why someone has an auditory experience as opposed to not having an auditory experience. But that surely isn't what we're after. It doesn't, for example, tell us why someone hears a voice as opposed to a non-voice sound (e.g. a klaxon, or even an unrecognisable sound). It doesn't tell us why we might be hearing a voice saying one kind of thing as opposed to another, or from a given individual and not another, or from an individual as opposed to no individual at all, etc. At the very least, the inner speech-based account gives us the vocal and linguistic aspect for free. Why is the experience linguistic and voice-like (as opposed to not)? That's because inner speech is linguistic and voice-like.

However, there are explanatory hurdles for the inner speech account, too (some of which can be derived from the contrast cases just listed), and attempts to overcome them have yielded interesting and ingenious theoretical developments. As to whether they are successful, I'll leave to the reader to judge.

In case you haven't read the previous chapter, allow me a brief recap of self-monitoring. In Chapter 7, this is used to explain the various symptoms of schizophrenia (and a fuller presentation can be found there), whereas here it is specifically used to explain voice hearing. The fundamental notion of self-monitoring is the same, however.

To quickly recap, the basic idea behind self-monitoring goes at least as far back as Herman von Helmholtz (1866). Helmholtz asked himself the following question pertaining to visual cognition: when you move your eyes and the image project onto your retina moves across it, how does your nervous system know that whether it is the world moving across our eyes, or our eyes moving across the world? Reflecting on the difference between these two states of affairs (viz. the eye moving and the world moving) he concluded that, when our eye moves there is a motor command, whereas there is no such motor command when it is the world moving. The nervous system must somehow make use of this motor command, over and above its use in generating behaviour, to infer what is going on and generate an appropriate visual experience. That is why when your eyes move across a scene, the scene itself doesn't seem to move. (As I said earlier, a nice illustration of these ideas is that, in contrast, when you move your eye with your finger, the world seems to tilt.)

To elaborate on this hypothesis, the mechanism of self-monitoring works as follows. Information about the motor command, which Sperry (1950) dubbed "corollary discharge" (also known as an "efference copy"), is used by the nervous system to predict the sensory consequences that would be brought about by the action. If the predicted and actual sensory consequences match, the action is deemed by the nervous system to be "you," and the percept is adjusted accordingly. If it doesn't match, as in the case of you prodding your eye, it's deemed to be "not you."

Eye movement, however, is a very special kind of action: it is an *epistemic*, information-gathering action. More canonical forms of action are *instrumental*: they are not about gleaning information, but about bringing certain states of affairs into being, given information you've already got. For example you see where the glass of water is, and you act so as to drink from it. In instrumental actions of this sort, the relevant adjustment is not a percept, as it is in eye movement, but what is known as *sensory attenuation*. In other words, the perceived intensity of sensory consequences of your actions is down-modulated. The evolutionary advantage of this is obvious enough: it is your nervous system's way of saying: "Don't pay attention to this, it's only you." This sensory attenuation explains, for example, why you can't tickle yourself. Some data suggest that some individuals with diagnoses of schizophrenia can tickle themselves, thereby fuelling speculation that self-monitoring is impaired in schizophrenia (Frith, Blakemore, and Wolpert 2000).

However, self-monitoring might be impaired by all manner of things, in all manner of contexts, and we can remain agnostic about the validity of the schizophrenia concept. What we can say, however, is that, to the extent that inner speech is an action (Jones and Fernyhough 2007), it too can be badly self-monitored, and as result it might be experienced as loud, intrusive, strange, alien and so on.

8.2.3 Arguments in favour of inner speech-based self-monitoring accounts

Even granting a degree of caution or agnosticism about schizophrenia, it structures research and (as discussed in Chapter 7) often provides the participant inclusion criteria. Several studies, using populations with schizophrenia diagnoses, seem to support the idea that self-monitoring of speech might be involved in voice hearing. For example as we've seen, Johns et al. (2001) got participants to read words aloud and played their voices back to them with mild acoustic distortions. The participants with schizophrenia diagnoses were significantly more likely than controls to claim that the voice they heard was someone else's. But, interestingly, among those with schizophrenia diagnoses (since of course many who get the diagnosis aren't voice hearers), the voice hearers were significantly more likely to misattribute their voice than those who did not hear voices. This suggests that misattribution of one's own speech is related to voice hearing, rather than schizophrenia *per se*. (Although, of course, certain manifestations of voice hearing attract the schizophrenia diagnosis by themselves (e.g. DSM-5 and commentary voices)). An important caveat is in order here, however. As I said in Chapter 7, this task is actually testing source-monitoring, rather than self-monitoring in the low-level, physiological sense appealed to by Frith and colleagues. Nevertheless, if the latter *mechanism* impacts performance on the former *task*, then this can be interpreted as a suggestive result.

You will also notice that this study is about the attribution of overt ("outer") speech. Overt speech, like bodily action, has the required ingredients for self-monitoring, most notably: motoric activity and sensory consequences. As you will recall, self-monitoring involves using motor commands to generate predictions about sensory consequences and seeing if actual and predicted sensory consequences match or not.

Although it is not immediately obvious that inner speech involves motoric elements, this has been empirically supported by several EMG (electromyographical) studies, which measured muscular activity during inner speech. This is a straightforward technique, with some of these studies dating as far back as the early 1930s (e.g. Jacobsen 1931, but see Rapin et al. 2013 for a more recent replication). Later experiments made the connection between inner speech and AVH, purportedly showing that similar muscular activation is involved in healthy inner speech and AVH (Gould 1948; McGuigan 1966). The involvement of motoric elements in both inner speech and AVH is further corroborated by findings from Gould (1950), who, using a highly sensitive throat microphone, showed that when his subjects hallucinated, subvocalisations occurred that are similar to those that occur in inner speech. Of course, correlation doesn't imply causation, so this did not definitively prove that these subvocalisations were causally responsible for the inner speech and, by extension, the AVHs: they could have been echoing it, as has been hypothesised to happen in some cases of verbal comprehension (e.g. Watkins, Strafella, & Paus 2003). However, this causal contribution was more robustly supported by Bick and Kinsbourne (1987), who demonstrated that if people experiencing hallucinations opened their mouths wide, thereby physiologically stopping vocalisations, then the majority of AVHs stopped.

There is other research that provides significant support, both for self-monitoring accounts in general and for inner speech-based self-monitoring in particular, and that is work pertaining to "voice hearers" in the congenitally deaf population. Joanna Atkinson (now Atkinson-Hearn) has done some extremely valuable work correcting the audio-centrism of mainstream clinical perspectives. While it has long been acknowledged that congenitally deaf individuals can have psychosis, receive diagnoses of schizophrenia and report voice hearing experiences, it was always assumed that reports of voice hearing should be taken at face value, namely, as genuinely *auditory* experience (e.g. Altshuler 1971; Critchley et al. 1981; du Feu and McKenna 1999). As someone who is congenitally profoundly deaf herself, Atkinson (Atkinson 2006; Atkinson et al. 2007) found this deeply implausible. These individuals have never had an auditory experience in their lives, and then suddenly, because of psychosis/schizophrenia, they suddenly do? So she asked congenitally deaf voice hearers (in British sign language) what their voice hearing experiences were like. However, unlike previous researchers, Atkinson and colleagues were sensitive to the fact that many congenitally deaf individuals operate with a different sensory semantic field (and that this could be responsible for the audio-centric misconceptions of past research) and may use auditory words to describe non-auditory experiences. So, "loud" might mean "intrusive" or "abrasive" rather than "auditorily loud." To get around this, the statements were further illustrated with picture cards (see Figure 8.3).

The responses were striking. Out of 17 participants, not a single one responded affirmatively to statements about literally being able to hear voices. What they reported instead was visual imagery of someone signing to them or lip reading or (more rarely) fingerspelling. They also reported somatic and kinaesthetic features of their experience (motor imagery).

What is the significance of this? First of all, this is bad news for audio-centric clustering (as per Figure 8.1) and good news for language or communication-centred clustering (Figure 8.2). In a related manner, it is bad news for spontaneous auditory cortex activation accounts. In contrast, Furthermore, it is in line with Shergill et al.'s (2000) finding that in AVH it is associative areas, not primary auditory cortex, that are predominantly active. Furthermore, it is very promising for inner speech-based self-monitoring accounts, since the functional equivalent of inner speech in congenitally profoundly deaf people (their daily "inner voice"/"interior monologue") plays out in the "mind's eye" not the "mind's ear" (McGuire et al. 1997) and bears a striking resemblance to the features of these "voice hallucinations."

8.2.4 *Problems and further developments*

There are at least four well-rehearsed objections to inner speech-based self-monitoring accounts. Logically speaking, you can either aim your criticism at the self-monitoring side of the account, or the inner speech side of the account, or a combination of both. Here are the four objections (in no particular order). I will rehearse responses to them, some of which are more successful than others. The responses to them also serve as useful elaborations on (or clarifications of) the inner speech-based self-monitoring account.

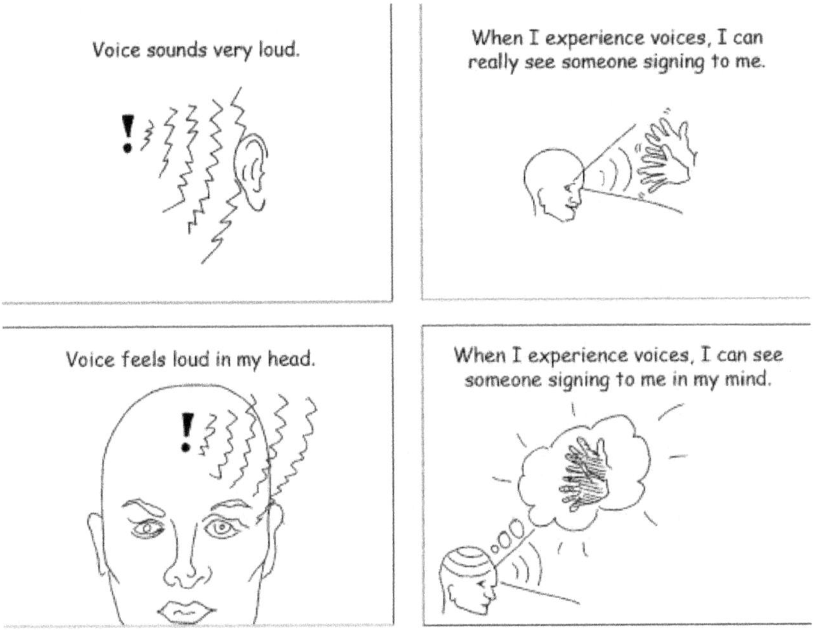

Figure 8.3 Examples of picture cards from Atkinson et al. (2007)

1. Inner speech isn't monitored at all (and therefore can't be badly monitored).
(Stephens and Graham 2000; Gallagher 2004)

2. Inner speech-based self-monitoring can't account for the auditory phenomenology of AVH.
(Wu 2012)

3. Inner speech-based self-monitoring can't account for the variety of AVH.
(Jones 2010)

4. Inner speech-based self-monitoring can't account for the sporadic nature of AVHs.
(Shergill et al. 2001)

Let's unpack these objections little bit and rehearse some responses to them.

Criticism 1 – *Inner speech isn't monitored at all (and therefore can't be badly monitored)*

This criticism has been put forward by several theorists, but perhaps most prominently by Stephens and Graham (2000) and Gallagher (2004). The grounds for these criticisms are subtly, but importantly, different. Gallagher (2004) uses the example of "unbidden thoughts," episodes of inner speech and other mental episodes that aren't intentional and therefore aren't actions. Self-monitoring, Gallagher claims, only operates on actions, since the operant mechanism involves a match or mismatch between intended consequences and actual consequences. This, however, is a misinterpretation of self-monitoring and gives us a useful opportunity to clarify it. The misinterpretation seems to stem from a difference between what "action" means to philosophers and to cognitive scientists. For philosophers, an action is behaviour that is intentional (or, more technically, "intentional under some description"). For cognitive scientists, "action" is simply a domain of cognition that is opposed to perception and emotion, etc. It basically is synonymous with motor cognition. Self-monitoring is postulated by cognitive scientists, and what the forward model is constructed from is a copy of a *motor command*, not an intention. What this means, crucially, is that something that is unintentional and motoric can still be self-monitored, and this can go well or awry.

Indeed, we can see that this is so if we look at a classic paper by Frith, Blakemore and Wolpert (2000). In this paper, two pathological phenomena are contrasted: Anarchic Hand Syndrome and Delusions of Control. With the Anarchic Hand, the hand in question will act in opposition to what the patient is trying to do. So, while the person is trying to get dressed, their right (unaffected) hand might button up a shirt, while the left one unbuttons it. And yet, the hand is not experienced as somehow alien. In contrast, in Delusions of Control, the behaviour can be in keeping with the person's overarching goals and intentions (e.g. they might be brushing their hair), but they will report experiences of passivity or, in more extreme cases, the experiences that their actions are being controlled by someone else. In other words, the former involves actions that run counter to intentions, but self-monitoring is still working fine. The latter involves actions that are aligned with intentions but self-monitoring has gone awry. The take home message is: self-monitoring is about – and operates on – motor commands, not intentions. And as we saw briefly earlier, there are reasons to think that inner speech (and indeed AVHs) involves motoric activity.

This, however, brings us to the second version of this same worry, which is expressed as follows:

> [What] could be the point of a monitoring system that keeps track of whether a thought [inner speech] occurring in my mind, presented to me in introspection, is *my* thought rather than someone else's thought?
>
> (Stephens and Graham 2000, p. 138)

Now, one can certainly criticise the introspectionist and infallibilist assumptions behind parts of this quote (see Carruthers 2009; Wilkinson 2020b; Deamer 2021); however, there are still two major worries here. The first is captured by talk of "the point of" monitoring and the second is captured by the focus of "occurring in my mind" and the two are related. In other words, there is a concern about *function* and also about *internality*.

To see the force of this worry, let's reflect not so much on how self-monitoring works, but on what it is for. In both the cases of epistemic action, as in eye movement (Helmholtz 1866) and practical action, as in reaching for my cup of coffee, the point of self-monitoring is to help me distinguish sensory changes that are the result of my impact upon the world or the world's impact on me. In a low-level and subtle way that we have no conscious appreciation of, this is involved (via sensory and proprioceptive self-monitoring) in online correction in what is known as "skilled reaching," a constant awareness of what is expected on unexpected as we move through three-dimensional space. Armed with these reflections, let's look at inner speech. Why would we need to distinguish between us and world, when it's happening in our minds? Perhaps even more worryingly, how can there be the necessary gap between the actual and predicted sensory consequences if what we are talking about is in our minds? How can what is in our minds frustrate our expectations, if it isn't a brute fact about the outside world?

Two quite different (put potentially related) kinds of response to this go some way (but in my view not all the way) to addressing these concerns. The first involves pointing to empirical research that demonstrates that, whether it seems intuitive or not, inner speech just does involve self-monitoring. For example Ford and Mathalon (2004) found (EEG) that during the production of healthy inner speech, corollary signals from speech production areas suppress activation in speech-processing areas. As Bentall and Varese (2013) nicely put it, "it might be said that when we talk to ourselves, the frontal language production areas of the brain tell the posterior areas not to bother listening" (2013, p. 71). This applies to inner as well as outer speech.

The second kind of response, which also fits this data, involves a broad theoretical shift. What I have in mind here is the move towards a Vygotskian conception of inner speech (see Fernyhough 1996; Jones and Fernyhough 2007). The first step is to ask: how does inner speech develop, and what purpose does it serve? Vygotsky's answer is that inner speech starts off as overt (viz. outer) speech (Vygotsky 1934/1987). That is to say, whatever function inner speech plays, once it has developed, is played by overt speech in children who have not yet developed the capacity to engage in inner speech. This capacity to engage in inner speech is usually seen as the capacity to inhibit the overt production of speech (see Alderson-Day and Fernyhough 2015).

According to this story, inner speech is the end product of a developmental trajectory that begins with private speech. "Private speech" refers to speech that is not produced for the benefit of anyone other than the speaker. So, according to Vygotsky (and neo-Vygotskians), young children will first, under the guidance of a caregiver, learn to reason verbally, but out loud, for the benefit of guiding their thinking and attention and for learning to "navigate" in light of basic social norms. Over time, they learn to inhibit its overt production, which is sometimes called (somewhat misleadingly since nothing actually "goes inwards") "internalization."

What does this mean for the criticism currently under consideration? Well, it means that to the extent that self-monitoring occurs in overt speech, one might expect it to exist *vestigially* in inner speech, its developmental descendent. Also, aside from addressing a functional worry, it points towards inner speech being an active and motoric phenomenon (Jones and Fernyhough 2007; Wilkinson 2020b), rather than merely being a case of conjuring up auditory mental imagery. Furthermore, self-monitoring can go further in explaining why, if the production of outer speech is inhibited, auditory phenomenology is still reported. It is because, as with any aborted overt action (motor imagery), the predictions of the sensory consequences of the would-be action come into play, activating sensory (and somatosensory) cortices.[2]

Criticism 2 – *Inner speech-based self-monitoring can't account for the auditory phenomenology of AVH*

This is a concern expressed by Wu (2012) as a criticism of what he calls the *strong identification thesis*, namely, "AVH *is* the subject's own inner speech misattributed to another" (p. 92). I'm not quite sure how we are deciding whether something that is not experienced as inner speech remains inner speech (since the term "inner speech" plausibly picks out an experience), but I have (Wilkinson 2014) expressed a similar concern under the guise of *the transformation worry*. In particular, if the inner speech-based self-monitoring theorist claims that inner speech is somehow implicated in the generation of the AVHs (which is a weaker claim than the *strong identification thesis*), then the onus is on them to explain how you get the *transformation* from the various phenomenological features of inner speech to the various phenomenological features of AVHs. And by "phenomenological features," I don't mean to focus, as Wu (2012) does, on the auditory qualities (which leads Wu to postulate spontaneous activation in auditory cortex as the better account), but on far more besides: the type of speech acts, the mode of address, the presence of other agents, the experience of communicative intention, etc. (see Section 8.1.2).

Some theorists (Moseley and Wilkinson 2014) have tried to address this by drawing on questionnaire data (McCarthy-Jones and Fernyhough 2011) and theoretical underpinnings (Fernyhough 2004; Alderson-Day and Fernyhough 2014) claiming that inner speech is rather more complex and contains dialogic elements and the presence of others agents (Fernyhough 2004). In other words, inner speech is more complex, in the first place, than these critics seem to appreciate. Ultimately, however, some may not be convinced. And, although the literal presence of auditory phenomenology is overstated (see Wilkinson 2019), it remains a difficult thing to explain. In short, yes, perhaps inner speech is more complex than many acknowledge, but that doesn't necessarily bridge the gap to the complexity of AVH, at least not in all its varieties, and this is where we get to the next criticism.

Criticism 3 – *Inner speech-based self-monitoring can't account for the variety of AVH*

As we saw before (in Section 8.1.2), there are major variations in AVHs. Given this variety, we might do well to entertain the possibility that there won't be a one-size-fits-all account for all varieties of AVHs. What's more, it is not just the phenomenological variety that may point to the need for a pluralistic approach (see Jones 2010 for an argument for inner speech and memory-based subtypes based on phenomenological patterns), it is also about varieties in context. Garwood et al. (2013) found in an interview study that, whereas some AVHs get worse in quiet contexts and better in noisy contexts, others exhibit the exact opposite pattern of aggravation. Why would that be if you're ultimately talking about the same basic phenomenon in both cases? This leads the authors to postulate the existence of an outwardly directed hypervigilance subtype of AVH in addition to an inwardly directed inner speech subtype. The hypothesis is that the former are made worse by noise but better by silence. For the inner speech subtype, however, external noise (and activity generally) draws attention away from inner speech, making AVHs less pronounced.

Criticism 4 – *Inner speech-based self-monitoring can't account for the sporadic nature of AVHs*

More specifically, if AVHs are badly monitored episodes of inner speech, then how come voice hearers can engage in normal inner speech (Langdon et al. 2009)? It doesn't seem plausible to say that there's a generic environmental or emotional trigger or stressor, since it seems that voice hearers can engage in normal inner speech even during episodes of high-frequency voice hearing. Indeed (from personal communication with voice hearers), some voice hearers respond to and converse with their voices by using inner speech.

If we want to hold on to the idea that inner speech is implicated in AVHs, we need to think about why some episodes of inner speech form the basis of AVHs while others do not. Environmental stressors might go some way towards explaining this, but the actual content of the episode of inner speech is bound to be of relevance (Wilkinson and Fernyhough 2018). In particular, factors like the representation in experience of features linked to personality, agency (Wilkinson and Bell 2016) and communicative intention (Deamer and Wilkinson 2015) are good candidates for playing such a role.

Taking stock of criticism and responses

Some of these criticisms seem more damaging than others. Or, rather, which comes to the same thing, some of the responses from defenders of the inner speech-based accounts are more successful than others. More specifically, I think Criticism 1 can be successfully put to rest. An inner speech-based self-monitoring account is at least viable and coherent. The three remaining criticisms suggest at the very least a degree of explanatory incompleteness to the account in question.

8.2.5 Self-monitoring without inner speech?

Most of these criticisms (namely, 2, 3 and 4) apply not to self-monitoring *per se* but specifically to self-monitoring as it applies to *inner speech*. But might there be another "raw material" that might fare better and, perhaps, if not in general, for different subtypes? Borrowing Frith's (1992) schema for accounting for the various symptoms of schizophrenia, then, perhaps AVHs are best explained in terms of self-monitoring, but the varieties of AVHs are down to varieties in what is failing to be self-monitored. On this hypothesis, some subtypes (e.g. soundless voices and inserted thoughts) might be badly monitored (hence misattributed) inner speech, while others might be something else.

What are the other candidates? Theorists have proposed episodic memories as the basis of AVHs (West 1962; Waters et al. 2006) and also auditory imagery more generally (McCarthy-Jones 2012). One of the problems here, however, is that it is not at all clear that self-monitoring, as sketched by Frith (1992), can operate upon episodic memories and auditory imagery. As we have already seen, the self-monitoring mechanism requires motor commands (so that copies of these motor commands can be used to generate forward models). Having said this, perhaps there are other ways in which these "raw materials" can give rise to AVHs in ways that don't rely on self-monitoring at all.

8.3 Alternative accounts

There are several other accounts that don't make use of self-monitoring, whether this involves inner speech of not. I present a selection of these now.

8.3.1 Spontaneous activation accounts

Motivated mainly by concerns about the phenomenological differences between inner speech and AVH (see the "transformation worry"), Cho and Wu (2013) propose an alternative to the self-monitoring account, which they call the spontaneous activation account, building on a fascinating study from the 1960s by Penfield and Perot (1963). Over the course of this study, they had access to over 500 patients with epilepsy who were underdoing neurosurgery. The surgery involved removing portions of their skull in order to directly access the brain. This gave them an amazing opportunity: they were able to use an electrode to directly stimulate the brain, all while the patients were wide awake and able to report on what they experienced. The findings, published in a 100-page report in *Brain*, are extremely detailed and varied and include reports on experiences in modalities other than simply audition. What is most relevant to us here is that stimulation along the temporal lobe resulted in auditory hallucinations. As Cho and Wu (2013) note as being of particular interest, "stimulations along the superior temporal gyrus elicited AVH with phenomenology typical of schizophrenia AVH" (p. 3). For example in one of their case reports, the participant claims that "They sounded like a bunch of women talking together" (Penfield and Perot 1963, p. 622, quoted in Cho and Wu 2013). In another case report, the experience is less distinct: "like someone whispering, or something, in my left ear" or "a man's voice, I could not understand what he said" (Penfield and Perot 1963, p. 640, quoted in Cho and Wu 2013).

How plausible is this proposal? I would also like to ask: how *explanatory* is it? Let's take these in turn.

First of all, let's reflect on the fact that Penfield and Perot's subjects had no psychosis or diagnosis of schizophrenia. The aim of their study was to correlate brain activity, produced by the experimenters, with experiential (phenomenological) reports. Now, just because you can recreate something that is like voice hearing in psychosis, it doesn't mean that voice hearing in psychosis is neurally or, more importantly, causally like that recreation. There are plausibly many ways of generating experiences that may be reported similarly. Second, verbal report is such a blunt instrument, as we've seen so many times (most strikingly in Atkinson et al. 2007). And note that the participants in Penfield and Perot's study can't themselves speak to similarities between what they experience and voice hearing in psychosis or schizophrenia, since they have no experience of the latter. Lastly, even if we take the reports at face value, we may question whether they really are typical of AVHs. Interestingly enough, McCarthy-Jones (2012), in examining the findings of the study, makes the exact opposite claim: he is struck by *just how different* the experiences of the Penfield and Perot study are, phenomenologically, from AVHs (pp. 193–195).

How explanatory is this proposal? Even if these lab-induced experiences closely resemble AVHs (and that is up for dispute), one thing we know is that voice hearers haven't got electrodes in their heads. It's unsurprising (indeed trivial) that, *if* they are experiencing something auditory (and many voice hearers aren't), you'll get activation in auditory cortex. The question is, *why* do you get it? And why do you get that activation and not relevant other kinds? Remember that you don't simply explain something by *describing* what is going on when that thing happens. You explain why you get that thing and not something else. So spontaneous activation accounts need to say more about this spontaneity. They also lack explanatory power in that they tell us why someone has an auditory experience rather than not, but they tell us nothing about why they have a particular kind of auditory experience rather than any other (e.g. a voice, rather than a klaxon; a voice saying this, rather than that; a voice of an abusive stepfather, rather than a stranger, and so on).

Before moving on, it is worth mentioning that this proposal arises from a criticism of inner speech-based self-monitoring accounts. Inner speech cannot be the raw materials of AVHs since AVHs are nothing like inner speech. They are auditory and too complex. But the motivation for this account can be undermined by responding to this criticism. First, as Moseley and Wilkinson (2014) have said, inner speech is far more complex than Cho and Wu (2013) acknowledge. Second, they may be overestimating the auditory nature of voice hearing, or at least its centrality. Not only are there, as we've seen, voice hallucinations in congenitally deaf people, but there are also reports (among hearing individuals) of soundness voices (Wilkinson 2019). All of this also motivates a theoretical shift away for audio-centric accounts. Following Deamer and Wilkinson (2015), we can say that these are communicative experiences that happen to be auditory, rather than auditory experiences that happen to be communicative. This casts it in a different, social and interpersonal light.

8.3.2 Memory-based accounts

Waters et al. (2006) present the view that AVHs are the result of an inability to inhibit memories. Their account comprises two factors. The first concerns a problem with "conscious inhibition" (as it happens, there is a vast literature on "inhibition" and problems with "inhibitory control" in schizophrenia more generally (see Westerhausen, Kompus and Hugdahl 2011, for a nice overview), and it means subtly different things in different contexts). In this context, inhibition simply means the ability to filter your mind, to keep irrelevant or unwanted "cognitions" (i.e. thoughts, episodic memories, imagery) out of your conscious experience. This first factor alone, however, won't account for AVHs, since, as the authors recognise, there are individuals (e.g. with various forms of obsessive disorder) who have similar intrusions but who don't report these intrusions as hallucinations. This is where the second factor comes into play. Voice hearers in addition to these inhibitory deficits also have a "context integration" or "context memory" deficit. Episodic memories, especially if they are intrusive, can be imagistically vivid and don't carry their status *as a memory* on their sleeve. How we recognise them as a memory, it has been hypothesised, is that we tie them to a particular context in the past. The core idea is that a failure to recognise the context of the memory will likely result in us failing to recognise it as a memory. But if it's not a memory, an experience related to the past, then what is it? The answer could be: some kind of experience of the present, which (in spite of potential phenomenological differences to normal perceptual experiences) might result in reports of hearing voices, hallucinations and the like.

Is this account plausible? As McCarthy-Jones (2012) points out, this account would predict phenomenological similarities between obsessive compulsive disorder (OCD) and AVHs. This, he argues, is somewhat supported. Thoughts in OCD, like some AVHs, are repetitive and also compelling (i.e. they drive action). This is especially seen in command hallucinations. One might think it's a fine line between the thought "I must go home and check I haven't put the gas on" and hearing the command "Go home and check!" Having said this, OCD and AVHs have different aetiologies, and one would expect different contents and different themes in each case. What's perhaps even more convincing, as McCarthy-Jones (2012) also points out, is that Stephane, Polis & Barton (2001) report two cases where fluvoxamine, an OCD drug, helped reduce the severity of AVH. As Stephane et al. themselves acknowledge, this only appears to be effective with a particular subtype of AVH but nonetheless is highly suggestive.

Waters et al. (2006) have garnered support for each of their two factors. For example in an earlier study, Waters et al. (2003) investigated inhibition in voice hearers with a diagnosis of schizophrenia, using the Hayling Sentence Completion Test. This task tests inhibition by asking participants to complete a stock sentence with a word that is not the obvious word. So, for example, you might get "The cat sat on the . . .," and then the word to avoid is "mat." Voice hearers with diagnoses of schizophrenia were less good at avoiding the completion with the obvious word than "controls."

After supporting the idea that there might be deficits in intentional inhibition, the same group then explored the second factor, namely, context integration deficits. Participants had to assess two things: (i) *who* did something and (ii) *when* it happened. In this task, there were cards with pictures of household items, and the action in question involved pairing two cards together. There were two sessions, 30 minutes apart. Participants either watched an experimenter put two cards into a pair, or they themselves made a pair. Thirty minutes later, this was repeated with different cards, but also randomised as either observer or agent. Five minutes after the second session, participants were presented with a sequence of card-pairs and had to answer a number of questions for each. First was: was this one of the pairs that had been made in the experiment? If yes: was this pair made in the first or second session? (This was supposed to ascertain the *when*.) Then: was it they themselves or the experimenter who had made the pair? (This was supposed to ascertain the *who*.) Fascinatingly, in 2004 and 2006, this team ran this study for importantly different populations. Waters et al. (2004) simply contrasted schizophrenia participants with controls and found that patients did worse than controls at both the *who* and *when* tasks. But 2 years later, Waters et al. (2006) contrasted participants with schizophrenia who were voice hearers from those who were not. The aim of this is to ascertain what might be associated with schizophrenia *per se* or more specifically with voice hearing. For the *when* task, there was no difference between the voice hearers and non-voice hearers within the schizophrenia group. However, performance on the *who* task was worse in voice hearers and correlated negatively with severity of voice hearing. This can be interpreted as showing that voice-hearing is not caused by context-integration more generally but by the recognition of one's own agency, potentially underpinned by self-monitoring.

The following is worth mentioning. The two factors in the memory-based account, namely, the inhibitory deficits that yield intrusive memories and thoughts and contextual integration deficits that lead to misattribution, though presented together in Waters et al. (2006), are in fact logically distinct. Indeed, context-integration has been seen as a potentially central feature of schizophrenia since at least Cohen and Servan-Schreiber (1992), with a focus on attentional and linguistic tasks (i.e. not specifically AVHs). Furthermore, the hypothesis that inhibition is central to schizophrenia goes back at least to Frith (1979).

8.3.3 Social deafferentation accounts

There is a somewhat marginal account that I find extremely interesting and worthy of mention, and that is Hoffman's "social deafferentation" account (2007). But, first, a caveat: this is presented as an account of schizophrenia and is supposed to account for both hallucinations and delusions. Nevertheless, I would like to examine it specifically as an account of AVHs in particular.

Hoffman brings together the following two observations. The first is that sensory deprivation ("deafferentation," the total interruption of afferent, namely, incoming, signals) over sufficient periods leads to hallucination, sometimes highly vivid hallucinations, as in the case of Charles Bonnet Syndrome (Berrios and Brook 1982). The second observation is that psychosis is often accompanied by social isolation. Granted, there is a "chicken-or-egg" issue here regarding whether the isolation is a result of psychosis or a cause of it. Nevertheless, running with the hypothesis that social isolation is causally implicated in psychosis, the two observations yield the following explanatory hypothesis. Just as sensory deafferentation eventually results in the nervous system generating sensory experiences for itself, so might social deafferentation (i.e. a lack social stimuli) lead to the self-generation of social experiences.

Two things are worth saying about this. First, this is clearly not going to work as an account of all AVHs, since they certainly do not all occur in the context of deprivation of social stimuli. Second, and this is why I find it worth mentioning, it presents the core phenomenon in a diametrically opposed way to how, for example, the spontaneous activation accounts present it, namely, as primarily a social phenomenon (that might happen to be experienced as auditory) rather than an auditory phenomenon (that might happen to be social). My opinion, for what it's worth, is that, regardless of the plausibility of this account, this is certainly the more useful way of carving up the AVH conceptual territory.

8.3.4 *Taking stock*

What are the relationships between these alternative accounts? Are these alternative accounts mutually exclusive, either with each other or with self-monitoring accounts? One could reason as follows. Perhaps the spontaneous activation account might work upstream from the intrusive cognitions account. What's more, perhaps the social deafferentation could be one possible explanation of why you might get spontaneous activation in the first place (although, as we've seen, this could only work for a subset of cases). Lastly, it could be that, in the face of the wide variety of AVH experience, these accounts work sometimes in tandem and to varying degrees with different subtypes.

One particularly interesting question concerns whether episodic memory, with its rich and dynamic imagery, is also subject to similar forms of self-monitoring as inner speech. Might there be two forms of self-monitoring account, one according to which it operates on inner speech, the other according to which it operates on episodic recollections. If the latter account is feasible, then do we need to postulate the inhibition deficits (which give rise to the intrusive memory) and context-integration deficits (which results in misattribution)? If the self-monitoring deficit could explain both of these things, then one might think we wouldn't need to.

8.4 From self-monitoring to predictive processing

One might think that, although there are some seemingly insurmountable criticisms to the inner speech-based self-monitoring account, there is an important kernel of truth to it. Furthermore, putting its accuracy aside, it has some explanatory virtues that other accounts lack. For example although it struggles to explain some of the finer details of the voice-hearing experience, it does at least explain why it is linguistic and vocal, since, arguably, inner speech is too. For example spontaneous activation in auditory cortex or intrusive memories (accurate though they may turn out to be, as far as they go) can't explain why the hallucinating subjects wouldn't just experience indiscriminate noise (in the former case), or memories that aren't voices (in the latter). So how can some of the important insights of self-monitoring be kept and built on?

8.4.1 *The problem of hypervigilance hallucinations*

Whatever problems there may be for the finer details of self-monitoring accounts, whatever the "raw materials" may turn out to be, one thing that self-monitoring fundamentally involves is the misattribution of something self-produced. It involves wrongly experiencing as passive, external, exogenous, something active, internal, endogenous. So, one thing that self-monitoring cannot account for is any phenomenon that is passive. In other words, it cannot account for "passive symptoms" (Wilkinson 2015). This is where a subtype of

AVHs, known as hypervigilance hallucinations (Dodgson and Gordon 2009), presents a problem.

What are hypervigilance hallucinations? Delespaul, DeVries and Van Os (2002) investigated the contexts in which AVHs most commonly occur, and they found that they are most likely either in the presence of lots of other people or when alone. Building on this observation, Dodgson and Gordon (2009) suggested, on grounds of clinical case studies, a kind of AVH called a "hypervigilance hallucination." These experiences tend to occur against a backdrop of heightened arousal, anxiety, threat, and in contexts of drug abuse, gang violence and so on.

In a further development of subtypes, Garwood et al. (2013) showed that AVHs tend to occur when: (i) attention is directed inward in quiet contexts and (ii) attention is directed outward in noisy contexts. They took (i) to point towards an inner speech subtype and (ii) a hypervigilance subtype. Now, unlike inner speech hallucinations, for hypervigilance hallucinations, the basis stimulus is not internally generated but comes from the outside world: they involve the boosting a shaping of an environmental signal.

Two things arise at this point. First, one might claim that, since hypervigilance hallucinations do, strictly speaking, have some kind of worldly stimulus, then they are not, hallucinations; they are rather illusions. It is a distortion of something that's there, rather than a creation out of nothing. I see this as a mainly terminological point. You can think of a hallucination not causally, but in terms of representational content, and how far removed it is from what is actually there. By the latter conceptualisation of hallucination, hypervigilance hallucinations are indeed hallucinations.

The second, more important point is that because of this, self-monitoring cannot account for hypervigilance hallucinations. Self-monitoring exploits mechanisms (e.g. motor commands) linked to self-generated events, and it is this that goes wrong and is misattributed. Thus, it must remain silent about hypervigilance hallucinations, which are precisely not a case of a self-generated stimulus being misattributed, but rather an external stimulus being distorted.

8.4.2 Prediction is pervasive in all cognition, not just action monitoring

One way of salvaging what's good about self-monitoring accounts, while making room to account for hypervigilance hallucinations, is to see self-monitoring in the context of a radically different way of thinking about brain function. For example, Pickering and Clark (2014) differentiate auxiliary forward models (AFMs) and integral forward models (IFMs). Self-monitoring accounts operate with the former, but, as Wilkinson (2015) argues, commenting on Pickering and Clark (2014), they would do well to operate with the latter. What are integral forward models? To answer this question, we need to turn to a way of thinking about the brain where everything is anticipatory and predictive, not only self-generated phenomena.

You will recall from Chapter 7 that the predictive processing framework (PPF) is a general way of thinking about cognition, which can be applied to schizophrenia and psychosis (and other changes to typical cognition). I won't repeat another detailed introduction. For those who haven't read Chapter 7, I'd recommend looking at Section 7.3 for a primer on the PPF.

In short, there is a big explanatory payoff from adopting an IFM which, as Pickering and Clark (2014) are very much aware, falls naturally out of the PPF. This also allows us to explain passive psychotic symptoms, like hypervigilance hallucinations.

At the most general level, approaching voice hearing from the perspective of predictive processing changes the central explanatory question. Theorists have tended to ask themselves, within a standard framework: where does the erroneous stimulus come from? And, of course, this has led to hypotheses about inner speech, memory, spontaneous activation in auditory cortex and so on. In contrast, with the PPF, experience is much less about stimulus, much less about the contribution of the outside world. And so the question instead becomes: why did this person's nervous system generate a hypothesis that corresponds to that experience (namely, that someone is talking to them)?

Predictive processing provides a backdrop from which to provide subtypes, of both active and passive varieties. First, I will present the predictive processing account of hypervigilance hallucinations and then show how inner speech and memory might both generate other subtypes.

8.4.3 A predictive processing account of AVH

The simplest way of explaining hypervigilance hallucinations with the PPF is in terms of Bayesian inference. You will remember from Chapter 7 how hypothesis selection is Bayesian within the PPF, so the hypothesis selected is determined through a trade-off of fit with input and prior probability of the hypothesis.

To re-borrow an example from Pezzulo (2014) that we saw in Chapter 7, suppose I hear a window squeak. Suppose for simplicity that there are two potential hypotheses: burglar and wind. Both have an adequate fit with the data, which is to say that, if they were correct, they would produce that sound. But suppose it's a windy night and I live in a low crime area. Then the prior probability (the probability of the hypothesis before/independently of the evidence) will be higher for the wind hypothesis: and so that's the hypothesis that I ought to select.

However, now let's modify the example a bit. Suppose I have just watched a scary film, and I am in a residual state of emotional arousal. This means that the two hypotheses now have a different fit. Why is that? Because, whereas the wind hypothesis explains the sound, the burglar hypothesis explains not only the sound, but my emotional state too. This is why, as a result of my emotional state, I might become convinced by the burglar hypothesis. This, of course, is not a good inference. My state of fear is not in fact a good indicator that there is something to be afraid of – indeed, it has nothing to do with a burglar or anything else that ought to frighten me. Nevertheless, it certainly rings true that we often engage in these sorts of "circular inferences," where we "freak ourselves out" in a dark alleyway, ratcheting up our baseless fears.

Now this example involves my personal-level inferences based on experiences. But the PPF is about the nervous system's "inferences," which themselves generate experience (see Chapter 7 on the predictive processing hierarchy). This is the basic idea behind hypervigilance hallucinations. People are in a state of extreme hypervigilance (fear, roughly speaking) and when presented with a noisy sensory input, like a boiling kettle or the sound of neighbours through a partition wall, their nervous system boosts and shapes the input into a voice-hearing experience. The person's nervous system infers (circularly) "I am afraid, therefore there must be something I am afraid of."

We would expect such experiences to have a particular aetiology and contexts of aggravation (Delespaul, deVries and Van Os 2002). More specifically, we would expect this to occur in context of trauma and threat and also (separately or in conjunction) in the context of certain forms of drug abuse (especially stimulants like amphetamines). We would also, with these background neurological/psychological states in place, expect contexts of complete silence to give the hair-triggered nervous system nothing to go on, while also allowing a de-escalation of the arousal and hence to reduce incidences of hypervigilance hallucinations. In contrast, we would expect noisy contexts, especially where the noises are "noisy" in the informational sense, namely, where an unambiguous signal is hard to extract, to be triggers of voice-hearing experiences. Indeed, one area where you would have a "perfect storm" is in a prison setting: prisons are deafeningly noisy, like an inchoate soup of shouting voices. And yet individuals with a history of gang violence and drug abuse are likely to be over-represented within correctional facilities, not to mention the threats (perceived or actual) within the prison setting itself. To my knowledge (and I may well be wrong), there is no recent detailed study on voice hearing in prison. Having said that, Launay and Slade's original study (1981), which gave rise to the widely used Launay–Slade Hallucination Scale, was conducted on a prison population, but has since been used elsewhere. In line with the Bayesian interpretation we've just seen, they found that the "aggressive-paranoid tendencies" (Launay and Slade 1981, p. 221) that they found to be far more prevalent in a sample of prisoners correlated strongly with scores of hallucination-proneness.

Hypervigilance hallucinations are an example of the passive symptoms that self-monitoring accounts can't account for. They are outwardly directed and don't involve motor commands. So this is one thing that a predictive processing account can accommodate that self-monitoring accounts cannot. But how does the PPF fare when it comes to explaining those kinds of hallucinations that self-monitoring is better at accounting for? In particular, the question arises: what does a predictive processing account of inner speech hallucinations look like?

In Wilkinson (2014), I acknowledged that more work was needed to go into a predictive processing account of inner speech itself before it could be in any way deemed to be the "raw materials" of AVH. In Wilkinson and Fernyhough (2017,

2018), we went some way (although not far enough) towards developing that. In order to understand what inner speech looks like within the PPF, we first need to appreciate how it deals with action and motoric activity. Unlike in the standard framework, within which the self-monitoring account operates, action and perception operate according to the very same principles of prediction error minimisation. According to the PPF, the brain is trying to select models/hypotheses that generate accurate predictions. But there are two ways of making accurate predictions: one involves making your predictive models fit the world, the other involves making the world fit your predictive models. The former is known as perceptual (or passive) inference and the latter as active inference. In short, this means that what plays the functional role of motor commands with the PPF are *predictions* (see the aptly named paper "Predictions not Commands" by Adams, Shipp and Friston (2013)).

What is the significance of all of this for inner speech and AVH? To simplify, inner speech involves generating the prediction of speaking but without actually "going through with it." This inhibition involves turning down the precision in motoric regions. However, what remains is imagery, since imagery is simply simulated sensation, a prediction that is decoupled, that the world doesn't need to live up to (Jones and Wilkinson 2020), and, in fact, you get both the sparse and degraded auditory imagery as well as the motoric imagery (or visual and motoric imagery for deaf people (see Section 8.2.1)).

So much for inner speech, what about AVHs? The devil would be in the details, but the basic idea would be that the generated prediction (though decoupled at the motoric level and not giving rise to overt speech) would be richer, powerful and unpredictable in the context of excessive dopamine-weighted prediction error in sensory areas.

We could also hypothesise a mechanism for memory-based hallucinations. This is less well worked out, but what episodic memory might look like within the PPF would be like multimodal imagery that is somehow recognised as being tied to the past. Again, we are more likely to successfully tie it to the past when we are in control of its content and its intensity. If the conjuring up of this imagery that is part of normal recollective process gets disrupted, we can easily see that someone may misidentify past recollections as present sensory experiences.

Indeed, there might be a fine line between "flash backs" in post-traumatic stress disorder (PTSD) (Wilkinson, Dodgson and Meares 2017) and memory-based hallucinations. In the former, an intense recollective experience (a re-living of a traumatic event) is recognised as such, in spite of its unusual intensity. It may phenomenologically resemble a multi-modal hallucination but is not experienced or interpreted as perceptual. This recognition of the vivid experience as a flash back must, one would think, be related to the individual being able to remember and hence recognise the similarity between the traumatic event and the flash back. But what if the trauma is childhood trauma? What if it is childhood sexual trauma with all of the associated confusion and guilt that accompanies that? Then we might have something that resembles a flash back, but that isn't recognised as such because the trauma is either too far in the past and/or has been buried by guilt-driven processes of repression (Otgaar et al. 2019).

8.4.4 *Remaining challenges and future directions*

The PPF helps to account for the varieties of AVHs (including the three subtypes of hypervigilance, inner speech, memory-based hallucinations) (Wilkinson 2014). It also helps to account for the phenomenology of AVHs or what we might call the "transformation worry." How can AVHs not be like inner speech, if they are misattributed inner speech? The PPF renders this talk of "raw materials" senseless, and the explanatory focus is not on where the erroneous stimulus comes from but rather on why the person's nervous system adopts the hypothesis that corresponds to the experience. In short, the notion of a fully formed stimulus plays no role within the PPF. So, for the inner speech subtypes, some of the processes that give rise to inner speech give rise to AVHs as well, but by that point there is no reason to think that there would be much similarity between AVHs and inner speech. Part of what makes inner speech the experience that it is, is that it is controlled and hypothesised to come from me. You need only disrupt that slightly for the hypothesis-generation to spiral off that course and generate an experience that's very different.

However, we must not be complacent, and there are many remaining challenges. Furthermore, it may well be that addressing these challenges will require us to move away from thinking in terms of the PPF. This is not necessarily to say that it will conflict with it, but rather that certain questions will need answers to be couched in different terms (see Chapter 6; you don't explain stock-market crashes with nuclear physics).

Among these challenging explanatory questions are the following. Why – and how – are auditory verbal hallucinations experienced often as coming from particular agents (Bell 2013; Wilkinson and Bell 2016)? What do experiences of felt presence have to do with this phenomenon (Alderson-Day et al. 2022)? How are we to understand the claims concerning "hearing soundless voices" (Wilkinson 2019)? How do we make explanatory accounts of this phenomenon (the PP hierarchy might come in useful here)? These sorts of questions pose challenges for all concerned.

Ultimately, AVHs are a lens through which we can better understand the human mind. In other words, we might ask: What sort of a thing is a human being, such that it experiences voice hearing in the way in which it does (i.e. the prevalence, variety, etc., that it does)? A partial and preliminary answer: one that is deeply social and communicative.

Summary

- Hearing voices (a.k.a. auditory verbal hallucination or AVH for short) is often, but unhelpfully, associated with schizophrenia.
- There is a significant variation in voice-hearing experiences, and we need to be open to the possibility that, although they are grouped together, this variety may represent what are fundamentally distinct phenomena, with different phenomenologies, causes, mechanisms and so on.
- We examined the view, which has some claim to being the mainstream view, that AVHs are the result of self-monitoring deficits or abnormalities, which

result in the individual failing to recognise that something that they have produced comes from them.

- The most prominent candidate for the misattributed phenomenon is inner speech, so you end up with the mainstream view that AVHs are misattributed episodes of inner speech.
- There are, however, several other accounts, but they ultimately fail to have the same explanatory power as the inner speech-based self-monitoring orthodoxy.
- However, the orthodox view does have some insurmountable challenges.
- One way of retaining the power of the mainstream view, while addressing these challenges, is to re-frame the self-monitoring account in terms of the predictive processing framework.
- Many other exciting challenges lie ahead.

Questions

- Do you experience inner speech? If so, are you tempted to talk about it in auditory terms? (Is it like "hearing a little voice inside your head"?)
- Have you ever heard a voice when no one was speaking? Why would you think that this is a sure sign of mental illness?
- When something pops into your head, how do you know it comes from you? And what's the relationship between something coming from you and not being real?
- If someone told you that you are hearing voices because you have too much dopamine in your brain, or because you have spontaneous activation in your auditory cortex, would that satisfy you as an explanation? Under what circumstances might it not?

Notes

1. Indeed as Wilkinson and Krueger (forthcoming) point out, the word "voice" in clinical contexts can express two very different concepts (an auditory and an agentive one).
2. See Jeannerod 1995 for a classic presentation of this, and Adams, Shipp and Friston 2013 for the same within a predictive processing framework.

Further reading

McCarthy-Jones, S. 2012. *Hearing Voices: The Histories, Causes and Meanings of Auditory Verbal Hallucinations*, Cambridge: Cambridge University Press.

Wilkinson, S. & Bell, V. 2016. The Representation of Agents in Auditory Verbal Hallucinations. *Mind and Language* 31 (1):104–26.

Woods, A., Jones, N., Alderson-Day, B., Callard, F. & Fernyhough, C. 2015. Experiences of Hearing Voices: Analysis of a Novel Phenomenological Survey. *The Lancet Psychiatry* 2 (4):323–31.

9 Delusion

Overview

We say that people have delusions – or are delusional – when they make highly implausible claims that seem to lack the support of evidence. Delusions are of central diagnostic importance in psychiatry. However, in being related to notions like belief, justification, evidence and rationality, delusion has also drawn the attention of philosophers. This double role in theorising, as both a diagnostically vital concept, while also being arguably an epistemic notion, is what makes delusions both an interesting and at times perplexing topic. I will try to navigate through this theoretical landscape in a way that does justice to complexity and nuance, while hopefully not muddying the waters too much.

We start by asking: what is delusion? As we've seen at several junctures, this question can be interpreted in two quite different ways, and we will address both. First, what is the concept of delusion? In other words, what are we doing when we call something a delusion or someone delusional? What grants something the status of "delusion"? Second, what things that actually exist in the world (if any) correctly/appropriately fall under that concept or have that status? Then, we go through various explanatory accounts of specific delusional phenomena. Finally, we examine a particular philosophical debate that centres around the question: "are delusions beliefs?" To put the question more intuitively: "do delusional people genuinely

DOI: 10.4324/9781315450124-11

believe the things they claim?" One's answer to this question depends on a catalogue of other theoretical commitments about the nature of belief and the way specific delusional patients really are.

9.1 The concept of delusion

The concept of delusion, namely, the concept expressed by uses of the word "delusion," varies depending on the context and purpose of use. There are more fine-grained variations, but as a rough starting point, it is important to distinguish delusion as a concept that has significance – in particular diagnostic significance – for clinicians, from the concept that has significance for philosophers.

9.1.1 Delusion as a diagnostically important psychiatric concept

As Jaspers put it, "Since time immemorial, delusion has been taken to be the basic characteristic of madness" (1913/1963, p. 93). Delusion remains, to this day, an important psychiatric concept, not least for the purposes of diagnosis. Delusions are one of a number of symptoms that may lead to a psychiatric diagnosis, most famously, schizophrenia, but plenty of others besides (e.g. schizo-affective disorder, bipolar disorder, major depressive disorder) including also neuropsychiatric conditions (caused by traumatic brain injury, carbon monoxide poisoning, encephalitis, meningitis, Alzheimer's disease).

As we saw in Chapter 5, the DSM characterises a delusion as follows:

> A false belief based on incorrect inference about external reality that is firmly sustained despite what almost everyone else believes and despite what constitutes incontrovertible and obvious proof or evidence to the contrary. The belief is not one ordinarily accepted by other members of the person's culture or subculture (e.g. it is not an article of religious faith).
>
> (APA 2013, p. 819)[1]

Many theorists (e.g. David 1999; Coltheart 2007; Bayne and Fernandez 2009) have noted that, although this definition is the "official" one, it is, if taken to present necessary and sufficient conditions, problematic and subject to several counterexamples. For a start, delusions needn't be false; they could be accidentally true. Consider the dishevelled self-proclaimed prophet (perhaps who even has a schizophrenia diagnosis) who declares on the basis of no evidence that the end of the world is nigh, and (thanks perhaps to a helpful asteroid) just so happens to be right! Furthermore, do delusions have to be based on incorrect inference? And what exactly do we mean by that? (Surely if someone has powerful experiences that lead them astray, we can think of this as involving correct inferences, or even no inferences at all.) Do they have to be about external reality? (Couldn't I have delusions about my mind?) To what extent should the beliefs of others in one's community play a role in determining delusional status? (see the discussion in Chapter 5.)

While some theorists have stated the inadequacy as a matter of fact (e.g. Marshall and Halligan 1996), others have actively bemoaned it. For example the prominent psychiatrist Anthony David (1999) speaks of the "Impossibility of Defining Delusions" in a paper of that title:

> Most attempted definitions begin with "false belief," and this is swiftly amended to an unfounded belief to counter the circumstance where a person's belief turns out to be true. Then caveats accumulate concerning the person's culture and whether the beliefs are shared. Religious beliefs begin to cause problems here and religious delusions begin to create major conflicts. . . . The beleaguered psychopathologist then falls back on the "quality" of the belief – the strength of the conviction in the face of contradictory evidence, the "incorrigibility," the personal commitment, etc. Here, the irrationality seen in "normal" reasoning undermines the specificity of these characteristics for delusions . . . as does the variable conviction and fluctuating insight seen in patients with chronic psychoses who everyone agrees are deluded. . . . Finally we have the add-ons: the distress caused by the belief, its preoccupying quality, and its maladaptiveness generally, again, sometimes equally applicable to other beliefs held by non-psychotic fanatics of one sort or another. In the end we are left with a shambles.
>
> (pp. 17–18)

This lament (and it surely is a lament with strong negatively laden terms like "beleaguered" and "shambles") is revealing of two interesting things. First, when theorists talk of "defining" delusion, they seem to want a clear presentation of necessary and sufficient conditions. In other words, they want a strict recipe for picking out all and only delusions. This is seen by the fact that counter-examples are seen to be damaging to such a definition. What provide counter-examples to any definition (e.g. the DSM definition) are cases where delusional status (and indeed pathological status) has already been recognised. That's why they are counter-examples! This means, though, that any definition (the DSM definition included) isn't what is guiding our judgements about delusional status. We've already made these judgements intuitively.

This raises the question: so what is the DSM definition doing? If it is seen as providing a strict definition in terms of necessary and sufficient conditions, then it clearly fails. If, however, it is seen as a description of paradigmatic cases of delusion, then it seems to do the job. Most delusions do fit the DSM description. That this is what the "definition" is really doing is, after all, suggested by the acronym DSM: it is a *Diagnostic* and *Statistical* Manual. If somebody didn't know how to recognise a delusion, it would be helpful in enabling him or her to do that in most cases. It is highly unlikely that such a person would encounter one of the counter-examples first. Similarly, the delusions that clinicians are likely to encounter will tend to fit the DSM "definition."

Since theorists can recognise delusions when they encounter them (see, "patients who everyone agrees are deluded" in the aforementioned quote), when people criticise the DSM, it is not the fear of misdiagnosis that is motivating them. It is that the DSM definition is failing to "capture the concept" of delusion, whether this is what it is trying to do or not. In the DSM's defence (although I don't want to defend it too much), I don't think that's what it's doing.

The second revealing thing is that David is assuming that delusions must, as a matter of conceptual necessity, be pathological. This can be seen from the observation that he takes irrationality in the normal, healthy, population to undermine a definition that might be based only on irrationality, rather than allowing that delusions in healthy people might not be a contradiction in terms.

Contrast this with, for example, what philosopher Kengo Miyazono (2015) writes in a paper explaining what makes some delusions pathological: "I do not assume that all delusional beliefs are pathological. . . . I only discuss typical delusional beliefs that are pathological" (Miyazono 2015, p. 561, fn.1). This contrast reveals two ways of approaching delusion. First, you can think of it as a diagnostically important psychiatric concept, a Jaspers-style "marker of madness," in light of which non-pathological delusion is indeed a contradiction. Second, you can think of it as a concept that has to be carefully defined in terms of epistemology, or at least, not as a medical phenomenon, but as a (more abstract, perhaps) mental state that is subject to various normative evaluations. One can easily understand why clinicians will tend to see delusions as pathological by conceptual necessity: because delusions that are pathological are the ones that are likely to come to their attention. Suppose that, instead of thinking of delusion as a marker of mental illness, and hence of delusion as necessarily pathological, we thought of the relationship between mental illness and delusion as less direct. This can play out in a couple of ways.

In line with discussions in Chapter 2 about the nature of mental illness, one can either be an objectivist and think that mental illness is about some kind of underlying dysfunction, or one can be a values-based theorist and think that mental illness is more about certain kinds of interference with human happiness and flourishing. Either way, the account would continue, there is conceptual independence between mental illness and delusion. Within an objectivist framework, delusions might well be caused by underlying dysfunction but aren't conceptually tied to this. Delusion, on such a view is only causally, but not conceptually, tied to mental illness. Thus, delusions that are pathological might be indicative of, for example, underlying harmful dysfunction (Wakefield 2007; Miyazono 2015).

The values-based theorist can have a similar degree of causal dependence, coupled with conceptual independence. Since it is often disruptive to be delusional, delusional subjects will *tend* to be deemed mentally ill. However, even if it were always disruptive to be delusional, it would be fallacious to think that this should make pathology *enter into the very concept of delusion*. (By analogy, being blind might make you more likely to bump into things, but bumping into things is not part of the concept of blindness.)

In general, it's perfectly possible that the concept of delusion can be analysed without reference to pathology, and yet the phenomena that fall under that concept resultantly have the kinds of properties that will always make them pathological (whether that be about underlying objective pathology or a more values-based interference with daily functioning and happiness). This conceptual independence is sometimes overlooked because pathological (or at least clinical) delusions are the paradigm cases, since (trivially and by definition) they come to the attention of clinicians.

Having said all of this, it is worth noting that there are a several terms in David's grievance that are epistemological ("unfounded," "irrational," "incorrigible"). So the following question seems to arise quite naturally. Might it not be the role of the philosopher in general and the epistemologist, in particular, to "define" delusion?[2]

9.1.2 Delusions as epistemically defective mental states

The aim of such a philosophical definition is typically thought to be that it should enable us to take any case, real or imaginary, and put it into the category of "delusional" or "non-delusional," and to do so for its own sake, not for the sake of clinical diagnosis. As Murphy (2013) rightly says, we can even draw an illustrative parallel with the concept of knowledge:

> A delusion is a false belief, just as knowledge is true belief, but, as with knowledge, philosophers do not rest there. Knowledge is true belief plus something else. So too, philosophers try to find that extra property of the false belief that converts it from a mere false belief into a delusion.
>
> (Murphy 2013, p. 115)

The point is that just as epistemologists have looked for the extra ingredient you need to add to true belief in order to get knowledge, so have recent theorists looked for the extra ingredient you need to add to false belief in order to get delusion. And as with analyses of knowledge, real or imaginary counterexamples can be appealed to in criticism of analyses of delusion. We will look at potential attempts at doing this, before considering an abandonment of the overall approach.

Whatever we take delusion to be, one thing that seems fairly obvious is that delusions (and the subjects who have them) are breaking norms. In other words, delusion involves belief or belief-formation that is "bad" or has "gone wrong" in some way. When you say that something is a delusion, or that someone is delusional, you are performing a negative evaluation; you are claiming that they are somehow deviating from some standard or benchmark.[3] So, what kinds of norms might help us define delusion?

We need to distinguish between *deontic* and *epistemic* norms, with a view to putting aside the former. Deontic norms are norms that don't only set the standard for our evaluations of certain things: they entail some kind of obligation. You don't just say that murder is wrong in the sense that it is merely defective (as you might for a belief that is false or a heart that fails to pump blood); you expect to have normatively binding consequences for the way people act (and evaluate, reprimand and punish actions). These kinds of norms, therefore, seem to apply only to actions, namely, to phenomena that an agent has voluntary control over. It is safe to say that doxastic involuntarism, the view that you don't have direct control over your beliefs, is widely accepted. You can't believe that I'm 10 feet tall even if I offered you a million pounds to do so. In any case, regardless of whether one takes belief in general to be subject to voluntary control, it seems clear that delusions (or at least most of them) do not arise as the product of anything that is under the direct voluntary control of the subject. We should therefore put aside deontic norms altogether as being of little use for defining delusion. The delusional subject (at least in most paradigmatic cases) has not obviously neglected an obligation of any sort. The delusional subject is often, and perhaps always, a victim to be pitied, rather than a wrongdoer to be blamed.

An *epistemic norm* is a standard or benchmark that specifically concerns epistemic matters, matters concerning the phenomena of knowledge and belief, and that will guide evaluations of these phenomena. These kinds of norms are commonly thought to be clearly definable by theoretical work. Indeed, one might even see the main task of epistemology as being one of mapping out the normative landscape for belief and cognition. Epistemologists ask some variety of the question: when and in what way is a given belief good (and, more traditionally, good enough to warrant the accolade of being called knowledge)?[4] We might call the norms that arise from answering this question in terms of clearly definable conditions, "traditional" epistemic norms.

So, what are the ways in which belief and belief-formation can be good or bad? One obvious way is by being true or false. True beliefs, all other things being equal, are better than those that are false. However, this is not usually what we mean by an epistemic norm since it is in principle independent from the epistemic agent. An epistemically good process, where the agent is doing things right, can get unlucky and yield a false belief. Conversely, one that isn't good might get lucky and yield a true belief. The point is that epistemically good belief-formation is (to use the epistemological terminology) truth *conducive* but not infallibly truth generating.

What might such a truth-conducive process look like? At this point, we risk delving into many decades of core research in epistemology (e.g. between internalists and externalists, reliabilists and virtue theorists). Let us focus on the sort of processing that has tended to be appealed to in the delusions literature: *epistemic rationality*.

Epistemic rationality is to be contrasted with practical rationality. The norms of *practical* rationality tell us when *action* is rational (or irrational), whereas the norms of *epistemic* rationality tell us when a *belief* is rational (or irrational). Rationality is best thought of in terms of the attainment of *goals* or *aims*. Practically rational action is action that maximises our chances of fulfilling our own aims (or "desires"). So an irrational action is an action that does a bad job of fulfilling your desires. For example a reluctant addict's behaviour is a prime example of practical irrationality. So what is epistemically rational belief? Whether *we* can aim at anything while believing is controversial, since it is not directly under voluntary control (Williams 1973); however, the idea that belief *itself* aims at truth is seen by many as highly plausible (Williams 1973; Velleman 2000; Wedgewood 2002). In light of this, a simple characterisation of epistemic rationality is belief-formation that has maximised its chances of achieving its goal, namely, *truth*. But how is that achieved?

Here is one answer: correctly using certain evidential inputs. Irrationality would then involve *incorrectly* using certain inputs (implementing biases) or forming beliefs on the basis of no evidential inputs at all. Here is the core question we come to: can delusion be defined in terms of epistemic irrationality thus construed?

This is far more promising than defining it in terms of falsity. It also fits nicely with our intuition that someone who is being elaborately misled (like Neo in the Matrix) is not delusional (and that the true-out-of-luck street prophet remains delusional). However, it is crucial to see why this proposal won't work. Irrationality is not sufficient in itself. There are many cases of irrationality that fall short of being delusional, and so where do you draw the line? More worryingly, perhaps, irrationality might not be necessary either. In other words, there might be instances of epistemically rational delusion. Let us elaborate on these two worries in turn.

Not sufficient: non-delusional irrationality

Stating the obvious, there are instances of irrational beliefs that we would be unwilling to call delusions. There are heuristic biases that stray from epistemic ideals, but that are adaptive shortcuts most of the time. We may say that this is not ideal rationality but rather "bounded rationality" (Gigerenzer and Selten 2002): there are temporal constraints or conflicting norms (perhaps even various epistemic norms) that are weighed against one another (you might, e.g. sacrifice accuracy for exhaustiveness or vice versa (see, e.g. Proust 2013)). On a different note, as we have mentioned, there are various "healthy biases" in the normal population. These include, for example, the fact that it is considered healthy to have an unrealistically positive self-image (i.e. "Positive Illusions," see Taylor 1989). People who are realistic, who approach what would be epistemic rationality in these domains, are more likely to suffer from clinical depression (Alloy and Abramson 1979).

In any case, it should come as no surprise that there are cases of irrationality that aren't delusional. No one expects to be able to say that *any* degree of irrationality is a sufficient condition for delusion: we are all guilty of irrationality at times (see Bortolotti 2009). But perhaps a certain degree of irrationality may suffice. Perhaps somebody needs to be irrational *enough*, and then they will count as delusional.

However, not only is it difficult to know where to draw the line, degree of irrationality might not directly correlate with "delusionality" as well. Stated plainly, Person A might be *more* epistemically *irrational* than Person B but in fact turn out to be less *delusional*. Consider, for example (from Nozick 1993, cited in Murphy 2012) a mother whose son has been convicted of murder. We can understand that she will be *highly* resistant to evidence that suggests that he is guilty. We will not, however (I would suggest), be tempted to call her delusional. People in these situations are believing in ways that are *epistemically* deeply irrational (they are far from being truth-conducive), but they are intuitively not delusional.[5] Why is this? I have suggested elsewhere (Wilkinson 2020a) that it is because we can recognise their motivations, and we can recognise the influences that these can have on belief-formation and maintenance. This means that we find their epistemic irrationality unsurprising and *understandable*. We might even recognise (implicitly or explicitly) that in similar circumstances, we would do similarly. We might even be repulsed by someone who calmly and dispassionately evaluated evidence pertaining to their son's guilt. We model other human beings (and ourselves) as understandably biased and emotional creatures.

This, I would suggest, marks a major change of approach. However, before moving on to the new approach, let's look at the other reason why delusion cannot be defined in terms of epistemic rationality, namely, that it might not even be necessary, let alone sufficient.

Not necessary: rational delusion?

Might there be cases of delusion that don't involve *any* irrationality in the sense we have just sketched? There are two very different kinds of grounds one might have for claiming this. One is on the basis of already existing (and in principle empirically testable) theories about how certain cases of delusion come about. We will examine these very shortly. The other is a conceptual argument that can be supported with thought experiments.

With advancements in cognitive neuropsychiatry, we have moved beyond the observable behaviour of delusional individuals to some understanding of what might underpin the formation of these delusions (more on this soon). In particular, there has been increasing support for the view that these delusions are in fact formed on the basis of some kind of anomaly at the level of experiential input. To put it in more intuitive terms, if you or I were to experience what these patients experience, then we too would form the delusions that they form. As Brendan Maher presciently put it, at a time before neuropsychological theories of delusions were available, "The delusional belief is not being held 'in the face of evidence strong enough to destroy it,' but is being held because evidence is strong enough to support it" (1974, p. 99).[6] The point is that we can think of these delusions as arising from the correct use of very bizarre input (what Maher calls "evidence"), instead of from a misuse of normal input.

As we are about to see, most philosophers and neuropsychologists in the field agree that many paradigm cases of delusion have at least some experiential grounds. The main source of contention is whether this experiential anomaly is strong enough (carries enough epistemic weight) to explain why the delusion is maintained for so long, or whether we need to postulate a bias of some kind. In the latter case, the delusional patient would be charged with epistemic irrationality.

However, whether or not there actually are biases at work is an empirical question, and our aim is to ascertain, regardless of whether certain real-world delusional patients are epistemically irrational or not, whether, *if* there *were* people who believed these bizarre things on the basis of fully adequate private grounds, and hence are plausibly epistemically rational, we *would* still rightly consider them to be delusional. To put it another way, if Maher's theory happened to be correct (regardless of whether it actually is or not) would these patients still count as delusional?

Jennifer Radden (2010) calls these, rather aptly, "perceptual delusions." Her view is that "As reasonable inferences from misleading perceptual experiences, 'perceptual delusions' are not epistemic lapses of the sort by which delusional states are identified" (2010, p. 28). This amounts to us retrospectively revising our delusion attributions in the light of a *stipulation* that delusions are tied to epistemic irrationality and the *discovery* that a significant proportion of delusional states aren't irrational in the required way. In other words, it would turn out that some paradigmatic cases of delusion aren't delusional after all. There's nothing incoherent about such a position, but I find that such an overarching revision of what we deem to be a delusion needs to be thoroughly motivated, and I fail to see such motivation.

In contrast, I am tempted to say: "Yes, these patients still count as delusional," for reasons related to the change of approach I mentioned earlier, and I am about to elaborate next. Notice that this is in line with Maher's implicit view. It certainly wasn't his intention to show that these patients that we previously had taken to be paradigm cases, were not, after all, really delusional. Rather, the question is: *granting* that they are delusional, how can we explain their delusional state? This is precisely the question we will address in the next section. But first, what is this change of approach that I've mentioned?

9.1.3 A change of approach

The change of approach I have in mind is as follows. We need to reflect on the kind of term that "delusion" is (or the kind of concept that delusion is), how it came about and the role that it plays when delusion is attributed to.

In Wilkinson (2020a), I argue that delusion attribution, for example saying "This person is delusional," is not (fully) fact-stating. It is not like saying "This person is 6ft tall" or "This person has brown hair." Picking up on themes we encountered in Chapter 2, it is fundamentally an evaluation. Some evaluations involve failing to adhere to an objective benchmark (e.g. "Anyone who is under 6ft doesn't make it into my basketball team"), while others are more fundamentally, irreducibly evaluative ("That's bad," "That's ugly" etc.). As we saw in Chapter 2, one way of thinking about these fundamental kinds of evaluations is as *expressive*, rather than descriptive.

In other words, delusion attributions are more akin to expressions of folk-epistemic bafflement. It's like saying "You can't believe *that*!" Why think that this is so? And what are the consequences of this?

First of all, it *seems* like delusion expresses a negative evaluation. Consider a hyperbolic use among friends ("You're delusional if you think Manchester United will win the Premier League this season!"). Now, one might object that, while this is an expressive use of "delusion," this is not at all how psychiatrists use the term. In response, though, I would insist that the calm, institutional use of the term only masks its true nature. The word has been appropriated and used in cold, pseudo-descriptive contexts in a way that is (as Gilbert Ryle would say) systematically misleading (Ryle 1932).

Second, as with expressivists in meta-ethics, who are motivated by a metaphysical unease with postulating moral properties, we might be similarly reluctant to countenance properties of "delusionality." In any case, as the foregoing discussion shows, nailing down a property of delusionality that everyone can agree on is elusive. I would simply add that it is elusive for good reason. Furthermore, as we saw with Anthony David's lament, part of the problem is that we are already intuitively deciding upon delusional status and lining up our intuitions with the definitions. The definitions do not guide our judgements of delusional status: they attempt to capture them. But this is in vain, since why assume that our intuitive (folk-epistemic) reactions track anything that can be captured by a neat definition?

Now we ask: what are the consequences of the view? First of all, an inability to define delusion is to be fully expected. As just mentioned, we can't assume that all of the many things that arouse negative reactions from our folk-epistemic sensibilities can be regimented under a strict definition. In a related manner, if you try to characterise the sorts of things that get called delusions, you get a sort of *disjunctive norm pluralism*. What do I mean by this? Put simply, there are different kinds of folk-epistemic badness, but once a (not very well-defined) threshold is reached, however it is reached, you call that thing a delusion.

We can see this by contrasting two cases that cross that threshold but in very different ways. Contrast, for example, Reverse Othello syndrome (the delusional belief in the fidelity of a romantic partner) with a bizarre delusion in the context of a psychotic disorder diagnosis. You might, for example, think about the implausibility of the content. Here, the bizarre delusion ("I am the left foot of God") scores highly, *so* highly in fact that you don't even need to ask yourself about the individual's grounds for the claim. And even if you established that there were experiential evidential grounds, though an interesting discovery, this couldn't remove delusional status. The belief content is simply too implausible.

In contrast, if somebody said "My wife is being faithful to me," this is true (one hopes) of many people, and it could only be an examination of their personal situation, their bafflingly extreme resistance to evidence, etc., that would then tip them into delusional territory.

Another consequence of this view is as follows. Since delusion attribution is an expression of bafflement, understandability takes precedence over rationality. What I mean by "understandability" is an empathetic understanding of "where someone is coming from." Returning to this example: the mother who is reluctant to believe that her son is a murderer exhibits profound levels of epistemic irrationality, but, to reiterate, we understand that people are reluctant to admit the guilt of loved ones, and so, while we may charge her with understandable levels of bias and wishful thinking, we wouldn't call her delusional. Indeed, we'd find an impartial, fully rational, acceptance of the evidence rather more troubling.

9.2 Examples of clinical delusions

Although delusions need not be pathological by definition (granted, some may insist otherwise), they are certainly paradigmatically encountered in clinical settings, and most specifically in psychiatric contexts. It is quite striking how varied these clinical contexts are. The orthodox way of talking about these different clinical contexts is in terms of diagnostic category. For ease, I will do the same, but note that, for reasons expressed elsewhere (especially Chapter 3), I do not accept these categories uncritically. More problematically, the presence of delusions is often what leads to the diagnosis, rather than the diagnosis itself having any real explanatory power.

There are three main contexts in which delusions occur: psychotic disorders, mood disorders and brain injury.

9.2.1 Schizophrenia spectrum and other psychotic disorders

In DSM-5, under the large overarching chapter "Schizophrenia spectrum and other psychotic disorders," you will find a plethora of categories that have delusions among their diagnostic criteria. Although interesting, it would be too time consuming to go through them all. Aside from the obvious presence of "schizophrenia," there is just plain "delusional disorder," as well as "psychotic disorder due to another medical condition" where you are then, as a clinician, supposed to specify whether it is with delusions (code: 293.81) or with hallucinations (293.82). More relevant for our purposes is what these delusions look like. What is their *content* like? What are they about? And how, broadly speaking, is the person who holds the delusions? This will become clear as we progress.

Without any expectations of uniformity, we can still see that certain patterns emerge. In particular, we see that delusions that occur in psychotic disorders are often parts of *delusional systems*, and often, although not always, paranoid delusional systems. This means that you don't simply have one free-standing delusional belief but a network of implausible beliefs. Another feature of these delusions is that they can be bizarre, with content that is either impossible or hard to comprehend. For example you might get delusional claims of self-identity, such as "I am Napoleon," "I am Jesus," "I am Wayne Rooney." (What could the claim to be someone else actually mean?) We might also, as we've already touched upon, see delusions of thought insertion: "someone else is putting their thoughts into my head." This is often considered an experiential, rather than cognitive, disturbance; but if the experience is taken seriously, it leads to delusional belief (rather than mere experiences of passivity). Then you get the very bizarre, and physically impossible, claims like "I have a power plant in my ear" and then ones that are hard to comprehend like "I am the left foot of God" (attributed to the Nobel prize-winning mathematician John Nash (Nasar 1998)). I understand each word in that sentence (except perhaps "God"), and it is syntactically well formed. However, I have no idea what it would be for it to be true.

One very important feature of beliefs, and also of delusions, is that they are relatively stable mental states (we will also see that delusions may be especially stable and fixed in a particular way). The point is that beliefs are supposed to outlast their linguistic expression. I believe that Paris is the capital of France, even when I don't say it, even when I'm not thinking about it or even when I'm asleep. Furthermore, I wouldn't suddenly change or abandon that belief for no reason, or on a whim. The relevance of this for delusion in schizophrenia and psychotic disorders is that we need to be mindful, when we attribute delusions, that we are attributing something that is held over a period of time and is not simply a bizarre thing that a psychologically troubled individual just happened to say, once, as a fleeting expression of a momentary and extreme state of mind. The trouble is that there is a huge amount of variety among patients. Some will hold tenaciously to a particular claim, while others will say one thing and you won't be able to get them to either affirm or deny it immediately afterwards (so insulated they may be from rational engagement). These latter cases most closely resemble "classic" schizophrenia of Bleuler and Kraepelin, and it can sometimes be hard to tease apart various other "symptoms" such as "thought disorder" and "disorganised speech" from bizarre delusions. Indeed, this is so much the case that Berrios (1991) claimed that delusional claims were just empty speech acts, with not meaningful content at all. In other words, these patients are "just saying things," without it being an indication of what they really believe.

9.2.2 *Delusions and mood disorders*

Again, there is a lot of fluidity between diagnostic categories. It is widely acknowledged that delusions (and the softer term "delusional ideation" or "delusional thinking") occur in mood disorders. However, many patients who have mood disorders with delusions ("with psychotic features") could often just as easily be labelled with a psychotic or delusional disorder with bipolarity or depressive features. The dynamic temporal progression is probably what would lead to one diagnosis over another (Is this a depressed individual who became delusional, or a delusional individual who became depressed?). In any case, the overarching feature of delusions in the context of mood disorders is *mood congruence*. This means that the content of a delusion is in keeping with a given mood. There are a great many examples, but we can illustrate this with the following four representative ones.

1. Bipolar disorder (especially "mania" or a "manic episode")
2. Major depressive disorder (unipolar depression)
3. Anxiety disorder
4. Postpartum depression/psychosis

So, to simplify enormously, we have in these examples, states of feeling up (1), feeling down (2), feeling anxious (3) and feeling guilty (4). In "up" episodes of bipolar disorder, people can present with grandiose delusions. Sometimes, these will have supernatural content, like the belief that one is immortal or God, but more commonly it will involve implausible self-evaluations (e.g. that they are writing something of supreme importance, like a paper that will revolutionise the whole field). In down episodes in bipolarity, or in unipolar depression (major depressive disorder), some subjects can present with delusions of guilt and inadequacy; in other words, they don't merely have feelings of inadequacy but will endorse and verbalise them. Third, in anxiety disorders, experiences of anxiety and hypervigilance can give rise to persecutory delusions (Freeman, Garety and Phillips 2000). Indeed, Docherty et al. (1978) noted strong patterns of heightened anxiety prior to the onset of delusions.

Finally, a very striking example of how delusions can take on not only emotion and affect, but also thematic context is in postpartum psychosis. There are varieties of presentation, but often anxiety about the safety of the child, fear of and guilt about being a bad mother, all in the context of the emotional vulnerabilities accompanying the physical exhaustion of childbirth, lead to delusion-like claims (e.g. the mother's belief that the father is poisoning her food; the expressed belief that she is a terrible mother and doesn't deserve her family (Sit, Rothschild and Wisner 2006)).

What all of these have in common is that we can draw a straightforward line from the intense emotions and moods experienced, to the beliefs expressed. What makes these interesting as delusions (to the extent that they can be considered such) is that they tend to be as short lived as the moods and emotions to which they give expression. Perhaps, because of this, they occupy a conceptual space between delusion and delirium. Having said this, the moods and emotions can be long-lasting.

9.2.3 Delusion and brain injury

Delusions can occur in the context of brain injury, although only a very small percentage of brain injury survivors have them. Brain injury (as a result of stroke, trauma or disease) is immensely varied but is usually associated with neurological rather than psychiatric difficulties, most commonly fatigue, memory problems, irritability, hemiplegia, speech and language difficulties and so on. Although a very small proportion of those with brain injury have delusions (less than 1 per cent estimated in Deb et al. 1999), and conversely a tiny proportion of those with delusions have them as a result of brain injury (especially if we put aside dementia), delusions in the context of brain injury are (whether explicitly or *de facto)* the focus of most of the philosophical work on delusions.

Given some of the features of these delusions (namely that they are, as I am about to explain, monothematic, circumscribed and relatively tractable), this is perhaps unsurprising. However, it does mean that we need to be wary of cases where philosophers make sweeping claims about delusions (e.g. "Delusions are not beliefs," "Delusions are endorsements of experience," etc.). If the focus of these claims is delusions in the context of brain injury, they may only apply to a relatively small (although fascinating) subset of the things that we call delusions. Furthermore, some theorists count as a point in favour of their explanatory models that they can generalise across different delusions. But unity of explanation should reflect unity of the phenomenon explained. In the face of all of this variety, seeking and promising unity could be a misleading enterprise. Indeed, a similar point could be made to the one Jones (2010) made about hearing voices: with a heterogeneous phenomenon, we would do well to expect different accounts for different types.

Enough of the caveats: what are these delusions that have captured the attention and imagination of philosophers? The most theorised about, undoubtedly, is the Capgras delusion, named after Joseph Capgras, the clinician who first identified the condition (Capgras and Reboul-Lachaux 1923). This is often (although somewhat misleadingly, for reasons we will touch upon later) characterised as the delusional belief "that a loved one has been replaced by an identical-looking impostor." This is a particular form of delusional misidentification. Delusional misidentification, as a generic term, picks out the inability to correctly identify someone that you would expect the person to be able to correctly identify. You would expect them to be able to because they have no sensory or perceptual impairment (e.g. they are not blind, nor face-blind) nor do they have an impairment of memory (e.g. they do not have dementia). Indeed, the contrast with dementia is particularly illustrative. Whereas someone with dementia might fail to retrieve names and information pertaining to a loved one, in the canonical Capgras case, all is remembered, but it is simply that the person perceived is deemed by the delusional individual to not be the remembered individual. As a concrete example, a Capgras patient might say: "that man looks just like my father, but he is not him." Such an individual may well remember every relevant detail of their lives up until the accident that damaged their brain. Having said this, it is worth noting that Capgras-style delusional misidentification does indeed occur in the context of dementia.

There are other forms of delusional misidentification. There are cases of objects, pets and places being misidentified in ways that are logically analogous to Capgras (i.e. it looks just like the individual in question but somehow fails to be that individual). These things will all have to be things that are known to the individual because *identification* as a task only makes sense with singular individuals who have been encountered before and are known. I cannot ask you to identify a total stranger, to answer the question "Who is that?," since that is asking you to draw a connection of identity between the perceived individual and someone that you have previously encountered (whether in person, or through the media, books, etc.), where there is no previous encounter. More on this topic follows in this chapter.

These cases are Capgras-analogous in that the misidentified objects/places/pets look to the subject just like the objects/places/pets in question (that they in fact *are*!) but somehow fail to actually be them. The same can be said for mirrored self-misidentification, namely, the delusion that "the person I see in the mirror is not me."

In contrast, other examples of delusional misidentification involve hyper-identification, namely, the conviction that an individual is a given individual in spite of looking nothing like them. This amounts to the belief that a particular salient individual is taking on the appearance of strangers around one. This is called the Fregoli delusion (named after Leopoldo Fregoli, an Italian comic actor who was known for quick changes of appearance on stage (Courbon and Fail 1927)).

Confusingly, individuals with no clinical record or evidence of brain injury exhibit delusional misidentification of a kind that is often called Capgras delusion but occurs in the context of broader psychotic or mood disorder. At this point, it is crucial to disambiguate something. "Capgras delusion" could simply pick out the delusional misidentification of a significant other, regardless of the underlying cause. In this case, this labelling would be fine but should be handled with care. In particular, we should expect aetiology, clinical presentation and associated behaviours of brain injury Capgras and psychosis Capgras to look very different from one another. One particularly relevant pattern is the extent to which the delusional belief is acted upon. Pandis, Agrawal and Poole (2019) report a 38 per cent prevalence of aggression among "functional" (i.e. non-brain-injured psychotic disorder) cases, as opposed to 23 per cent in "organic" (brain-injured) cases.

Alternatively, however, one could take Capgras-resembling delusional misidentification in the context of a psychotic disorder, but in the absence of brain injury, to be just that: something that resembles but that is not actually Capgras. Having said that, to avoid confusion, as a useful (but not authoritative) stipulation, I use "Capgras syndrome" to pick out Capgras delusion in the context of brain injury (partly for ease, but also because it is common to distinguish "organic syndromes" from "psychiatric disorders"). In any case, it has some unique and striking features that delusional misidentification in psychosis lacks. These are also the features that make it so interesting to philosophers. So what are they?

First of all, in contrast to delusions in the context of psychotic disorders, they tend to be about one thing (it is "monothematic"), rather than involving a broad, expansive delusional system. Also, and here is a particularly striking feature, aside from the delusional belief, individuals with the condition seem remarkably normal and rational otherwise (it is "circumscribed"). In a related manner (although we will see that some aetiological accounts, namely, causal stories about their genesis, appeal to affect), these delusions, again in contrast to delusions in psychotic and mood disorders, seem not mood-congruent. Indeed, here is one area where the distinction between neurological and psychiatric delusional misidentification proves especially important.

Compare two cases. One individual, who is deeply paranoid and is prone to misreading social cues because he is overwhelmed by negative affect in the presence of social ambiguity. This person might, for example, have developed a friendly passing relationship with a cashier in the local supermarket. One day, he encounters the cashier, but he is being a bit less friendly than usual. In response to this, a cascade of negative social emotions needs explaining. The cashier must still like me. Perhaps he has an identical twin. That's it. It must have been his identical twin.

Contrast this emotionally grounded (in this case, paranoid) misidentification with, say, DS, from Hirstein and Ramachandran (1997). DS claims that his father has been replaced by an identical-looking impostor but has no concerns for the welfare of whereabouts of the real father. And, quite the opposite of the being the expression of paranoia, or negative emotions towards the father, DS actually likes the "impostor." DS says to the experimenter: "he's a nice guy, but he's not my father, Doctor" (Hirstein and Ramachandran 1997, p. 438). This ties in with another more variable and disputed feature of these kinds of delusions, which we have already touched upon, namely, that they tend to be agentively *inert*. They do not drive behaviour in the way one would expect. Would someone who really believed that they had lost a loved one really not worry about them? This issue will be highly relevant in Section 9.4.

9.2.4 *Taking stock: taxonomies of delusion*

There is a lot of variation among the delusions that we encounter in clinical settings. In spite of this, they can be broadly taxonomised.

First, we might distinguish monothematic from polythematic delusions. The former centre around one narrow theme and tend to occur in the context of brain injury. The latter can be about many different things. A related distinction is between circumscribed and elaborated delusion. Circumscribed delusions are insulated inferentially from other beliefs, whereas elaborated delusions spread out into delusional systems. (I'm feeling paranoid, something that I was expecting to get in the mail hasn't arrived. The CIA is stealing my mail. They might also be reading my thoughts. That neighbour opposite just looked at me through a narrow opening in his curtains, he works for the CIA too. Etc.) To re-iterate a very important and related distinction, we can distinguish mood-congruent from non-mood-congruent delusions.[7] The former involve delusions that are in keeping with current moods and emotions, whereas the latter are not so clearly continuous.

There are exceptions to this, but, in general, delusions in the context of brain injury (and certainly those that philosophers have spilt most ink over) tend to be monothematic, circumscribed and not mood-congruent. Delusions in the context of psychotic disorders occurring in the absence of brain injury have tended to be polythematic, elaborated and mood-congruent.

9.3 Explanatory accounts of delusion

Why do delusional individuals believe the things that they do? Notice that, following themes from Chapter 6, there are two perfectly valid senses of "why" at play here. We could be asking the justificatory why-question, which would amount to asking: on what evidential grounds do they believe what they do? Alternatively, we could be asking: how do we causally explain why they come to believe what they do? As we saw in Chapter 6, there might well be a satisfying account that makes use of both: that gives us a causal story about the generation of a particular experience, which then is taken to produce grounds for the belief.

Since the majority of the work on these explanatory models has been done either explicitly on the Capgras syndrome (i.e. in the context of brain injury) or makes general claims about delusions with those kinds of monothematic and circumscribed delusions in mind, I will focus largely on these. This focus is understandable, since the delusions, as we will see shortly, are in principle more theoretically tractable. We can, for example, make hypotheses about how certain kinds of localised brain damage give rise to certain experiences, and in turn these give rise to certain beliefs. However, I will first say a few words about explanatory accounts of delusions in the absence of brain injury, namely, in the context of psychosis and, most canonically, schizophrenia diagnoses.

9.3.1 Delusions in psychotic disorders

In discussing the sorts of delusions that might occur in the context of psychosis (e.g. the sorts of individuals who might attract a diagnosis of schizophrenia or similar), it is helpful to separate delusions of control, and delusions of thought insertion, from other kinds of delusions in psychosis (e.g. delusions of reference, grandeur and persecution, as well as florid delusions).[8]

As we saw in Chapters 7 and 8, self-monitoring accounts might help us to understand why an individual's nervous system might have difficulty distinguishing self-generated phenomena from externally generated phenomena. In particular, this has been hypothesised (e.g. by Frith 1992) to be the result of a self-monitoring mechanism that compares the predicted and actual sensory consequences of one's actions (see Chapter 7). When these fail to match, the nervous system signals "this is not you!." Interestingly, Frith took the self-monitoring deficit to unify different symptoms: the symptoms differ to the extent that the phenomenon (the action) that is failing to be properly monitored differs. For delusions of control ("Somebody is controlling my actions") it is bodily actions, whereas for auditory verbal hallucinations (AVHs) ("Somebody is talking to me") and for thought insertion ("Somebody is putting thoughts into my head"), it is inner speech. What is fascinating about this is that, even though hallucinations and delusions are traditionally thought to be very different, several theorists (e.g. Frith 1992; Stephens and Graham 2000; Ratcliffe and Wilkinson 2015) have claimed that AVHs and delusions of thought insertion are fundamentally the same phenomenon interpreted, and hence reported, differently. If this is correct, it is no surprise that reports of inserted thoughts get branded as delusions, because thought insertion seems like an outlandish belief, whereas claims of hearing a voice look like the report of a hallucinatory experience.

In any case, one popular explanatory account of delusions of control and thought insertion is that self-monitoring deficits causally explain a strange experience, which is then taken at face-value, yielding the delusional belief (that someone is controlling my actions, or putting thoughts into my head).

In stark contrast, a similar story cannot be told for other delusions like "I am Napoleon," "I am the left foot of God," etc. This is not simply because it is not clear what self-monitoring deficits might contribute here, but rather because there is less of a clear connection between the experience and the delusion. There is a degree of embellishment, and how do we explain that?

Here one promising strategy might be to look at how people with psychosis process language, and in particular non-literal language (i.e. metaphor). Deamer et al. (2019) showed that participants with diagnoses of schizophrenia were significantly more likely to interpret metaphorical expressions literally. Now, if this is the case, perhaps the gap between very powerful experiences (e.g. of grandiosity) and strange, florid claims can be bridged. A poet might conjure up the expression, "I feel so powerful, it's like I'm the left foot of God." Someone with metaphor comprehension deficits might generate such an idea (perhaps in inner speech), take it literally and assert it. This is all very speculative, but it is a potential future direction (Deamer and Wilkinson 2021).

In stark contrast, attempts to explain Capgras delusion have been rather different, since the delusional claims are not nearly so separate from what might be conceivable experiential grounds.

9.3.2 The classic bottom-up account of Capgras delusion (Ellis and Young 1990)

Ellis and Young put forward the classic "bottom-up" account of Capgras, in the sense that they present the delusion as a reaction to an unusual experience (hence "bottom-up," where the "bottom" is experience/sensation, and the "top" is judgement/cognition). Borrowing Bauer's (1984) model, whereby there are two streams for processing facial information – one covert, affective and anatomically dorsal, the other overt, semantic and anatomically ventral – Ellis and

Young (1990) put forward the influential proposal that Capgras syndrome can be understood as a sort of "inverse prosopagnosia."

In cases of prosopagnosia, patients have difficulty in the overt recognition of faces, even though their ability to recognise other classes of stimuli can remain intact. Show them a picture of a familiar face (i.e. a friend, family member or famous person), and they will not be able to tell you whose face it is. However, fascinatingly, they appear to have differential autonomic responses to these faces, as measured by heightened skin conductance response (SCR). In other words, although they themselves cannot tell you whose face they are looking at, their affective system seems at the very least to be able to "tell" that it is someone familiar.

Ellis and Young hypothesise that Bauer's two streams can be selectively impaired, leading to (that Holy Grail of neuropsychology) double dissociation. According to them, whereas with prosopagnosia the affective stream for "covert recognition" is intact and the semantic stream for "overt recognition" is impaired, with Capgras it is the other way around. The Capgras patient, on their view, is presented with someone who, thanks to intact semantic processing, looks to them exactly like a loved one, but there is a lack of affective response. The patient therefore concludes that this person cannot be the loved one in question. This model was then given experimental support when it was discovered that, in contra-distinction to prosopagnosia, Capgras patients show diminished SCR when presented with familiar faces (Ellis et al. 1997; Hirstein and Ramachandran 1997).[9]

This model is highly attractive for a number of reasons. First, it builds on and is consistent with past research, in particular on prosopagnosia and face-processing (e.g. Bauer 1984; Landis et al. 1986). Second, it presents a hypothesis and shows how it might be falsified. If Capgras patients did not show diminished affective response then their theory would be *prima facie* falsified. Third, it presents an account for understanding the formation of the delusion that not only renders the phenomenon causally tractable, but also gives us an idea of what it might be like for the subject. It therefore makes the delusion at least in principle intelligible: you can put yourself in the patient's shoes by asking yourself, "What would it be like to see your mother, but to not feel that affective warmth, or to experience her as deeply unfamiliar?" In a related manner, it enables us to answer the why-question: "why does the subject believe that this person isn't her mother?" We can answer something like: "because she feels unfamiliar to him." This provides a personal explanation in the sense sketched in Chapter 6. Finally, it seems to explain the statistical and behavioural data. It explains why Capgras patients don't come across as deeply psychotic or globally irrational. It also has the tools for explaining the fact the Capgras patients only misidentify loved ones (however, we will later see that this is not so with all cases). As Ellis and Young (1990) make explicit, it is only with people who look like loved ones to the subject (namely, the people we onlookers know *are* their loved ones) that one *expects* a strong affective response. When this expectation is frustrated, misidentification ensues. That explains why those who are not loved ones are not misidentified.

With all of this going for it, it is hardly surprising that the Ellis and Young model, at least in rough outline, has been widely accepted and constitutes an orthodoxy or at least a starting point for other more recent theories. Accounts that rely on the basic idea of the Ellis and Young model include, but are not restricted to Stone and Young (1997), Davies et al. (2001), Bayne and Pacherie (2004) and Coltheart (2007). In the next chapter, we shall see that I have some suggestions of my own to make in improving this model.

9.3.3 Explanationist versus endorsement models

It is worth noting that Brendan Maher foreshadowed the Ellis and Young model in spirit if not in detail. As far back as 1974, he claimed that delusions are unusual beliefs that are held on the basis of unusual experiences. However, if we accept this, there is an important issue to resolve concerning the precise nature and role of the anomalous experience. What content is carried or encoded in the experience? In other words, what does the experience "tell" the subject?

Following Bayne and Pacherie (2004), we can isolate two options, which correspond to what they call "explanationist" and "endorsement" models. For the explanationist, the experience only encodes a sparse content (something like "unfamiliar"), and the delusional judgement is a hypothesis that has the role of *explaining* the anomalous experience. Roughly speaking, the subject reasons: "This man, who looks like my father, feels unfamiliar (or fails to invoke a feeling of warmth or familiarity in me). What could explain this? It must not *be* my father." In other words, the delusion is formed through an abductive inference on the basis of sparse content. Maher (1974) seems to hold something like this view, as do Ellis and Young (1990) (who refer to a "rationalization strategy") and Stone and Young (1997).

Bayne and Pacherie (2004), on the other hand, want to replace the explanationist model with their own "endorsement" model. Endorsement models claim that the content of experience is rich: the content of the delusional judgement ("that's not my father") is encoded directly in the unusual experience, and all that suffices, namely what the delusional judgement consists in, is *endorsement* of that content. To put it another way, what the subject's experience is "telling them" is something rich like "This person is not your father" rather than something sparse like "unfamiliar."

How are we to adjudicate between the two? As Pacherie (2008) neatly puts it:

> Experience-based accounts of delusions involve (at least) two components: (a) an explanation of the delusional patient's experiential state, and (b) an explanation of the delusional patient's doxastic state. Endorsement and explanationist models face distinct challenges in providing these explanations. Explanationist models appear to have an easier job of (a) than endorsement models: the less one packs into the content of the perceptual experience, the easier it is to explain how the experiential state acquires its content. . . . But what explanationist models gain with respect to (a) they

> lose with respect to (b). The explanationist holds that delusional beliefs are adopted in an attempt to explain unusual experiences. The problem with this suggestion is that delusional beliefs are typically very poor explanations of the events that they are supposedly intended to explain.
>
> (2008, p. 107)

In other words, the downside of the explanationist model is that it fails to explain why the subject adopts what we might call the "doppelganger hypothesis" and not some other, more plausible hypothesis, whether this be self-constructed ("maybe I'm in a bad mood," "maybe I don't like dad anymore") or actively suggested by their doctor, like the "brain-damage hypothesis" ("the reason why this person feels unfamiliar is because I've damaged my brain in such a way that they will feel unfamiliar").[10] As Fine, Craigie and Gold (2005) ingeniously point out, if, as the patients sometimes admit, the doppelganger is an *exact* replica of the original loved one, then the patient ought not to expect them to feel odd. If a perceived individual is perceptually indistinguishable from one's father, one would expect them to feel like one's father too, namely, one would expect the deception to be complete. After all, what, other than perceivable properties, can the disguise, indeed any disguise, or indeed *any* conscious judgement of identification, be based on? (Later, we will see what my response to this question is.) Therefore, the doppelganger hypothesis is not only a bad explanation compared to competing hypotheses: it is also a bad explanation in that it almost seems to defeat itself.

Conversely, the main selling point of the endorsement model is that it can explain this. There is no hypothesising: only endorsement. It seems to me that P, therefore P. However, as Pacherie points out, endorsement models "pack more into the content of the experience" and therefore have more explaining to do on that front. Later in this chapter, we will see some options for how to achieve this.

At this point, it is interesting to note, given what we established in Chapter 6, that the "explaining" that is done at (a) and (b) correspond to subpersonal and personal explanation, respectively. Explaining how the experience has the content that it has requires us to understand what has been disrupted, and how that will impact on experience. Experience itself is a starting point for personal explanation, an input for inference or belief-formation. It is not *itself* explainable by personal explanation. You can't ask someone: "why (on what grounds) did you have that experience?" In contrast, explaining how one gets from the experience to the belief is paradigmatically personal. On the endorsement model, you get:

> Q: Why did the subject judge that P?
>
> A: Because it perceptually seemed to her that P.

Alternatively, on the explanationist model, you get:

> Q: Why did the subject judge that P?
>
> A: Because it seemed to her that Q, and the best explanation for Q was P.

It is extremely important to note, as Bayne and Pacherie emphasise, that both the explanationist and endorsement models count as bottom-up accounts in the sense that the delusional judgement is grounded in experience. They merely differ in their views about what the content of this experience is and, therefore, what kind of reasoning process the subject has to go through in order to form the delusion.

9.3.4 One-factor versus two-factor accounts

Note, however, that delusions are not only formed, but also tenaciously held in the face of contrary evidence. Following Coltheart (2007), we might call these two features "formation" and "maintenance."[11] Although these bottom-up accounts explain the formation, neither the explanationist, nor the endorsement theorist provides an account of why it is that the delusion is held so tenaciously in the face of contrary evidence. In particular, regardless of whether the experience carries sparse or rich content, it still constitutes evidence on the basis of which the delusional judgement is formed, and presumably the subject can refrain from making this judgement or can reject it later. On the endorsement theory, the experience tells the subject something like "this man is not your father," but the subject could in principle reject the testimony of his senses. After all, we can do this when we are presented with illusions that we know are illusions, but which have conceptually rich content: we may see the stick as bent, but we don't *believe* that it is. Bayne and Pacherie themselves are aware that they need to say more to account for the tenacity of the delusion. The richness of the experiential content cannot account for this alone. What else might be appealed to is what we will look at now.

As the name implies, a one-factor theory is one that claims that the patient has one deficit and that that is sufficient to explain both the formation and the maintenance of the delusion. The deficit is usually taken to produce an anomalous or bizarre experience on the basis of which the delusion is formed. A two-factor theory claims that the bizarre experience, though very much necessary, is not sufficient to explain the formation and maintenance of the delusion. It needs to be supplemented by some abnormality, either a deficit or a bias, at the level of reasoning or belief-formation.

The first theorist to clearly express a one-factor view (which was also the first clearly formulated bottom-up view) was Maher (1974), who, as we have already mentioned, foreshadowed the experimental work that, over a decade later, came to support his bottom-up theory. Maher claimed that delusions are hypotheses that the subject forms and adopts in order to explain an unusual experience. What makes this a one-factor theory is that the delusions are developed as a result of "the operation of normal cognitive processes" (Maher 1974, p. 103). There is no reasoning abnormality: only an experiential one. The disagreement between one- and two-factor accounts is usually seen as a disagreement about whether the delusional hypothesis is a "normal" response to an anomalous experience. Assuming rough agreement on what the delusional hypothesis is, they disagree, first, about the nature of the anomalous experience, in particular its evidential weight, and second, about what constitutes a "normal" response. This disagreement is very important since it is a dispute about whether deluded patients have a more global pathology of thought and reasoning. One-factor theorists claim that they haven't; two-factor theorists claim they have (see Sullivan-Bissett 2020 for an excellent clarification of the one-factor/two-factor distinction).

In direct opposition to the official DSM definition's claim that delusions are held in the face of contrary evidence, Maher states:

> The delusional belief is not being held "in the face of evidence normally sufficient to destroy it," but is being held because of evidence powerful enough to support it. Where the patient may differ from the normal observer is not in the manner of drawing inference from evidence but in the kinds of perceptual experience that provide the evidence from which the evidence is being drawn.
>
> (Maher 1974, p. 99)

Two-factor theorists agree that there is an unusual experience (cf. Davies and Coltheart 2000) but think it is implausible that the experience alone can account for the delusion. They think that, without postulating some post-experiential abnormality, the one-factor theorist cannot explain, first, why such an implausible hypothesis is formed and adopted, and, second, why the patient doesn't reject the hypothesis once she is presented with better explanations. In other words, assuming that the delusion involves an inference to the best explanation of the anomalous experience, one would expect the delusional hypothesis pretty swiftly to cease to be the best explanation. For example one would expect the following assurance on the part of the doctor to provide defeating evidence and hence to cure the Capgras patient: "The reason why this person doesn't feel like your mother is not because she isn't your mother, but because you have had an accident that has damaged your brain and will make her feel unfamiliar." But this doesn't cure the patient; hence the need to postulate a second-factor, according to the two-factor theorist.

What can the one-factor theorist say in response to this claim of implausibility? Since the criticisms that the two-factor theorists launch rest on the idea that the delusional hypothesis is a bad explanation of the anomalous experience, there are two clear and complementary strategies open to the one-factor theorist. First, she can claim that the delusional hypothesis is not as poor an explanation as it appears to be. Second, she can claim that the experience is stranger, and hence in need of a stranger explanation, than the two-factor theorist realises. I say these are complementary because the two strategies help each other out. The less implausible you take the delusional hypothesis to be, the less strange the experience needs to be in order for it to be a good explanation of it. On the other hand, the stranger the experience is, the stranger the explanation is allowed to be that is constructed to explain it. Reimer (2009), for example, protests that the nature of the anomalous experience makes all the difference. We just cannot underestimate what these patients are going through, experientially speaking. I think there is a great deal of truth to this, but it will require us to give up on some fairly ingrained presuppositions about experience and judgement and how they relate to one another. We will see more along these lines later.

However, it is important to see that there are views that don't fit the bottom-up orthodoxy at all. In a sense, they avoid these issues altogether.

9.3.5 An alternative approach: top-down accounts

Though popular, it is not universally accepted that delusion formation is the "bottom-up" process that the Ellis and Young model suggests. In other words, it is not universally accepted that the delusional belief arises as result of an anomalous experience, an experience that (i) precedes the delusional judgement and (ii) serves as evidential input or grounds for it. According to alternative "top-down" theories, the delusion is certainly the result of brain damage (obviously, since we know that these patients weren't delusional before their lesions and became delusional subsequently), and there is an anomalous experience, but it is not the experience that is caused first by brain damage. Rather, the way things appear to the subject, experientially speaking, relies on their expectations and beliefs, which can be prior to the experience and colour it (or alternatively are a crucial component of the experience). As Campbell 2001 puts it, delusion "is a matter of top-down disturbance in some fundamental beliefs of the subject, which may consequently affect experiences and actions" (Campbell 2001, p. 89). Therefore, applied to the Capgras delusion, the subject believes that this person is not their mother, and as a result this person will not be experienced as such or will be experienced as unfamiliar. It is not the experience that drives the judgement ("bottom-up"), but the judgement that alters the experience ("top-down"). Of course, this provides an alternative explanation of the reduced skin conductance. It is a consequence, not a cause, of the delusional misidentification.

This has been less popular than the bottom-up theories that have been built more directly on the Ellis and Young model. More specifically, top-down theories seem to lack many of the explanatory advantages mentioned earlier. First, they cannot explain as easily why it is only loved ones that are misidentified. Second, they do not enable one to understand the subject's point of view: how can you simply judge on the basis of nothing, that this person, who looks just like your mother, is not your mother? Lastly, and in related fashion, there is something perplexing about the idea that a knock on the head can directly induce a change to the subject's beliefs. It is far more understandable to think of the damage causing some deficit to the experiential input of a belief-formation mechanism.

Implausible as top-down theories may seem, there are valuable insights to be extracted from them, especially concerning the way in which belief (or belief-like states) need not always be held for epistemic reasons. Furthermore, they have the *prima facie* advantage of accounting for the tenacity of the delusion, without having to appeal to a second-factor reasoning deficit. For the top-down theorist, the delusion is held for so long because it is not rationally held on the basis of evidence, experiential or otherwise, which the subject could reject. Rather than being the product of a rational process, the delusion is directly caused by brain damage, by something pre-evidential, pre-rational. Some have even claimed that it has the epistemic role of a premise or a Wittgensteinian "hinge proposition" (Campbell 2001; Eilan 2000).[12] Perhaps, for these reasons, namely, that it doesn't have any evidential grounds (let alone bad ones), it doesn't even deserve to be thought of as a belief at all. We will examine this issue in Section 9.4.

9.3.6 *Tracking accounts*

My own contribution to this literature is what might be called a "tracking account" (Wilkinson 2016). The best way of explaining a tracking account is as follows. If you reflect on what identification is, it concerns *singular cognition*. That is to say, it is about individuals, which are represented in formal logic by singular terms, like a, b, c, etc. They represent unique, unreplicable individuals in the logical domain. Identification is about drawing connections of identity between two individuals. It is not about predicating a property of an individual. It is not what is represented in formal logic as "Fa." Of course, this gets obscured by the fact that we use the word "is" (the copula) to express both predication and identity. We use it in both "Batman is tall" and "Batman is Bruce Wayne," even though in the former we are attributing a property to an individual (Fa), and in the latter, we are drawing a connection of identity between two individuals (a = b).

The relevance that this has for delusional misidentification is that previous theorists, most prominently bottom-up experience-based accounts, have failed to fully appreciate this and its consequences. Ultimately, the acknowledgement that identification (mis-identification) is of the form a = b, and not Fa, means that it involves rationality-bypassing, singular cognition, otherwise known as *tracking*. Nothing needs to rationally ground my judgement that a = b, because my appreciation of singularity is brutely causal (and so when it goes wrong, I am bound to have a tenacious and unsupported belief – something that people are understandably likely to call a delusion).

I haven't the space to go into the details here, but the main point is that properties, however complex (e.g. the property of looking exactly like my father), are multiply instantiable (can be instantiated, can exist, more than once). Logically, more than one individual can bear the same properties. What's more, a given individual can radically change its properties.

One useful way of thinking about this is through the metaphor (or analogy) of "mental files" (Recanati 2012). When I encounter an individual, and they are sufficiently salient, I open a mental file for them. The file corresponds to a bare singular term. I then fill the file with information (roughly, predicates) that I take to pertain to that individual. It is the file, and not its content, that makes it refer to that individual. This aligns with our intuition that if you meet Jane-1, and someone thinks you met Jane-2 and starts telling you lots of things about Jane-2 ("Did you know she's climbed Everest?") which you take to be about Jane-1, what you end up with is false beliefs about Jane-1, not true beliefs about Jane-2.

At last we get to the tracking account. As deeply social creatures who live in groups where we need to keep track of conspecifics as individuals (not simply as the bearers of certain properties) so that we can identify friends, foe, kin, people with shared histories, etc., we have developed a highly specific, low-level, intuitive mechanism for tracking individuals over time. (We can appreciate the extent to which our tracking of individuals outruns our explicit recollective and justificatory capacities when we re-encounter a long-forgotten acquaintance and identify them immediately.) Delusional misidentification in the context of brain injury involves neurological disruption to this mechanism. In terms of mental files, Capgras involves an inability to retrieve the correct mental file when presented with a particular individual ("hypo-identification"). This amounts to a neurological inability to believe that this individual I now perceive is that same individual from my past. No amount of recognition that they look the same will get around that. In contrast, the Fregoli delusion involves the erroneous retrieval of mental files ("hyper-identification"). Here, you have the judgment that this person is someone from my past in spite of looking nothing like them. Aside from its simplicity, why think that this account might be onto something?

First of all, there are two different ways of interpreting the tracking account: one is as experiential, the other as doxastic. In other words, does the failed tracking mechanism generate an inability to *experience* the perceived individual as a specific individual or an inability to *judge* them to be that individual? For reasons that are similar to those put forward by Hohwy (2004), I'm not sure if a clear distinction is necessary. In any case, I am sceptical of the "testimonial view of experience," according to which your experience "tells" you something and you calmly decide whether to believe it or not. It seems clear to me that experience carries with it a doxastic pull (in this respect I am sympathetic to the view held by the likes of Armstrong (1968), Pitcher (1971) and Dennett (1991) that experience is a disposition to believe/judge). Let's suppose that the tracking failure has a doxastic force: this would explain the tenacity of the belief. What is more, it would explain it in the presence of epistemic rationality elsewhere, as well as an apologetic acceptance on the part of the delusional subject that it is deeply implausible, both of which are evidenced in the case studies. Attempts to convince the patient otherwise would be futile: they are neurologically incapable of judging that individual to be the individual that they really are.

As I emphasised in Wilkinson (2016), the standard bottom-up accounts, which rely on lower-than-expected affective response as a cause rather than a consequence of misidentification, struggle to explain reported instances of:

1. Global Capgras
2. Reduplicative Capgras and
3. Fregoli delusion

Global Capgras involves misidentification of more than just loved ones: for example acquaintances like hospital staff and nurses. Here, a strong effective response would not be expected since they aren't loved ones. Reduplicative

Capgras (which is actually present in the original case report, Mme M. in Capgras and Reboul-Lachaux 1923) involves the impostor themselves being replaced (Mme M. reduplicated her husband an estimated 80-odd times!). Again, if the reason for taking him to be an impostor in the first place is a lack of affect, why would that hypothesis not be stabilised? In other words, you shouldn't expect affective warmth from seeing the impostor, since it was lack of warmth that, according to the standard view, would have led to the impostor hypothesis in the first place. Finally, although the classic account can appeal to hyper-affective response in its account of Fregoli, the tracking view does a better job of accounting for the specificity of the wrongly retrieved identity. On the standard, affect-based view, you would expect a nebulous feeling of familiarity ("Haven't we met before?") rather than something so specific ("You're James, my childhood friend!").

9.4 Delusion and belief

Although delusions are typically thought of as beliefs, there have been some who have questioned this. One's answer to the question, "Are delusions beliefs?" will obviously depend on what one takes beliefs and believing to be and on what one takes delusions to be. Furthermore, unless one takes the features that make a phenomenon delusional to be the very same features that prevent it from being a belief, one's answer may also depend on what cases or kinds of delusion one focuses on (as we've seen they are tremendously varied). We can't assume at the outset that the belief-status will be constant across the wide array of phenomena that we call delusional. It will then also depend on the nature and aetiology of the delusion that one has chosen to focus on.

9.4.1 Why might (some) delusions not be beliefs?

On what grounds have theorists denied that delusions are beliefs, and what kinds of claim have typically been made in response to such views? Following the terminology of this debate (Bayne and Pacherie 2005), I call those who deny that delusions are beliefs "anti-doxasticists" and those who insist that they are "doxasticists."

It is useful to see this debate as raising a particular instance of a general problem (see Wilkinson 2013), namely, the problem of distinguishing between evaluative and constitutive norms; in other words, distinguishing between doing ϕ badly or well (evaluative) versus doing ϕ and not doing ϕ at all (constitutive).

To get a grasp of the distinction between evaluative and constitutive norms, consider an analogy with chess (a real favourite among philosophers). If, during a game of chess, I move so that my queen can be captured by my opponent with impunity (and as a result she gains an enormous material advantage and goes on to win the game), that is a *bad* chess move. An evaluative norm (the norm of playing chess well) is broken. If I move my bishop like a rook, however, that's not a chess move at all, let alone a bad one. A constitutive norm (dictated by the rules of chess) is thereby broken. The doxasticist will say that delusion is like the former. The anti-doxasticist will say that it's like the latter.[13]

So why have some theorists denied that delusions are beliefs? The main proponents of the anti-doxastic position are Gregory Currie and his collaborators (Currie 2000; Currie and Jureidini 2001; Currie and Ravenscroft 2002). Their view comprises two claims: a negative claim and a positive claim. The negative claim tells us that delusions aren't beliefs (anti-doxasticism). The positive claim tells us what delusions are, namely, they are imaginings that are mistaken for beliefs (what Bayne and Pacherie 2005 call "The Metacognitive View"). For now we focus exclusively on the negative one.

Currie thinks that delusions ought not to be counted as beliefs because, although they have the superficial trappings of belief, namely, sincere assertion, they lack the right kind of functional role. Since, according to Currie, delusions

(i) are not supported by evidence in their initial formation,
(ii) are not open to review in the face of contrary evidence and
(iii) do not fully guide action and reasoning,

they should not be counted as beliefs at all. (And, as mentioned, should be counted rather as imaginings that are mistaken for beliefs.) The claim that delusions aren't beliefs can, in principle, be opposed on the basis of two different kinds of considerations:

A. Conceptual – you can disagree with the constitutive norms of belief that are put forward by Currie and Co.
B. Empirical – you can disagree that, as a matter of fact, patients infringe these norms.

B would ideally require us to pick a particular kind of delusion (see Section 9.2) and choose a fleshed-out account of the delusion in question (see Section 9.3 for some options). Note also that a critique could clearly make use of both A- and B-type considerations, both disagreeing with the norms taken to be constitutive of belief and with the portrayal of the deluded subject.

People with doxastic leanings might accept Currie's constitutive norms of belief and respond with exclusively B-type considerations. To (i) they respond that delusions may be based on evidence of a sort, namely, on strange experiences (e.g. as we saw, lack of affective response to familiar faces in the case of Capgras, cf. Maher 1974; Ellis and Young 1990).[14] In response to (iii) they claim that, although delusions often fail to generate the kinds of actions and emotional responses one might expect, the Capgras delusion (for example) leads to violence against the impostor in 18 per cent of cases, and sometimes of a particularly gruesome sort (a caveat is in order here – these tend to be in the context of psychosis rather than brain injury (de Pauw and Szulecka 1988)).[15] Granted, it is much harder to explain away (ii) since the delusions are (and some, including the

DSM, say *by definition*) highly resistant to correction in the face of contrary evidence. However, perhaps one can claim that the experiential evidence in favour of the delusion is so strong that this resistance to correction is not irrational since the experience trumps all possible testimony; we just don't know how weird these subjects' experiences are (see Reimer 2009).

But, in any case, these are all descriptive, B-type issues about what these patients are actually like. Furthermore, there is too much variety in the set of those we call "delusional" to make easily generalisable claims. As a result, we focus on more tractable A-type considerations, namely, on the constitutive norms of belief. We are concerned with this question: what would *any* case have to be like in order to qualify, or fail to qualify, as a case of belief?

And of course in order to answer this, we need to ask: "what is belief?"

9.4.2 What is belief?

What kinds of entities are believers? One might ask a related question: what kinds of entities *need* something like belief? Perhaps not every organism is a believer, but the claim that organisms believe and, say, rocks don't, is at least a start. Then we might want to narrow this down a bit further to organisms that move. Entities that have no means of operating in the world have no need for an informational view of it. Frank Ramsey (1929) put belief precisely in these terms. He called belief "a map . . . by which we steer" (p. 146). This idea of beliefs as being like maps that help us through the world in the service of our goals is extremely useful. In particular, it picks out an opposition between the phenomenon of belief (viz. the map) and the phenomenon of desire (viz. the goal). The former involves *taking* the world to be a certain way, whereas the latter involves *wanting* it to be a certain way. This picks out a fundamental distinction between information and motivation.

This opposition between belief (information) and desire (motivation) has sometimes been put in terms of "direction of fit." Elizabeth Anscombe's distinction between a shopping list and a detective's record illustrates this nicely:

> [i]f the [shopping] list and the things that the man actually buys do not agree, and if this and this alone constitutes a mistake, then the mistake in not in the list but this man's performance (if his wife were to say: "Look, it says butter and you have bought margarine", he would hardly reply: "What a mistake! We must put that right, and alter the word on the list to "margarine"): whereas if the detective's record and what the man actually buys do not agree then the mistake is in the record.
>
> (Anscombe 1957, p. 56)

So, the shopping list, which has a "world-to-word" direction of fit, is analogous to a desire, which has a "world-to-mind" direction of fit. The detective's inventory, which has a "word-to-world" direction of fit, is analogous to a belief, which has a "mind-to-world" direction of fit. It is in the nature of belief to fit with how things stand in the world. If there is a disparity, it is the "fault" of the belief, so to speak. Conversely, it is in the nature of desire to motivate one to behave in such a way that will make the world fit with it. If there is a disparity, it is the "fault" of the world, so to speak.

So, belief is the basic informational state, believing is the basic informational phenomenon and this involves taking the world to be a certain way. The question that interests us here is: what does one have to do (in the weakest sense of "do") in order to count as taking the world to be a certain way?

At this point, it is useful to draw attention to two things that we have already mentioned and to see the connections between the two. First, recall our distinction between constitutive and evaluative norms, namely, the distinction between norms that dictate doing ϕ badly or well (evaluative) versus doing ϕ and not doing ϕ at all (constitutive). Second, recall the "functional idiosyncrasies" that lead Currie and Co. (and Egan (2008)) to claim that delusions aren't beliefs.

Now, let us think of the functional idiosyncrasies as being instances of breaking norms, namely, norms dictated by how a good belief ought to (evaluative) or any belief must (constitutive) function. The norms that (i–iii) break, divide into "upstream" and "downstream" norms. That delusions are not supported by evidence in their initial formation, that is (i), and not responsive to review in the face of contrary evidence, that is (ii), both break upstream norms, namely, norms that are related to causal or evidential precursors of the belief.[16] That delusions don't guide action or subsequent reasoning, that is (iii), breaks a downstream norm (namely, norms that govern the consequences of the subject's belief). Note that any such belief-like state can be divided in this way. We can talk of how the subject came to be in such a state (upstream) and the use that the subject makes of the state (downstream).

Which of (i–iii) break constitutive norms, and which break evaluative norms? Elsewhere, I have suggested that there is a very simple pattern (Wilkinson 2013). Upstream norms are evaluative, whereas downstream norms are constitutive (I call this "the downstream-only view"). In other words, whereas upstream considerations are relevant for telling us whether the subject is believing well or badly, they are not relevant for telling us whether the subject is believing at all. In other words, to move from talk of norms to talk of functional role, these upstream issues about sensitivity to evidence describe the functional role of *good* belief, not belief *full stop*.

The relevant notion of belief for our purposes is concerned with whether an organism (in this case a delusional subject) really takes the world to be a certain way. It is not strictly relevant (although it is clearly an interesting question) how it is that the organism came to have this belief. Indeed, the idea of a mad scientist manipulating your brain so as to "implant" a certain belief is not (or at least should not be) ruled out by our philosophical concept of belief. This would be a prime case of irrational (or perhaps non-rational, depending on how you cash out epistemic irrationality) belief, but nonetheless it would count as belief, for example if it were acted upon in the right way. One might think that one doesn't need far-fetched thought experiments to illustrate this point: the indoctrination of extremist cults, for example, is action-driving in the worst ways and gives rise to strongly held beliefs, but the belief-formation process is far from rational.

Returning to the map metaphor, if we want to know whether someone is using a map-like drawing as a map, we need to look at her dispositions to act in accordance with the drawing. The presence of "disposition" is important, for there may be parts of the map, say distant parts, that are irrelevant to the person's goal-directed navigation, but she can still take them to be accurate, and this is shown by the fact that *if* she *were* to be in those distant parts, she *would* use the map to navigate (e.g. would expect a river here, a mountain there). If, for some reason, the map was drawn using a bad method (analogue: unjustified belief) or a good method but which unfortunately went awry and was consequently inaccurate (analogue: unluckily false belief), then the map is a *bad* map, but still used as a map if the person is inclined to navigate by it.

The consequence that the downstream-only view has for the question of whether delusions are beliefs is clear. If we want to claim that delusional patients don't really believe what they assert, it is *not* to resistance in the face of contradictory evidence, nor to lack of supporting evidence during formation, that we should turn. This tells us whether they are believing *badly* (and arguably presupposes that they are *believing*). Rather we need to show that the delusional patient is, all things considered, failing to act or reason in accordance with her professed beliefs. Only then should we say that she doesn't truly believe what she appears to.

9.4.3 So do delusional patients fail to be believers?

So, do delusional patients display the downstream functional idiosyncrasies that we have just mentioned? Do they, in a perplexingly direct manner, fail to act in accordance with their claims in such a way that they need to be re-interpreted? As mentioned, this is bound to depend on which delusion we look at. Let us look again at the philosopher's delusion of choice, and the one that this debate has tended to centre around, namely, Capgras. It is, after all, often claimed and widely accepted that Capgras patients often fail to show concern for the replaced loved one. Let's look at some examples.

Alexander, Stuss and Benson (1979) report a 44-year-old man, who misidentified his entire family. "At no time after the injury were suspiciousness, paranoia . . . noted and he never displayed agitation or anger to anyone, including his wife" (p. 334). The patient "described positive feeling toward 'both wives', showed no anger or distress about the first wife's desertion" (p. 335). Another example is patient DS (Hirstein and Ramachandran 1997) who exhibits a striking lack of concern for the welfare or whereabouts of his real father. Speaking of his father's impostor, he calmly says: "He's a nice guy, doctor, but he's not my father" (p. 438).

It is not always the case that *no* concern is shown for the replaced loved one. However, even in some cases where there is concern, there is still the problem of how fleeting this concern is. Lucchelli and Spinnler (2007) give us the case of patient "Fred" and his wife "Wilma" (both, of course, fictitious names):

> On another occasion, he urged her [his wife, Wilma] to go with him to report Wilma's disappearance. *Most of the times*, however, he was quite pleased to see her as the "double" Wilma and addressed her in a very gentle way. His wife described his manner as "courting as when we were dating."
>
> (p. 189, emphasis added)

So, although Fred occasionally acted as if his wife was missing, he was mostly perfectly happy with the double Wilma. But if the double Wilma isn't the "real" Wilma, then where is the real Wilma? Surely these are not the actions one expects from a man who asserts that his wife, whom he loved very much, is missing.

Stone and Young (1997) sum up what they can glean from these and similar case studies as follows:

> So, although in some cases of Capgras delusion patients act in ways that seem appropriate to their beliefs, in many other cases one finds a curious asynchrony between the firmly stated delusional belief and actions one might reasonably expect to have followed from it.
>
> (p. 334)

Let us note, in passing, that there is here mention of "the firmly stated delusional belief" as if the delusional claim, firmly stated, guarantees belief, and the task is then to explain why it fails to have the consequences that it has. But putting this to one side, the main point can be rephrased unproblematically as follows. Just as it is partly constitutive, for example, of your believing that something is deadly poisonous that you should refrain from eating it if you want to stay alive, so it seems to be partly constitutive of your believing that a loved one has been replaced by an impostor that you should be concerned for the welfare or whereabouts of the loved one in question.

From here there are two options for reinterpretation. First, one might say that if you are not concerned then it can't be a "loved one" that you believe is missing (and here there are two possibilities: either you don't really believe that it is *that person* that's missing, or you can't have *loved* them very much). Second, you can't really *believe* that they are missing (i.e. it is some other attitude that falls short of belief, like imagining). In other words, you can locate the deviance at the level either of content or of attitude. Currie does the latter and that is why he proposes that delusions are actually imaginings mistaken by the subject for beliefs (his "metacognitive view"). In my own work, I have done the former (Wilkinson 2013b).

In general, it seems to be assumed that if an utterance doesn't give voice to a belief, it must be giving voice to something else. But on a view where the relationship between an utterance (or belief attribution, including self-attribution) and what is believed is much less direct, there is less pressure to think in this way. Granted, insofar as it is action (and not mere behaviour, like a nervous vocal tick), it will be in principle explainable in terms of what the subject believes and desires, but neither of these need, by conceptual necessity, correspond in content to the utterance itself. People are motivated to make certain utterances, including assertions, by all sorts of things. Someone might say to me, "You're looking well" merely as a matter of politeness. Does this mean that, since she doesn't truly have a belief that I'm looking well, she has something else, like an imagining, a hypothesising, a considering, which has the content that I'm doing well? Clearly it is enough to say that she said it because the context motivated her in some way to say it. We generally try to do something (however small or vague) with our actions, and our verbal actions are no different. Of course, I am not saying that the assertions of Capgras patients are anything like this example of harmless insincerity. In fact, I think that they are genuine and accurate attempts to communicate how they take the world to be. I am just using this example to illustrate a different perspective on the general relationship between utterances and what is believed. The challenge, as I see it, is not to reflect on what kind of mental state the patient is in, with the asserted content, if not belief (imagining, bimagining, or whatever), but rather of seeing if we can get a handle on what it is that is believed, if not that which is asserted.

Summary

- The concept of delusion plays an important role, both as a diagnostic tool in psychiatry and as a part of philosophy of psychiatry.
- There have been different attempts to define delusion, but they all fail.
- I think that this is unsurprising once we see that delusion is a folk-evaluative term that expresses our folk-epistemic bafflement.
- As a result, we shouldn't expect our reaction of such bafflement to track something that can be neatly defined.
- We then examined examples of clinical delusions and focuses in particular on how varied they are in a number of respects.
- Some occur in the context of brain injury, whereas others in the absence of brain injury where there is instead a diagnosis of a mood disorder or psychotic disorder.
- A variety of different explanatory accounts were presented. They mainly focus on the relatively rare instances of delusional misidentification in the context of brain injury.
- Finally, we examined the debate concerning whether delusions are beliefs.

Questions

- What are your linguistic intuitions about the term "delusion"? When did you last use the word or encounter it?
- Can you imagine the wide variety of contexts in which someone might say something that strikes you as delusional? How might they behave on either side (i.e. before and after their claim)? Might they be calm and matter-of-fact, or agitated?
- Can you imagine what it might be like to see a loved one, but to experience them as unfamiliar, or as threatening, or as someone who doesn't actually have the shared history with you that they in fact have?
- What is belief? People can certainly say things that they don't believe, but can they *sincerely* say things that they don't really believe? If so, under what circumstances?

Notes

1. This is the definition in the glossary. There is a much longer description under "Schizophrenia spectrum and other Psychotic disorders" (APA 2013, p. 87).
2. I put "define" in scare quotes because, strictly speaking, you define *terms*, not phenomena themselves; you *characterise* phenomena. And furthermore, you don't define all terms. Some escape adequate definition, and, as we will see, delusion may be one of these.
3. To put it another way, being non-delusional is normal – at least with regard to being delusional or not; there are obviously plenty of other ways of being abnormal.
4. Indeed, in the narrow tradition of epistemology as theory of *knowledge*, the central question is typically: when is a belief *good enough to constitute knowledge*?
5. Consider (from Bayne and Fernandez 2008) the active young man, who is diagnosed with terminal cancer, but who is in denial, refusing treatment and planning adventurous trips that go well beyond his life expectancy. This is more of a borderline case. An informal opinion poll I conducted through social media suggested that significantly more than half (32 out of 50) of those I asked do not think of this as delusional. In contrast, no one thought of the mother as delusional.
6. This is perhaps a non-standard use of evidence in recent philosophy, which can construe evidence as being *de facto* truth or knowledge conducive. However, Maher's use of "evidence" is coherent, intuitive, and points to an important point. Namely, that when belief-formation "goes wrong" it can be down either to the process or to the input (or a combination of both).
7. Note that not mood-congruent does not mean mood incongruent.
8. The term "bizarre delusion", which has low inter-rater reliability (namely, different clinicians rate the same delusions differently on the scale of bizarreness), is usually taken to mean delusions that cannot possibly be true. So, delusional jealousy is not bizarre (it is "mundane", its content could be true), whereas delusions of control or thought insertion are typically rated as bizarre. I am using florid to mean delusions that not only could not be true, but that you don't know what it would be for them

to be true as well ("I am the left foot of God", being an example we've already encountered).

9. However, it must be noted that this is inconclusive evidence since the lack of SCR could be a consequence rather than a cause of the misidentification, that is it could be that there is a lack of affective response because you judge them not to be a loved one, rather than vice versa. A normal SCR, however, would have presented seemingly falsifying evidence.
10. Reimer (2009) calls the delusional hypothesis "the impostor hypothesis," but that is a less neutral term than "the doppelganger hypothesis" and implies a malevolence on the part of the double that is not always perceived by the subject (e.g. patient DS in Hirstein and Ramachandran (1997) who says: "He's a nice guy doctor. He's just not my father.").
11. Coltheart goes on to make the rather strong claim that "if we can find the answers to just [these] two questions [about formation and maintenance] in any case of delusional belief, we will have arrived at a possible explanation of that delusion." In fact, I take these two questions to potentially subdivide into two further questions, namely, ones asking for personal and subpersonal explanations. Both the formation and the maintenance can, at least in principle, be addressed in mechanistic terms and in terms that ask for grounds. The Ellis and Young model potentially addresses both for formation (viz. damage to affective processing and feelings of unfamiliarity, respectively) but, as it stands, is silent on the issue of maintenance, both in terms of the mechanisms that cause it, and in terms of the subjects grounds for holding the delusion for so long.
12. Wittgenstein describes hinge propositions as follows: "the *questions* that we raise and our *doubts* depend upon the fact that some propositions are exempt from doubt, are as it were like hinges on which those turn. That is to say, it belongs to the logic of our scientific investigations that certain things are *indeed* not doubted. But it isn't that the situation is like this: We just *can't* investigate everything, and for that reason we are forced to rest content with assumption. If I want the door to turn, the hinges must stay put" (Wittgenstein 1969, §§341–3).
13. This analogy with chess, however, has limitations. For example chess is a game, the playing of which is utterly deliberate, whereas belief is not. A move in chess is an *action*, rendered intelligible by beliefs and goals (by informational and motivational states). Given this fact about chess, you can deliberately play badly if your goal is to do so. As we will see, it is not obvious that this can be said of belief, at least not directly.
14. Recall that Maher hypothesizes that "delusional belief is not being held 'in the face of evidence strong enough to destroy it,' but is being held because evidence is strong enough to support it" (1974, p. 99).
15. De Pauw and Szulecka (1988) tell of a young man who decapitated his stepfather, taking him to be a robot, in order to look for batteries and microfilm inside his head. This looks like a misidentification that is suffused with paranoia.
16. (ii) may seem to be breaking a downstream norm, since it is something that is exhibited after the belief has been formed, but since in attempting to correct the delusion, you are providing *input* for new judgement, it is in fact upstream.

Further reading

Bayne, T. & Pacherie, E. 2005. In Defence of the Doxastic Conception of Delusions. *Mind & Language* 20:163–88.

Bortolotti, L. 2009. *Delusions and Other Irrational Beliefs*, Oxford: Oxford University Press.
Currie, G. & Jureidini, J. 2001. Delusion, Rationality, Empathy: Commentary on Davies et al. *Philosophy, Psychiatry & Psychology* 8:159–62.
Tumulty, M. 2012. Delusions and Not-Quite-Beliefs. *Neuroethics* 5 (1):29–37.
Wilkinson, S. 2013. Beyond Believing Badly. *Teorema* 32 (3):105–19.

10 Depression

Overview

Depression is not only one of the most common examples of mental illness; it also seems to be one of the easiest to empathise with (alongside, perhaps, anxiety disorders). Unlike, say, psychosis, many (perhaps all) of us have experienced low mood. To understand what it's like to be clinically depressed, one might think that all you need to do is to imagine an extreme and tenacious version of that low mood with which we are familiar. One key question is: is this an accurate construal of depression? If it is: How do we delineate depression? When does an individual tip from everyday low mood into diagnosable depression? And yet, if this is not an accurate construal, how *should* we understand depression instead? What is depression if it's not simply an extreme and tenacious low mood with which so many of us are familiar?

More generally, this requires us to try to make sense of depression. This can be interpreted in at least two ways. First, we might examine what causes and underpins depression. Second, depression gives rise to highly problematic behaviours; an inability to get on with and flourish in life, to uphold commitments to others and so on. How do we make sense of these behaviours, and what is it that needs to be put right? In other words, how are we to make sense of the relationship between depression and motivation? In this chapter, we answer these questions in turn.

DOI: 10.4324/9781315450124-12

10.1 Introducing depression

At first glance it can seem like depression, at least to some degree, is a rather common experience. Many of us may have said, at some point, things like "I'm feeling depressed today." This is intended as an expression of low mood or, perhaps, when used more specifically, an expression of a low mood that is hard to shake and/or is seemingly unrelated to a specific rational cause (the sort of "I just got out of the wrong side of bed" phenomenon).

One might wonder whether this use of the word "depressed" is hyperbolic or literal. It doesn't seem obvious (to me at least) that there is a clear answer either way, but one thing that *is* clear is that there are manifestations of something called depression that are of clinical severity and which, indeed, constitute arguably the most widespread form of mental illness and the second leading cause of disability years worldwide (Vos et al. 2017; Ferrari et al. 2013).

In this opening section, I introduce the clinical, psychiatric phenomenon that is often referred to in daily discourse as depression. We start by looking at the relevant official diagnostic categories in the DSM and then move on to the phenomenology of depression. Then, we examine the epidemiology of depression: who gets it and are there interesting patterns in the prevalence of who gets it?

10.1.1 Depression in the DSM

Within the DSM-5 (APA 2013), the presence of low mood of clinical severity is clustered in a chapter under a broad category of "Depressive Disorders." These include: "disruptive mood dysregulation disorder, major depressive disorder (including major depressive episode), persistent depressive disorder (dysthymia), premenstrual dysphoric disorder, substance/medication-induced depressive disorder, depressive disorder due to another medical condition, other specified depressive disorder, and unspecified depressive disorder" (APA 2013, p. 155).

The most well-known of these diagnostic categories is major depressive disorder (MDD). This is the category that best aligns with what people tend to have in mind when they refer to "depression" or "clinical depression." Unlike in DSM-IV, "Depressive Disorders" are kept separate from "Bipolar and Related Disorders" in the fifth edition. One reason for grouping these together is that some of the most extreme and damaging forms of low mood occur in the context of bipolar disorder. However, reasons for keeping them apart (which won out in DSM-5) are that the extreme upward mood swings associated with bipolar disorder, which are sometimes accompanied by psychosis and grandiosity and which are notably absent in depressive disorders, suggest that we are talking about a very different "underlying condition."

Either way, our focus is on MDD. In line with much of the literature, going forwards, by "depression" I will tend to mean MDD unless it is clear that I mean something else.

The DSM-5 criteria for MDD can be found in Box 10.1.

Box 10.1 Major Depressive Disorder

Diagnostic criteria

A. Five (or more) of the following symptoms have been present during the same 2-week period and represent a change from previous functioning; at least one of the symptoms is either (1) depressed mood or (2) loss of interest or pleasure.

Note: Do not include symptoms that are clearly attributable to another medical condition.

1. Depressed mood most of the day, nearly every day, as indicated by either subjective report (e.g. feels sad, empty, hopeless) or observation made by others (e.g. appears tearful). (**Note**: In children and adolescents, it can be irritable mood.)
2. Markedly diminished interest or pleasure in all, or almost all, activities most of the day, nearly every day (as indicated by either subjective account or observation).
3. Significant weight loss when not dieting or weight gain (e.g. a change of more than 5 per cent of body weight in a month) or decrease or increase in appetite nearly every day. (**Note**: In children, consider failure to make expected weight gain.)
4. Insomnia or hypersomnia nearly every day.
5. Psychomotor agitation or retardation nearly every day (observable by others, not merely subjective feelings of restlessness or being slowed down).
6. Fatigue or loss of energy nearly every day.
7. Feelings of worthlessness or excessive or inappropriate guilt (which may be delusional) nearly every day (not merely self-reproach or guilt about being sick).
8. Diminished ability to think or concentrate, or indecisiveness, nearly every day (either by subjective account or as observed by others).
9. Recurrent thoughts of death (not just fear of dying), recurrent suicidal ideation without a specific plan or a suicide attempt or a specific plan for committing suicide.

B. The symptoms cause clinically significant distress or impairment in social, occupational or other important areas of functioning.

C. The episode is not attributable to the physiological effects of a substance or to another medical condition.

There are a number of things worth mentioning with regard to this entry. The first is that five out of nine symptoms need to be present, leaving a significant amount of presentational variation. Granted, it is not any five of the nine. One of them has to be either depressed mood or lack of pleasure. This means that, in principle, you can get a diagnosis of MDD without having depressed mood but by exhibiting lack of pleasure and four of the other symptoms. A second thing to notice is that depressed mood can manifest itself as irritable mood in children and adolescents. Third, several of these symptoms are categorically different kinds of thing: some are feelings, others are behaviours, others are physiological consequences (e.g. weight loss), others are somatic symptoms (fatigue) and so on.

All three of these observations point to one very general overarching point which is that MDD, contrary to what one might think, is not presented in the DSM as a phenomenological phenomenon by definition. Depressed phenomenology (mood) is neither necessary nor sufficient. Why is that? There are two related reasons. First (as gestured towards in Symptom 1), we are purportedly on epistemically shaky ground if we rely entirely on subjectivity or subjective report. Second, MDD is a disorder not an experience. As such, it is (conceived of as) an underlying pathological condition that *manifests itself* phenomenologically in potentially many different ways.

What we see here, in effect, is a tension between the intuitive idea that depression is an experiential problem, *defined by* its phenomenology, *problematic because of* its phenomenology and the biomedical diagnosis-driven orthodoxy as enshrined in the DSM. The latter position stands and falls on our scientific understanding of the objective, underlying condition (brain disease?) MDD. We will examine progress on this front later. Nevertheless, however well we manage to understand, and pick out, such a condition, depression, some of which may or may not be "MDD," of the sort that leads people to seek psychiatric help, is ultimately held to a phenomenological not a neurological benchmark. In other words, if someone presents with depression, you cannot, as a physician, overrule their testimony and claim to have cured them of their depression based on a blood test or brain scan. None of this is to say that there isn't an underlying and clearly biomedically defined condition called "MDD" that is responsible for some, potentially most (or we could even stipulate *all*) of the things we call depression or clinical depression. Nevertheless, what bothers people about having depression is what it is like to have depression, from the first-person perspective. With that in mind, let's look at the phenomenology of depression.

10.1.2 The phenomenology of depression

Matthew Ratcliffe's *Experiences of Depression* (Ratcliffe 2014) is to my mind the most comprehensive and compelling text on the phenomenology of depression. And we need phenomenological investigation because there is a worrying circularity in explaining depression in terms of "depressed mood." What is depressed mood, other than the mood one is in when one is depressed? One might try to escape the circularity by talking about low mood, but what does *that* mean? "Lowness" of mood is clearly a metaphor, but what does it really capture? Metaphors are so often used in depression precisely because it is so hard to describe. The novelist William Styron (quoted in Ratcliffe 2014) writes that depression is like "being engulfed by a toxic and unnameable tide that obliterated any enjoyable response to the living world," something "close to, but indescribably different from, actual pain" (Styron 2001, p. 14).

This is where skilled phenomenological investigation comes into its own: it helps us to unpack these ways of speaking about hard-to-describe experiences and allows us to understand and even empathise with them. We will see more on this later, in criticism of more mainstream approaches, but one of Ratcliffe's central insights is that depression is (or at least often can be) a far more profound and alien experience than most people recognise. It is not simply a quantitatively stronger version of something we all know well: it is qualitatively different. In terms of his specific proposals, depression involves a change in pre-reflective ways of finding ourselves in a shared world, a disturbance to what Ratcliffe calls "existential feelings" (Ratcliffe 2008). These changes can result in a wide variety of other phenomenological shifts: changes to the sense of reality and belonging; changes to the sense of possibility and different kinds of possibility; changes to bodily feeling, feelings of hopelessness, guilt, changes to the perception of time and so on. In short, scratch beneath the surface, and depression is not merely (or not even) feeling very, very sad. We will return to Ratcliffe's phenomenology in Section 10.2.2.

Another important contribution to the phenomenology of depression is found in the work of Thomas Fuchs (2013). Drawing inspiration from less individualistic (i.e. "non-Western") cultures, he suggests that, instead of thinking of embodied and interpersonal features of depression as secondary to low mood, we should think of them as *primary*. Fuchs introduces the notion of "intercorporeality" and "interaffectivity" and presents his account of depression as a "detunement" of the lived body which interferes with our participation in "shared affective spaces," leading to profound detachment and isolation.

10.1.3 Who gets depressed?

The World Health Organization (WHO) estimates that in any given year, over 5 percent of the global population will experience depression. Fifteen percent of people will experience it at some point in their lifetime. The prevalence, however, is not uniformly distributed, and we can see clear patterns.

Geographically, for example, reported prevalence is particularly high in France and the United States and particularly low in China. It has often been commented that there is a rough correlation along lines of GDP per capita, with rates of depression being in general higher in developed, high-income countries. This in turn raises the question of what it is about richer countries that increases the reported rate. It could be a sampling bias with richer countries encouraging more awareness of mental health issues and/or having more extensive screening capacity (or, indeed, from a more cynical perspective, needlessly medicalising unhappiness). This might then loop back into the sampled reality itself. The more people become aware of the fact that they might be depressed (and perhaps have the luxury to reflect on it), the more likely they are to experience depression. In a similar vein, some cultures may stigmatise mental illness, leading to shame and self-stigma in a way that might reduce help seeking. Alternatively, another factor (potentially in addition) might be that rural societies are generally less stressed and more socially supportive than urban ones (we saw a similar factor regarding schizophrenia in Chapter 6).

There are other socio-economic variables, many of them distinctly unsurprising – for example rates of depression are three times higher among the unemployed. Unemployment not only is financially precarious but also fosters inactivity and lack of sense of purpose. In contrast, one can be unhappily overworked, but one is usually not depressed in such a circumstance. Another fairly robust finding in that rates of depression are about twice as high in women as in men. Perhaps this is because it is under-detected in men (because they don't talk about their feelings, and/or because women are more self-reflective). Alternatively, there may be other explanations that actually posit a higher prevalence in women (rather than a detection effect) that appeal, say, to the role of women in patriarchal societies. There are plenty of other epidemiological patterns, too. For example the loss of a life partner to either death or divorce increases rates of depression and does so more clearly in men than in women.

Although these patterns aren't of direct philosophical relevance, the ability of any account to accommodate and explain these patterns would certainly count in its favour.

10.2 Delineating depression

One area where philosophical attention has been directed is on the question of how we differentiate depression from other related phenomena. How do we delineate its boundaries? More specifically, we can divide this into four questions. These questions (especially the first two) are related, and answers to each may influence the others.

1. What is depression ontologically speaking? Namely, is depression simply an extreme version of something with which we are all familiar (e.g. low mood)?
2. How do we distinguish depression as a psychiatric disorder from non-pathological forms of low mood and sadness?
3. How do we distinguish depression from other pathological mood disorders (e.g. dysthymia)?
4. How does depression relate to historically related concepts? Does this suggest that it is at least partially constructed/influenced by the ways of thinking at the time?

It is very important to distinguish the first two questions, since they amount to two different interpretations of the question: how do we distinguish depression from more common and everyday forms of low mood? The first interpretation is a question about the nature of the phenomenon itself. Pick out a canonical instance of depression: what is it made of? Is it made of something familiar? The second interpretation (the aforementioned second question) is a question about how the picking out of something as clinical, pathological, depression is done in the first place, regardless of what it is constituted by.

The third and fourth questions are about the relationship between depression and related concepts like dysthymia and melancholia. Melancholia is a historic category, but there is an interesting question about whether it picks out the same thing as modern-day depression. Is depression simply the modern-day word for what was once called melancholia? "Depression" was certainly not in circulation as a psychiatric term, before the late nineteenth century (although Richard Baker's Chronicle, of 1665, refers to General Lambert, acting upon imminent capture "with great depression of the spirit"), and melancholia was used to pick out low mood since at least the time of Hippocrates. In spite of this, the word melancholia has been revived by some theorists, who have taken it to pick out either something different from (e.g. Radden 2003), or a subspecies of depression (see, e.g. the melancholic subtype of MDD in DSM-IV, and the "with melancholic features" specifier in DSM-5). This debate raises the interesting general issue of whether the phenomena that these categories pick out are objective and "stay still" as we describe and categorise them, or whether the phenomena themselves are shaped by our ways of talking and thinking about them (a phenomenon known as a *looping effects* (Hacking 1995)).

10.2.1 Is depression an extreme version of something familiar?

The central issue is expressed by Matthew Ratcliffe in the following passage.

> Although much has been written about the phenomenology of schizophrenia, there has been comparatively little phenomenological research on depression. Perhaps this neglect is due to *the assumption that depression involves intensification or proliferation of commonplace feelings, emotions and moods, such as sadness, hopelessness, and guilt*. So all one need do in order to appreciate the phenomenology of depression is imagine having unusually pronounced experiences of familiar kinds. Experiences of schizophrenia, by contrast, are utterly alien to most of us.
>
> (Ratcliffe 2014, p. 1)

This invites the question: is this assumption correct? As you can probably tell from the tone of the above passage, Ratcliffe's answer is: no. According to him, depression is far more extreme and exotic than people realise.

Let's examine this issue for ourselves. Is depression simply intensely low mood? In answering this question, we could start by clarifying what we mean by "low mood" and "mood" in general. As a first approximation, a mood is affective, but it is more vague and general than an emotion. Emotions are often directed at things in particular (fear of the spider), whereas moods are not. They are in the background, colouring our engagement with the world. This directedness of particular states is sometimes put in terms of "intentionality"

(Brentano 1874). Beliefs, desires and fears are about or directed at specific things. They have intentional contents or intentional objects. Thus, it is common to talk about moods as states that are not "intentional," not in the more common everyday sense of being deliberate or under agentive control, but in the sense of being directed at objects or states of affairs in particular. However, moods bear a relationship to intentional states in that they enable certain intentional states. A mood is a background state against which intentional states occur. Being in a generally cheerful mood, I'm *more* likely to laugh at your joke, *less* likely to get angry at you for not doing the dishes, *more* likely to get excited by the thought of an upcoming trip abroad. If I'm in a generally sad mood, although the mood itself is not directed at these things, it makes certain mental states and events that are directed at things, and with a certain valence, more or less likely.

This notion of mood is broad enough that it seems plausible that depression can be thought of in these terms. After all, depression (along with anxiety) is classified as a mood disorder, and that certainly doesn't feel like a misnomer. But even accepting that depression might be a mood disorder doesn't mean that it is continuous with more mundane experiences.

So, what do we mean by "*everyday* low mood"? Two features that may distinguish depression from everyday low mood not just quantitatively but *qualitatively* are: (i) phenomenology, namely, the subjective qualitative features of the *kind* of mood that we are talking about (and Ratcliffe argues for discontinuity here) and (ii) appropriateness.

Low mood of the sort that we are all familiar with tends to have identifiable causes. Such causes may not be identifiable by the subject, but an expectation that there is such a cause is reflected in our daily practices (often called "folk psychological practices"). We ask each other, and sometimes ourselves, *why* we are in a bad mood. Often, it seems, identifying why we are in a particular mood can help to dissipate that mood (anxiety in particular). Sometimes, however, there is no identifiable reason for a particular mood. In such a case, we might say that the mood is inappropriate.

Mood disorders (depressive and anxiety disorders) are characterised by inappropriateness. In certain contexts, certain levels of anxiety, or low mood, are appropriate. In mood disorders, the severity or even presence is deemed inappropriate. The criterion of appropriateness points to an interesting shift from DSM-IV to DSM-5. According to DSM-IV, anyone presenting with depressive symptoms needs to be screened for cataclysmic life events (responses to "significant loss"), such as bereavement, relationship or family breakdown, loss of livelihood, and are categorically excluded from the diagnosis of MDD. In contrast, in DSM-5, it is explicitly stated that:

> Although such symptoms [which are indistinguishable from those of depression] may be understandable or considered appropriate to the loss, the presence of a major depressive episode in addition to the normal response to a significant loss should also be carefully considered. This decision inevitably

> requires the exercise of clinical judgment based on the individual's history and the cultural norms for the expression of distress in the context of loss.
> (APA 2013, p. 161)[1]

So, you might think that clinical depression is not only phenomenologically discontinuous from normal sadness, but also contextually discontinuous. In other words, it seems that mood disorders involve inappropriate moods. MDD involves being depressed when there's nothing to be depressed about; an anxiety disorder involves being anxious when there is nothing to be anxious about. However, one thing that adds complexity to the clinical picture is that sometimes cataclysmic things happen to people, and they are in appropriately low mood. And yet, that low mood (though appropriate, given the circumstances) can disrupt day-to-day functioning to the extent that a clinician may feel the need to prescribe treatment (whether antidepressants or CBT). Picking up on themes from Chapter 3, this looks like it could be a case of prescription without diagnosis: of using treatment instrumentally without giving the diagnosis of MDD. This brings us on to the next topic.

10.2.2 How do we distinguish pathological depression from "healthy" sadness?

Whereas the first issue is about whether the phenomenon of depression is continuous with other more everyday experiences, the second is about when it should be rightly considered a pathology. The continuity claim is conducive to (but does not entail) a dimensional approach to depression as pathology. Conversely, the anti-continuity claim is conducive to (but does not entail) a clean divide between pathological depression and healthy sadness, since it wouldn't simply be a matter of degree but rather a qualitative difference.

An important book by Horwitz and Wakefield (2007) is dedicated to this question and, more specifically, to a critique of how depression is diagnosed in the DSM. In particular, they are highly critical of the symptom-based, decontextualised approach (which they call "descriptivist") adopted by DSMs from the 1980 third edition onwards. They claim that the symptoms as listed in the DSM can arise as the result of severely adverse life events. The reactions "even when quite intense due to the severity of the experience, are surely part of normal human nature" (Horwitz and Wakefield 2007, p. 9). To oversimplify a rich and complex book, Horwitz and Wakefield effectively want the diagnosing of depression to be holistic and contextual. Only by taking life events into account can we see whether what we are confronted with is pathological. This proposal certainly has historical antecedents. Indeed, there is a traditional distinction between "reactive" and "endogenous" depression (e.g. Watts 1956). Having said this, the claim that the reactive form is not an illness is a further claim, and the fleshing out with reference to Wakefield's own (1992) *harmful dysfunction* account of pathology is another innovation. Extremely negative psychological states can be very harmful and distressing, but, if they are normal (for

Wakefield, roughly, biologically adaptive examples of proper functioning), then they are not pathological, and hence the person shouldn't get a diagnosis.

It is worth noting in passing that a less objectively grounded account of pathology (see Chapter 2) focusing, for example, purely on harm, or distress, or "flourishing," might struggle to exempt these cases from classification as pathology. A rebuttal might be: it is not the psychological states themselves that are harming the individual, rather they are part of the individual, and it is the *negative events themselves* that are the source of harm. And yet, not all mental illness is endogenous. Traumatic events (arguably by definition) leave their mark and trigger canonical instances of psychiatric illness (voice hearing, PTSD). Who are we to say that these reactions are "dysfunctions"? Perhaps, to elaborate on the anti-objectivist rebuttal, when my own psychological states become the primary source of harm for me, rather than the events that cause them, then I am psychiatrically ill. Take, for example, PTSD. Being hypervigilant and easily startled in the comfort of your leafy suburb is highly disruptive. It is your psychological states now, and not your experiences in Vietnam, that are the causes of your distress. We can stipulate that these lingering responses are the result of biological proper function. Wakefield would then have to say that they are not true instances of illness. However, this seems implausible (to me at least). The extent to which they are disruptive is clear for all to see, but it is merely that this is a practical rather than a biological kind of dysfunction. It's about the practical context of getting on with daily life.

Enough of the interlude on Wakefield's notion of pathology: the core point, that the de-contextualised presentation of symptoms is insufficient to tell us whether something is a case pathology or not (regardless of the precise account we give of pathology), is clearly an important and plausible one. Sometimes, people end up in extremely distressing psychological states that may persist and interfere with daily life, because they have experienced something that just is very distressing.

10.2.3 *Depression and dysthymia*

How does depression differ from other related categories such as dysthymia? Dysthymia is typically distinguished from depression in virtue of its reduced severity. It is however, at least typically, more persistent. In being less severe, it is less clearly pathological and involves less of an interference with daily functioning. As a result, those who suffer from dysthymia can be mistaken for grumpy individuals. Nevertheless, the DSM does categorically recognise dysthymia as a disorder under the label persistent depressive disorder (PDD). Dysthymia raises interesting questions about the relationship between the severity and duration of psychiatric symptoms in general. It also raises interesting issues about the boundaries of personality and personhood (see also Chapter 4).

In an APA (American Philosophical Association) Newsletter, Jennifer Radden wrote a very short but important piece on "Classifying Dysthymia" (Radden 2009) in which she raises valuable concerns for the diagnostic category of dysthymia and draws general conclusions about the challenges facing diagnostic categorisation in general. As Radden points out:

> As unipolar disorders of mood, Dysthymia and Major Depressive Disorder share similar felt symptoms – what is described as "depressed mood." The diagnostic separation of these two conditions from one another must rest on differences in *onset, course, duration, persistence, and severity*.
>
> (Radden 2009, p. 9, emphasis added)

Roughly speaking, whereas MDD is *episodic* (although not exactly fleeting), dysthymia (a.k.a. PDD) is *chronic*. Also, as we have noted, whereas MDD is severe, PDD is relatively mild (although not unproblematic).

Now, it may seem as though severity and duration came apart entirely; you could in principle have something either severe or mild be either episodic or chronic. Indeed, these are *conceptual* possibilities. But what about the brute causal realities of human existence with these conditions over time? This imposes constraints. Severe depression is *unsustainable*. It tends to be ended by remission or death. Milder depression can be sustained for longer, and, if it is episodic, it is more likely to go unnoticed or at least un-reported. In light of this simple reflection, the patterns we see in chronicity and severity (and they are just patterns, not rules – there are exceptions) are unsurprising.

Having said this, chronicity is a bit more complicated. Yes, if a problematic state of mind lasts longer, it is more likely to be reported. And yet, if it lasts for long enough, it is more likely to become tolerated, to fade into the background, to no longer be seen as something to be cured and so on. This relates to a very general point about illness and its place in human life: it cannot be seen in isolation but has to be seen in context and over the dynamic temporal flow of someone's life. The very format of diagnostic manuals and the typically short-lived nature of the clinical encounter, both raise difficult challenges in and of themselves.

This relates, also, to some questions of categorisation. More specifically, in the DSM (going back at least to DSM III 1980), there are overarching categories that cluster disorders into very broad groups. Two of these groups are *Mood Disorders* and *Personality Disorders*. In the current DSM (fifth edition), and in earlier editions (III, where personality disorders were introduced, and IV), dysthymia (a.k.a. PDD) was classified alongside Major Depressive Disorder as a mood disorder. However, an interesting thing to reflect on (as Radden 2009 does) is whether dysthymia might be better construed as a personality disorder rather than a mood disorder.

Why might one think this? First, we need to ask: what is the difference between a mood and a personality? They obviously are very different, but getting clearer on their nature is informative. Intuitively, a mood changes: it is not part of who you are. In contrast, personality is something more fixed, conceptualised as part of who you are, at least to some extent. Having said that, personalities can change, but that is usually seen as the result of profound and transformative experience. We readily, on the other hand, talk of being in a better mood now than we were yesterday, or even an hour ago. Also, and in a related manner, moods and our narratives surrounding moods readily exculpate (see Chapter 4 for broader discussion of diminished moral responsibility). Yes, we may have certain obligations to control our moods, even to "snap out" of certain moods, but we nevertheless readily say things like: "I'm so sorry I shouted at you, I'm in an irritable mood." We far less readily offer the excuse: "I'm so sorry I got angry, it's just my personality." And this is no doubt for good reason: personalities are legitimate targets of moral disapproval in ways that moods are not, or at least are less so.

Now, in light of these comments about personality and moral evaluation, coupled with the observation that disorders exculpate, a "personality disorder" seems like a contradiction. The more something is attributable to your personality, the less it is attributable to a disorder that sits outside of and afflicts your personality. There is, in my view, some truth to this, but a better way of thinking about it is that it is *contradictory* but not a full-blown *contradiction*. In other words, the compounding of "personality" and "disorder" engages concepts, discourses, which pull in opposite directions. However, that is not to say that the compound is senseless. Indeed, "disorder" is appended to "personality" precisely as a reminder that, though we may be tempted to blame this individual, psychiatric wisdom tells us otherwise.

If we look in more detail at what makes something a personality disorder, for example according to DSM-5 (which doesn't differ significantly from earlier editions):

> A personality disorder is an enduring pattern of inner experience and behavior that deviates markedly from the expectations of the individual's culture, is pervasive and inflexible, has an onset in adolescence or early adulthood, is stable over time, and leads to distress or impairment.
>
> (APA 2013, p. 645)

It is rather striking that dysthymia seems to fit this description, apart from, perhaps, the somewhat surprising stipulation about age of onset (dysthymia can be early, namely, pre-21, or late, post-21, onset (APA 2013, p. 169)).

Another question is: why even think about dysthymia as a disorder at all? Why not simply think of people with dysthymia as diverse or different rather than disordered? One good reason for holding on to the "disorder" status is that people with dysthymia, i) don't want to be dysthymic and ii) feel that it is not their true, authentic selves. The counterargument is, of course, that i) also applies to undesirable personalities. People can certainly be aware of their character flaws and dislike them and want to be different. That doesn't make them disordered. These individuals may also make claims along the lines of ii). If we say that these claims aren't true and are just wishful thinking ("The real me is better than this!"), what makes them such? This opens tricky, but fascinating, territory about selves and authenticity (Tekin 2011).

Alternatively, you could provide an arch-objectivist solution to all of this and say: either dysthymia is the result of underlying biological pathology, in which case it is a disorder, or it is not, in which case it is not. There are two quite different concerns here. First, borderlines between "different but within normal range" and "abnormal" are precisely one of the major problems facing objectivism in general and notions of malfunction and pathology in particular. Second, even if these objective notions were viable, it is not clear why the discovery that dysthymia falls one way or the other should have the far-reaching consequences that its classification as disorder (or mere difference) would entail (see Chapter 2).

10.2.4 Depression, melancholia and cross-cultural variation

Another question concerning the delineation of depression is: what is the relationship between depression as diagnosed in contemporary Western psychiatry and similar conditions from the past, most notably melancholia? This may seem like a question for a historian, but there are a number of important philosophical issues that emerge from reflecting on this. I focus on the most prominent. This is a question of ontology and, more specifically, objectivity, as opposed to culture-dependence. To what extent is there an objective phenomenon in reality called "depression"? To paraphrase Radden (2003), is depression, in this sense, somewhat like gallstones? Or is depression as we know it a fragile and contingent feature of a particular epoch, culture or way of thinking? To address this, we can look at both the past and other contemporary cultures for illuminating comparison.

Let's begin with the past by reflecting on the sorts of things that the term "melancholia" was taken to pick out. Putting aside some quite significant fluctuation in what melancholia was taken to refer to (by different theorists and at different times), there are both notable similarities and differences between melancholia and the label of depression, or, MDD. Among the similarities we have the following. Both categories seem to involve:

1. A clustering of sadness and fear (in today's clinical terminology, depression and anxiety).
2. A degree of unreasonableness, objectlessness, lacking in cause or foundation.
3. A degree of self-centredness, over-reflectiveness and over-sensitivity.

In contrast to unipolar depression, melancholia was at times associated with periods of mania, characterised by high levels of energy and grandiosity. This fits more with bipolar disorder, which of course used to be commonly known as "manic depression." However, this points to another problem with a straightforward equating of the two categories: melancholia seems like a much broader, indeed almost a "catch-all," category. At times, melancholia also seems to pick out things that we might call psychosis, as well as obsessive compulsive disorder and anxiety disorders.

Other differences aren't just about how wide the net is cast, so to speak, but the nature of the net itself and, in particular, unavoidable inflections of current thinking at the time, the background ideas, the *zeitgeist*. For example it is almost too obvious to mention, but being melancholic (which, of course, etymologically comes from humoural psychology, in particular, a preponderance of the humour known as "black bile") is something that is seen as coming in degrees and constitutes an explanation not simply of an affliction (as in the case of melancholia, a disorder of black bile, which diverged from "melancholy" in the nineteenth century), but a stable personality type.

Although these days we don't think in terms of humours, except metaphorically, and we don't quite to the same extent think of depressive personality types (although we certainly accept that grumpy or sad people exist), there are analogues that arise from contingent institutional features surrounding the theory and treatment of depression in the twentieth century. In particular, there is the emergence of biological psychiatry where conditions aren't merely surface phenomena but underlying disease entities. As part of some of the pressures and mechanisms we examined in detail in Chapter 3, viewing psychiatry as a branch of medicine and the emergence of randomised control trials (RCTs) as the gold standard in medical research promoted psychiatry and psychiatric classification as something geared towards pharmacology. This has given rise to what Radden (2003) calls "drug cartography." In drug cartography, you have a (clearly flawed and circular) inference from the treatment to the presence of the disease. So, depression is not a particular cluster of symptoms, nor is it a clearly definable disease, with a well-understood pathogenic cause and course: it is simply "any condition that anti-depressants alleviate" (Radden 2003, p. 38). This has led to a mock-scientific folk discourse where depression (akin to the humoral discourse) is thought of as a "chemical imbalance." I don't mean to suggest that it isn't explained by something that might be described thus but rather wish to point out that the readiness with which the vague and poorly understood term "chemical imbalance" is thrown about in a way that is reminiscent of humoral imbalance.

Other differences concern socio-cultural associations. Melancholia not only was more prevalent in men, but also was so strongly associated with men that for some it would have been considered impossible for women to have it. Also, contrary to the stigma we see today with depression, melancholia actually had various positive connotations. For example it was often seen as an upper-class ailment and often associated with genius and creativity. This is something that is somewhat lacking in the supposedly medically purified notion of depression, which afflicts women just as much as men, and if anything would be considered more prevalent among the working classes (although this would be a causal rather than associative relationship).

There is more to say about the conceptual history of melancholy and related phenomena. For example in the classical period (by which I mean the time of the Ancient Greeks and Romans), thinkers like Hippocrates and (half a millennium later) Galen thought of melancholy, and moods in general, as physically caused (albeit by humours). In medieval times this became moralised and/or spun into religious narratives. Although there is a great deal more to be said, we should briefly move on to contemporary cross-cultural variations in depressive phenomena.

As Chentsova-Dutton, Ryder and Tsai (2014) point out, in a review of varieties of depression across cultures, different cultures draw boundaries between the normal and the deviant in different ways and also talk about their experiences and how to make sense of them in different ways. I will briefly present a few suggestive examples that illustrate differences in cultural norms. Eastern Europeans are more likely to view depressive symptoms as normal and are less inclined to avoid negative emotional states than European Americans (Jurcik et al. 2013). Another example (out of a great many) is that Chinese cultural contexts "discourage focus on internal emotional experiences and emphasise the monitoring and sharing of somatic symptoms and social references" (Chentsova-Dutton, Ryder and Tsai 2014, p. 343). As we saw in Chapter 5, we need to simultaneously distinguish "context of formation" (the sorts of phenomena that are made more prevalent in a culture) from "context of evaluation" (the way different phenomena are received by that culture), while acknowledging that these will loop into each other. The more prevalent something is, the more likely it is to be normalised in society while, conversely, societal approval can encourage the emergence of certain phenomena (and disapproval can stifle it).

10.3 Explaining depression

We have introduced the phenomenon of depression, how it gets diagnosed and who gets it. We have also looked at contrasting concepts and conceptions of related phenomena. How do we explain it, its emergence and its features?

As we've mentioned, depression is, first and foremost, an experience and arguably quite a heterogeneous one. What is that experience really like and does this give us important clues about what underpins it? In addition to this, however, depression is a mental phenomenon, and mental phenomena are importantly affected by (although not, perhaps reducible to) an organism's nervous system. As a result, we will examine phenomenological and neuroscientific approaches to understanding depression. It is worth noting at this point that these are by no means mutually exclusive.

10.3.1 Phenomenological explanations

As we've seen, Ratcliffe's *Experiences of Depression* is a comprehensive text on the phenomenology of depression. However, the aim of the book is not only descriptive (although we saw that it is useful in this regard, see Section 10.1.2): it is also *explanatory*.

A central claim of the book is that existing work on depression has made the mistake of thinking about depression in terms of ordinary intentional states, which are, roughly speaking, the states postulated by folk psychology, namely, beliefs, desires, emotions, etc. In particular, depression is not about experiencing intense negative emotion, nor is it about making certain types of judgements about the world, rather, it is about what Ratcliffe introduces in an earlier book (Ratcliffe 2008), namely, *existential feelings*.

Existential feelings are pre-intentional ways of finding ourselves in the world; they are not intentional states themselves but rather background conditions against which certain intentional states occur (as either possible, impossible, likely, unlikely, etc.). More specifically, Ratcliffe borrows Husserl's notion of "horizon," which is, roughly, about the sense of possibilities for both action and sensation. The analysis is too rich for me to do justice to here, but the basic idea is that our experience at any given time is shot through with possibilities. Subtle changes to these possibilities amount to changes in existential feelings, which Ratcliffe calls "existential change." Depression, according to Ratcliffe, is a particular kind of existential change, a shift or sometimes a loss of the sense of possibility. Most prominently, the experienced possibility of meaningful projects is reduced or lost entirely. There are other manifestations of this shift in the possibility space from changes to bodily experience, loss of hope, feelings of guilt, loss of agency, alterations to temporal experience (experience of the passing of time) and changes to interpersonal experience.

As we saw earlier, a related account is put forward by Thomas Fuchs (2013). He argues that, in depression, one's body becomes "corporealised." In other words, it becomes a burdensome object to be dragged around, rather than a lived body, something light and transparent that we experience life *through*. This interferes with our embodies intersubjectivity.

These phenomenological insights are valuable because they explain things that are phenomenologically complex and harder to understand in terms of more basic building blocks that are easier to understand, and easier to empathise with too. By reading the likes of Ratcliffe and Fuchs, if they are right in what they say, one better understands the experience of depression. One has the tools to imagine, to some extent, what it might be like in ways that are more subtle and accurate than simply being very, very sad. Furthermore, it can couple with the neuroscience of depression to give it the correct (or more useful) explanatory target. For example a neuroscientist looking to explain sadness is going to be doing something different to one who is looking to explain a basic sense of the loss of possibility (the latter, in particular, lends itself to cutting-edge predictive processing approaches, which we will look at shortly).

10.3.2 Neuroscientific and cognitive approaches

The phenomenology gives us a rich and nuanced picture of what it is like to be depressed and, to some extent, why one is depressed (namely, the lack of experiential possibilities in Ratcliffe's account functions as an explanatory primitive that accounts non-trivially for a wealth and variety of experiential symptomatology). But what is going on inside the brain, and, more importantly, what might be its explanatory significance?

There are several theories that can be thought of as falling broadly under the banner of neuroscience (and again, which are not necessarily incompatible, but rather operate at different levels). The first that we will discuss has become so well known among the general public that it has become, at least in simplified form, part of daily depression narratives. I am talking about "chemical imbalance" theories as they are publicly known (and have entered, as we saw, into public discourse). More scientifically, they are known as neurotransmitter theories and, more specifically, a subset of neurotransmitters, the monoamines (dopamine, noradrenaline and serotonin).

The monoamine hypothesis of depression goes back at least to the mid-1960s when it was discovered through various mechanisms that monoamine depletion can result in symptoms resembling depression and that increased synaptic concentrations seemed to alleviate them. The first version of the monoamine hypothesis, namely, that low levels of monoamines result in, or even constitute, depression, was widely seen to be inadequate. First of all, whereas antidepressants (e.g. selective serotonin reuptake inhibitors, SSRIs) operate in a matter of hours, successfully increasing serotonin levels (in the case of SSRIs by making sure it sticks around and isn't metabolised), it takes several weeks for the subjective antidepressant effects to be felt. This suggests that greater understanding of precisely how the brain reacts to increased levels of monoamines is needed. Having said this, there is no illuminating explanatory step that goes straight from, for example serotonin to improved mood, so even if the effects were immediate, we would need to understand how and why the effect takes place (see parallels in Chapter 6 with dopamine and psychosis). For this, we need to understand what certain neurotransmitters actually contribute to the mechanisms underlying cognition.

As a result, these neurobiological hypotheses need to be complemented by, rather than compete with, cognitive hypotheses. How might depression be underpinned by general domains of cognition, like perception, attention and memory? Roiser and Sahakian (2013) distinguish "hot" and "cold" cognitive impairments. This distinction between hot, affect-dependent processes, and cold, affect-independent processes goes back at least to Aristotle (although that terminology became especially prevalent in the motivated reasoning literature of the 1980s).

Among the "cold" cognitive impairments associated with depression are problems with attention and memory (Roiser and Sahakian 2013). Keller et al. (2019) recently hypothesised that abnormalities to attention could have all manner of knock-on effects and prove explanatorily enlightening for various symptoms of depression. Among the "hot" cognitive impairments associated with depression are negative, emotionally driven biases in the interpreting of ambiguous stimuli, especially social cues (Roiser, Elliott and Sahakian 2012; Roiser and Sahakian 2013). A glance might be interpreted as threatening or derisory, or a conversation might lead to the conviction that the interlocutor doesn't like, or has a low opinion of, the depressed individual. These negative biases are also, and importantly, self-directed: the depressed individual isn't only unhappy, but has a low opinion of themselves and a low sense of self-worth. Eshel and Roiser (2010) review research suggesting that another important hot abnormality involves the ability to use affective information to guide behaviour or, in other words, to respond normally to reward and punishment, the fundamental building blocks (at least within the framework of experimental psychology) of motivation. In the next section, we will view motivational issues from a more traditionally philosophical perspective.

Of course, a big question for any of these proposals is whether you have those impairments because you are depressed or depressed because you have those impairments.

10.3.3 Predictive processing accounts

Accounts that both capture the significance of neurobiological difference and tie this to phenomenology are predictive processing accounts (Barrett, Quigley and Hamilton 2016; Badcock et al. 2017; Fabry 2020; Kiverstein, Miller & Rietveld 2020).

We saw, in Chapters 7 and 8, how predictive processing (PP) can be used to help make sense of psychosis in general and voice hearing in particular. What is the scope for it to help us understand depression? First of all, it is worth remembering that PP is not simply about perception and exteroceptive processing (processing information from the outside world). It is also about proprioceptive (the position of your body, most notably limbs) and interoceptive processing (the state of your viscera, your insides). A central idea, then, is that, whereas your conscious perceptual experience (e.g. visual experience) is determined by the multi-level hypothesis (predictive model) that your brain has selected in order to best explain and predict exteroceptive information, so it is that your conscious emotional experience (e.g. a state of fear, or sadness) is determined by the multi-level hypothesis that your brain has selected in order to best explain and predict *interoceptive* information. If this is what emotionality fundamentally is, according to PP, then, to the extent that emotions and moods (different though they may be) are affective, emotion-like, it seems like a PP account of depression would do well to look at interoception. This is exactly what Barrett, Quigley and Hamilton (2016) do. They use predictive processing and, more specifically, active inference accounts of interoception, to propose that embodied cognition, and especially emotion, centrally involve pro-active metabolic regulation (i.e. allostasis) as well as an appreciation of the sensory consequences of that regulation (i.e. interoception). Within this framework, they then speculate that depression can be construed as a disorder of allostasis. In other words, the "myriad symptoms" of depression result from a "locked in" brain that is insensitive to its sensory context.

More recently, but in keeping with Barrett, Quigley and Hamilton (2016), accounts have added further complexity by incorporating second-order prediction error dynamics (Kiverstein, Miller and Rietveld 2020) in order to make sense of mood and affect. In other words, cognitive systems don't only strive to minimise prediction error in the here and now, but over the long-run, and this requires an estimation of how well you are likely to do in your prediction error minimisation. As they put it, moods "summarize how the organism is faring overall in its predictive engagements, tuning the organism's expectations about how it is likely to fare in the future" (p. 1). With regard to depression, they hypothesise that

> persistent low mood associated with major depression we propose to understand in terms of negative momentum in expected free-energy [viz. prediction error] minimization. We suggest low mood persists and rigidifies because the agent exhibits the opposite profile to healthy individuals – they exhibit a low learning rate for negative momentum in free-energy reduction.
> (p. 9)

Since positive momentum in prediction error minimization is, on their account, both the fundamental computational underpinning of positive affect and functions as a behavioural attractor, this has crucial implications for action and motivation:

> The person experiencing major depression expects that nothing good could happen. Affordances that would previously have been inviting to the agent (e.g. succeeding at work, eating a fine meal) begin to appear less inviting, and so cease to draw the organism into action in the same way.
> (p. 9)

You will undoubtedly have noticed the synergies between this approach and Ratcliffe's phenomenological account in terms of the loss of possibilities. How all of this impacts on motivation feeds into the next section.

10.3.4 Depression and motivation

There is a targeted explanatory puzzle that pertains to depression, especially with regard to traditional analytic philosophy of mind and action. People with depression don't simply feel different as a result of their condition; they also behave differently. In particular, depressed individuals find it difficult to engage in the behaviours that they need to, and often claim they *want* to. In extreme cases, this can be even the most mundane of daily tasks. How do we explain these levels of behavioural inertia? What does that tell us about both depression in particular and human motivation more generally?

This issue is perhaps most clearly presented in a paper by John Russell Roberts (2001), where he distinguishes "compulsives" and "depressives" (which is suboptimal labelling, but never mind). Let's examine the following passage.

> When people act out of compulsion, it seems that they act as the result of one of their desires. However, although the action is the result of one of their desires, they are alienated in some sense from the desire. They fail to identify with it, in the sense that if asked they would say that they would rather not act on the basis of that desire. This raises a problem: if agents act out of their own desire, how do we make sense of the claim that they are not acting as they "really" want?

> Depressives also find they are not acting as they want. For them, engaging in normal activities becomes extremely difficult. But depressives are no more happy with this situation than are compulsives. They too fail to identify with their deviant motivational states. Depressives' lack of action is commonly explained by saying that depression has the effect of removing or weakening one's desires. But if this is true, then how is it that depressives are failing to act as they "really" want?
>
> (Roberts 2001, p. 41)

To summarise, in Roberts's own words, "in the one case we seem to have a preponderance of desire and in the other a desire deficit" (ibid.). The problem is, why does this desire deficit then not simply mean that the person with depression doesn't want to do the thing in question? Why do they, and we, insist that they really do?

David Hume famously claimed that "reason is, and ought only to be the slave of the passions, and can never pretend to any other office than to serve and obey them" (Hume 1739, p. 416). This reflects what is known as a "Humean Account of Motivation." This account states that belief alone ("reason" in the quote given earlier) doesn't bring about action. You can believe anything, but until you have *desire* ("passion" in the quote given earlier) action will not happen. In other words, the Humean account can be interpreted as a claim about the fundamental building blocks of the human mind. You need more than simply belief. The relevant belief to appeal to, when you are explaining action, is, for example, that a particular course of action is good for you. However, the Humean will contend that no amount of seemingly action-mandating belief will *alone* bring about action. For example the belief that if you don't do something you will die has to be coupled, among other things, with the *desire* not to die if it is going to result in you doing that thing. Just because almost everyone has that as a background desire, it doesn't mean that it is not there.

In contrast, anti-Humeans think that you can explain motivation without appeal to desires, namely, just in terms of belief. According to the anti-Humean, sure, we sometimes act on the basis of desire, but we also sometimes do things that we have no desire to do. We simply believe, for example, that we ought to do them. (Then the Humean retorts that this requires desires too; the desire to fulfil your obligations.)

What has this Humean (desires are necessary) versus anti-Humean (desires are not necessary) debate about motivation got to do with depression? One of the most personally and socially damaging symptoms of depression is that depressed individuals fail to act in ways that they feel they ought to and, deep down, want to. One common manifestation of this is not turning up to things that they ought to – and claim that they want to – turn up to. The account we give of this will vary depending on whether we adopt a Humean or non-Humean view of motivation. But then, Roberts (2001) claims:

> We are left with a dilemma. We can follow the Humean and be in a position to explain why depressives are not motivated to act. But this comes at the

> cost of not being able to explain the continued presence of their values. On the other hand, we can follow the non-Humean and be able to explain the depressives' continued identification with their values, but only at the cost of not being able to explain their lack of action.
>
> (pp. 45–6)

Roberts' solution is to retain a Humean view but to deny that the lack of action in depression stems from a lack of desire. He starts his solution by reflecting on the phenomenon of desire satisfaction. What is it for a desire to be satisfied? Plausibly, two necessary conditions for a desire to be satisfied are (i) that the desired state of affairs should be brought into being and (ii) that you also believe that it has. (Arguably there is a thin and non-psychological sense of "satisfaction" where (i) is sufficient (and therefore (ii) is not necessary) and that, with the discovery expressed by (ii), it is not that your desire becomes satisfied, but that you realise that it already was.) Roberts then asks, (i) and (ii) are necessary, but are they sufficient? He thinks not, and he motivates this with the example of someone (Jill) who has OCD (obsessive compulsive disorder). Jill keeps checking that the door is locked, even though (i) the door is already locked, and (ii) she believes (indeed *knows*) that it is. What's going on here?

At this point, Roberts introduces an old notion found in Bertrand Russell's work, the notion of "quiescence."

> [T]o his credit, Russell puts the most obvious aspect of satisfaction at the centre of his theory: satisfaction's close connection with feeling. To the extent that a concept of satisfaction is at work in areas like logic, it tends to be understood as a purely semantic notion. But this use is a derivative use.
>
> (p. 48)

In short, in addition to the "semantic sense," where the state of affairs desired (the "content" of the desire) obtains, there is an additional condition, without which you don't get proper, psychologically robust desire *satisfaction*.

Now, usually, these two things, the semantic and psychological/phenomenological senses of satisfaction, pull in tandem, which is why when we talk about human motivation in general we just talk about the satisfaction of desire without drawing this distinction. However, in cases of OCD, like Jill, what we have is precisely the lack of the quiescence. Roberts suggests that this can be used to shed light on depression too.

How this "lack of quiescence" actually works is an interesting question. Roberts adopts a "feelings as information" view according to which feelings carry information about the world. In Jill's case, an overwhelming feeling of unease carries information about the locked door and "tells her" that it's unlocked. Only seeing that the door is locked eases this feeling. This easing is quiescence. As she goes away, because her affective system is misfiring, the feeling returns, even though, at a higher cognitive level, she knows that it's locked. In short, there is a conflict between high and low-level information. To make sense of depression, we need to move from emotional responses to one's immediate environment to what Roberts calls "inner environments." In other words, we don't only act, but we think about our actions, our plans for the day, etc.

Here is Roberts' "sketch" of depressive action. Depressed individuals want certain things (have certain desires), and they fully recognise that behaving in that way would be best. However, nothing that they do provides this *quiescence*, this feeling of desire fulfilment. As a result, they (or perhaps their nervous systems) learn that there is no point in doing anything. More precisely, in fact, when they contemplate a course of action (in their "inner environment"), it fails to entice, even though on a rational level they know that it would be best. So, they may want to go to a friend's birthday party, but when they think about it, simulate, to some extent, actually going, it seems unappealing. Whether this is accurate or not, a further step, rather obviously, is the explanation of what this "quiescence" is neuro-cognitively, and what it is about people with depression that means that they lack it. Nevertheless, it is a solution to the theoretical problems raised by the apparent incompatibilities between an independently attractive (and widely accepted) Humean account of motivation and what we see in depression.

In direct response to the dilemma that Roberts poses, Benedict Smith (2013) presents an account that, although it shares significant similarities with Roberts' account, questions some of the assumptions behind the dilemma. In particular, Smith finds that motivation is seen by mainstream philosophers of mind and action, and especially Humeans, in excessively psychological terms and opts instead for a far more embodied approach.

> In depression, the motivational pathology is not (or not just) "psychological"; it involves changes to the bodily articulation of what can or cannot be done, of what is taken or felt as possible or not possible. Changes to motivation can profoundly affect a person's sense of their bodily capacities and thus, I have suggested, their sense of being an agent in the world. Recognizing this role of the body can serve to correct rather than simply replace the mentalistic account of motivation. What it seeks to correct is the assumption that motivation, and thus its pathologies, must be understood in terms of mental states and their relations.
>
> (Smith 2013, p. 633)

To elaborate beyond what Smith himself says in the paper, one might think that post-Humean analytic (as opposed to, say, phenomenological) philosophy approaches motivation not as something primitive, but that emerges when you put beliefs and desires together. In contrast, Smith (alongside other philosophers like Elizabeth Anscombe (1957)) thinks of motivation, action, as something primitive and embodied. Human beings act, and we can read off their actions, in easy cases, what they believe or desire. However, sometimes, the world is just perceived as grey and devoid of possibility, and the body feels heavy or numb. Action does not ensue in line with what the agent acknowledges would be the right or desirable way to act. But to ask what is believed or desired by the subject, and to insist that there always is an answer, is to take this folk-psychological framework too seriously, and to misinterpret the nature of the situation.

Summary

- Many intuitively think that depression is simply an extreme version of something that we have all felt before, but we have good reason to think that it is (or at least can be) a rather more alien and qualitatively different phenomenon.
- A close and critical examination of the DSM diagnostic category of "Major Depressive Disorder" (MDD) is important for thinking about what we are referring to with the more everyday term "depression."
- If by "depression" we mean experiences of depression of clinical severity, then they don't only occur in MDD, but in other diagnostic categories as well, most notably, in Bipolar Disorder.
- An important question is: how do we delineate depression? This in turn subsumes four quite different questions: What is depression fundamentally? How do we differentiate it from "healthy" forms of sadness? How do we differentiate it from other mood disorders (like dysthymia)? How do concepts capturing a similar space of phenomena change over history and across cultures?
- How do we explain depression? Phenomenological accounts are not only descriptive, but also explanatory: they recast hard-to-understand phenomenological complexes in terms of easier-to-understand phenomenological primitives. Neuroscientific accounts try to explain depression in terms of neurotransmitter abnormalities and cognitive deficits. Another, very specific computational approach tries to explain depression in terms of predictive processing.
- Finally, there is a philosophical puzzle concerning motivation in depression, where depressed individuals seem to simultaneously want and not want to do what they miss out on doing. Roberts suggests that this is because when depressed individuals fulfil their desires and motives, they don't get the rewarding feeling of fulfilment ("quiescence").
- In contrast, Smith finds the entire way in which this puzzle is set up far too psychologistic, and we should think about depression in a far more embodied manner.

Questions

- How does depression differ from sadness?
- What is the difference between "being depressed" and "having Major Depressive Disorder"?
- Why might severe forms of depression be so hard to describe? Can you think of the sorts of metaphors that people use?
- Can you still be clinically depressed, if you are in that emotional state because something very, very bad has happened to you? At what point might you be justified in seeking psychiatric help (i.e. anti-depressant medication)?

- Is the "chemical imbalance" explanation of depression satisfying?
- Is it possible to simultaneously want to do something, and yet feel no motivation to do it?

Note

1. Then there is a fascinating footnote: "In distinguishing grief from a major depressive episode (MDE), it is useful to consider that in grief the predominant affect is feelings of emptiness and loss, while in MDD it is persistent depressed mood and the inability to anticipate happiness or pleasure."

Further reading

Fabry, R.E. 2020. Into the Dark Room: A Predictive Processing Account of Major Depressive Disorder. *Phenomenology and the Cognitive Sciences* 19 (4):685–704.

Radden, J. 2003. Is This Dame Melancholy? Equating Today's Depression and Past Melancholia. *Philosophy, Psychiatry, and Psychology* 10 (1):37–52.

Ratcliffe, M. 2014. *Experiences of Depression: A Study in Phenomenology*, Oxford: Oxford University Press.

Smith, B. 2013. Depression and Motivation. *Phenomenology and the Cognitive Sciences* 12 (4):615–35.

11 Addiction

Overview

Addicts can behave in ways that are at odds with our understanding of human rationality and motivation. They can do things that they fully recognise are extremely harmful to themselves and their loved ones, and that they often claim to not want to do. How do we make sense of a case of someone seemingly freely acting against their own will? Before being able to answer this, we need some idea of what addiction is. What is the concept of addiction? What sorts of phenomena in the world does this concept pick out? Some answers to the latter question are philosophical in nature, others more scientific. How the philosophy and science interact is not always clear, and we will examine different options in this regard. We end this chapter by examining some further issues raised by reflecting on addiction. In particular, we will look at how addiction interacts with philosophical accounts of rationality, moral responsibility and pathology.

11.1 Introducing addiction

With reference to its Latin etymology, George Graham states: "If I am addicted to something, then I am given over to it. I surrender to it" (2015, p. 45). Although commonly associated with the ingestion of substances, both illegal (e.g. heroin)

DOI: 10.4324/9781315450124-13

and legal (e.g. nicotine, alcohol), one can also be addicted to certain activities that don't involve putting substances into the body. Gambling is the most high profile of these, but there are other candidate activities, such as sex, exercise, computer games, Internet and social media use and so on. In the interest of consistency, it might be more accurate to say that we don't really get addicted to substances but to activities. Sometimes, these activities involve the taking of substances, sometimes they don't. The heroin addict is not addicted to heroin (except as a manner of speaking), but rather to the taking of heroin.

As we are about to see, coming up with a clear definition of addiction is remarkably difficult, and there are particularly problematic cases. Nevertheless, canonical cases – exemplars – which, if anything were to count as cases of addiction, these surely must, are easy enough to come by. We are all too familiar with those addicted to alcohol drinking themselves to an early death; those addicted to gambling getting deeper and deeper into debt; lifelong smokers stood smoking outside the cancer clinic and so on. The DSM-5 dedicates about 110 pages to a chapter entitled "Substance-related and Addictive Disorders" and clusters together issues associated with the substance or activity in question. Thus, addiction to alcohol is dealt with alongside medical issues that may arise from the excessive long-term consumption of alcohol. The same applies to opioids, hallucinogens, tobacco and so on. The purest treatment of addiction *per se* in the DSM is under the sub-category of "Gambling Disorder." The diagnostic criteria are as follows:

A. Persistent and recurrent problematic gambling behaviour leading to clinically significant impairment or distress, as indicated by the individual exhibiting four (or more) of the following in a 12-month period:

  1. Needs to gamble with increasing amounts of money in order to achieve the desired excitement.
  2. Is restless or irritable when attempting to cut down or stop gambling.
  3. Has made repeated unsuccessful efforts to control, cut back or stop gambling.
  4. Is often preoccupied with gambling (e.g. having persistent thoughts of reliving past gambling experiences, handicapping or planning the next venture, thinking of ways to get money with which to gamble).
  5. Often gambles when feeling distressed (e.g. helpless, guilty, anxious, depressed).
  6. After losing money gambling, often returns another day to get even ("chasing" one's losses).
  7. Lies to conceal the extent of involvement with gambling.
  8. Has jeopardised or lost a significant relationship, job or educational or career opportunity because of gambling.
  9. Relies on others to provide money to relieve desperate financial situations caused by gambling.

B. The gambling behaviour is not better explained by a manic episode.

(APA 2013, p. 585)

One of these criteria (i.e. 6, chasing losses) is specific to gambling, but the others are hallmarks of addiction in general (e.g. 1. Desensitisation/habituation/tolerance-building, 2. Withdrawal symptoms). Furthermore, whether we are talking about gambling, drug-taking or whatever, when dealing with something that has attracted the unfortunate counter-accolade of "addiction," we are usually dealing with "persistent and recurrent problematic . . . behavior leading to clinically significant impairment or distress" (APA 2013).

As is so often the case with the DSM, as well as with our folk understanding of addiction, what we have here is an adequate characterisation of an archetypal clinical phenomenon. What we do not have, however (and in the DSM's defence, it isn't purporting to offer this), is a good definition of the concept of addiction. It is not putting forward clear necessary and sufficient conditions that could be applied to any future, possible or imaginary case study we might encounter. For example it seems committed to the idea that a willing and high-functioning addict is not really an addict (this puzzle of the willing addict is a recurring example in the philosophical addiction literature).

However, the very fact that addiction features so prominently in the DSM is revealing. Addiction is a phenomenon that causes severe problems to people's lives, to the point where they end up seeking clinical, indeed psychiatric, help. Its prominence in the DSM may also suggest that addiction is thought of as a disease or illness. In any case, with this in mind, let's ask ourselves: What is addiction?

11.2 What is addiction?

As we have seen at numerous junctures in this book, questions of the form "What is x?" are ambiguous between asking about an abstract concept and asking about a concrete phenomenon. Another way of putting is as follows. We might ask: "what gives something the status of x (in this case, addiction)?" Or we might ask: "what are those things, which we know we already pick out as x (in this case, addiction), constituted by?"

Depending on the value of x, it may be entirely inappropriate to ask one question or the other. "What is water?" should not be asking about an abstract concept, but about a natural kind, a substance that occurs in the world. You don't define water: you point to it. "What is justice?," on the other hand, is asking about something abstract or ideal, not about stuff in the world. Unfortunately, addiction is sometimes treated as a concrete phenomenon that just happens to occur in human beings in this particular world; at other times it is treated as an abstract concept, something that any rational agent, of any kind, in any possible world, might exhibit. I don't want to be prescriptive (at least not in this instance) about which approach is better. However, it is important to be clear about which of these one is doing, since they are such radically different enterprises.

Here, I present a number of attempts at answering the question "What is addiction?" I start by presenting what we might think of as a "lay" (i.e. a-theoretical) view of addiction. It is a view that is rarely if ever defended in contemporary theoretical literature, but it acts as a useful piece of context, since it is against this view that more theoretically informed views (explicitly or implicitly) contrast themselves or even act as a corrective (to a greater or lesser extent). These theoretical views include, viewing addiction (i) as a particular kind of brain disease, (ii) as a form of mental illness, (iii) in terms of compulsive behaviour or lack of control and (iv) through a socio-pragmatic lens, as flagging certain individuals for the purposes of certain societal responses, treatments. But let's first present the lay view, since it is a useful dialectical starting point.

11.2.1 The Lay View of addiction

The term "Lay View" gets explicitly used by Foddy and Savulescu (2010) and is explained as follows.

> This view states that people use drugs because they are morally corrupt hedonists who value immediate pleasure above all else and who rely on others to handle their ensuing health and survival difficulties.
>
> (Foddy and Savulescu 2010, p. 2)

This reference to morality is central, and, indeed, Levy (2013) refers to it as "the moral model." This relates to a theme that we have encountered at numerous junctures, namely, that as a society we negotiate our way through different domains of discourse, domains that do different things and have different consequences. Two important such domains are "the moralising" and "the medicalising," and they function to control and regulate society, and behaviour within it, in importantly different ways. When people behave in ways that we don't like (for whatever reason, it may harm them, others, society, or simply fly in the face of what we are used to), we can evaluate them morally and enjoin society to punish or rehabilitate. Another more theoretically informed approach is to say that the individual is ill in some way, that something is blamelessly wrong with them. Here we enjoin society to help, pardon and cure.

A question that presents itself here is: why might we get a lay, moralising view for *addiction*, more than we might for, say, depression or psychosis? What is it about addiction that makes the lay view relatively appealing, at least at first glance? I think there is a cluster of features that encourage moralising thinking surrounding addiction.

1. First, and most obviously, addiction is (almost always) perceived as self-inflicted. Unlike quintessential mental illnesses, like schizophrenia or depression, where, whether it's a genetic predisposition, or triggered by bad things in your life, you are seen as a victim of the condition, addiction in contrast is seen as the result of poor or unwise decisions or, in the very best circumstances, the result of difficult life circumstances that have pushed you towards engaging in addictive behaviours. Ultimately, however, and unlike

other mental illnesses, unless you freely engage in addictive activities, you won't get addicted. The lay narrative involves the fact that you can look at people with depression or schizophrenia and say "There but for the grace of God go I," whereas the same cannot (as easily) be said of addiction.[1]

2. Quite aside from how addiction happens, once it has taken hold, it involves damaging actions that appear entirely un-coerced. Yes, Obsessive Compulsive Disorder involves actions that may be damaging (obsessive handwashing leads to dermatitis, or you might miss an important job interview, etc.), but these are more often clearly against the overarching will of the individual, and they don't seem deliberate and planned. In the most striking cases, addicts will, for example, engage in multi-step goal-directed actions, like stealing a TV, in order to sell it, so that they can buy their drugs.
3. Addiction is less obviously conceptualised as a binary disease or diagnosis. We may feel like we have certain dependencies that aren't yet problematic but could become so. In any case, facets of addiction like dependency and self-control come in degrees, and the threshold at which someone needs help (and to accept that they need help) is pragmatic and vague. Also, if we think of disease as somehow tied to malfunction, it seems like the human nervous system is supposed to learn about reward, and isn't just addiction a case of the brain doing exactly what it is supposed to do? On such a view (more on which shortly), the problem with addiction is that the environmental stimuli that it is learning about are embedded in human life in such a way as to be damaging and unhelpful impediments to human happiness. But the addict isn't *ill* (or so this narrative goes). Another way of putting it is that addiction is built out of the stuff of daily life: we all have strong appetites. (If you think that depression is relatable, addiction is even more so!)
4. Addiction allows itself very easily to be made sense of using folk notions of virtue and vice (self-control, wantonness, etc.). These may well be inaccurate and unhelpful narratives, but they fit the data for addiction rather better than they do for schizophrenia and depression (although even these can be given moral or quasi-moral interpretations, such as opening oneself through vice and sin to demonic possession or divine punishment).

In spite of these motives, the Lay View is deeply problematic for a number of reasons, which we might divide into theoretical problems and ethical problems, in other words, whether it is defensible as an *accurate* portrayal of addiction, and whether it promotes a *humane and ethical* approach to addiction.[2]

First of all, the Lay View is often criticised for being too narrow in scope and based on confusion. These two features are causally related. In particular, the Lay View, in its focus on careless hedonism, tends to focus on *recreational* activities, namely, activities that are fun, thrilling or pleasurable. For example it focuses on gambling or heroin and cocaine use. This is, certainly, a significant path to addiction, but it is not central to it. Although relatively rare, you can get addicted to substances and activities that, though perhaps pleasurable, aren't recreational. For example you might become addicted to painkillers after shattering your pelvis in a fall. You don't take the painkillers recreationally. You take them to cope with pain. Even more rarely, you might be entirely passive in your path to addiction. Some babies are born with substance dependencies, or (to take an imagined example) you might be the subject of a coercive experiment at the hands of an evil scientist. It doesn't matter how rare these non-hedonistic paths to addiction are. What matters is that it shows that the hedonistic path into addiction isn't central to addiction itself.

And this relates to the confusion that seems as the heart of the Lay View. The view conflates the path into addiction, with how we are to deal with, and reason about, the individual *once they are addicted.* Ultimately, whatever vices may or may not explain a particular individual's path into addiction, we need to reflect on what it is about their *current* state that makes them less in control of their actions, to the point, indeed, where they might need help. The Lay View, ultimately, does not think of addiction as a psychiatric, or more generally medical, issue. It therefore encourages the punishment of addiction. And, as Foddy and Savulescu put it:

> This is the view that is at the root of public outrage over harm-minimization strategies, such as safe injecting rooms, needle exchanges, and methadone clinics.
>
> (Foddy and Savulescu 2010, p. 2)

At the exact opposite end of the "moralising vs. medicalising" spectrum is a view that sees addiction as a medical issue.

Before examining different medical perspectives, it is worth clarifying the extent to which the medicalising of addiction pulls it out of a moral domain and opens up the way to more humane treatments and policies surrounding addiction. Is the disease or illness hypothesis necessary or sufficient for a humane approach towards addicts and addiction?

We might well argue that it is sufficient for a humane approach, namely, that it mandates one. Having said that, in 1996, US Congress voted to terminate access to certain forms of disability benefits if the disability in question is the result of drug or alcohol addiction, which demonstrates either that elements of the Lay View remain in political circles or that scientific research lacks the power to influence the minds of lawmakers.

Now looking at necessity, as Nick Heather (1992) rightly points out in his paper *Why alcoholism is not a disease*, the disease concept (and I would add the illness concept) is not necessary either to justify medical intervention or a caring approach to alcoholism in particular (and hence addiction in general). Here is the crux: as a society, we take caring and compassionate approaches to all sorts of problems that aren't diseases or illnesses. Being a victim of poverty, destitution, domestic violence, an unhappy marriage, geographical displacement from war or famine, all are reasonably approached with compassion, but they are not medical issues. Furthermore, we often intervene medically to alleviate problems that fall well short of being diseases or illness, either because they are short lived, or are part of normal life, or regular and common (e.g. paracetamol for a headaches or period pain, beta blockers for stage fright, sleeping tablets for insomnia, and so on). So, just because we may use methadone to ease someone off heroin, or disulfiram to encourage abstinence from alcohol, this is not to imply that these addictions, still less addiction in general, is a disease or illness.

Nevertheless, even if viewing addiction as a disease or illness doesn't strictly entail a compassionate de-moralising approach, and isn't the only way of securing one, it certainly *points us in that general direction*. There are different versions of this "up-medicalising, down-moralising" approach to addiction. Two such versions involve viewing addiction as a *brain disease* and, second, as *mental illness or disorder*. Often, it must be said, the "brain disease" view and the "mental illness" view are not distinguished. (It would take us too far afield to dissect exactly why that is so, but the relationship between mental illness and brain disease is dealt with at length in Chapter 2.) We examine these in turn.

11.2.2 Addiction as a brain disease

One might think that the "brain disease hypothesis" of addiction would be more historically recent than the "mental disorder hypothesis." After all, mental disorders are diagnosed on the basis of observable things, such as behaviours (broadly construed to include, e.g. weight loss) and first-person reports of experiences. Brain diseases, on the other hand, will require neuroimaging techniques only made possible by relatively recent technological advancements. Indeed, safe neuroimaging techniques like CT and MRI only emerged in the 1970s. (There were much earlier techniques, such as pneumoencephalography, but this involved draining all cerebrospinal fluid and replacing it with air, so as to enable the brain to show up better on an x-ray. Aside from being extremely painful, it also posed huge risks to the patient.)

To some extent, this chronology is accurate. While addiction was added to the DSM in 1968, the idea that addiction is a *brain disease* became fashionable in the wake of Leshner's seminal 1997 paper, which appeared in *Science*, entitled: "Addiction is a brain disease, and it matters." Although this paper certainly brought the brain disease hypothesis into the mainstream (where it remains to this day), this chronology overlooks the fact that, as far back as the late nineteenth century, the brain disease hypothesis was being put forward by the *Society for the Study and Cure of Inebriety*. In his inaugural presidential address, Dr Norman Kerr described addiction as "a diseased state of the brain and nerve centres characterised by an irresistible impulse to indulge in intoxicating liquors or other narcotics, for the relief these afford, at any peril" (Kerr 1884, p. 2). The evidential grounds on which this claim was made are unclear, but having added this slight modification to the accepted chronology, let's examine the brain disease hypothesis as it is presented in Leshner (1997) and beyond.

Leshner's paper certainly goes beyond what was known in the late 19th Century, but the argument is very simple. Neuroimaging (e.g. PET in Volkow et al. 1990) shows that there are robust differences in the brains of addicts when compared to non-addicts. Therefore, addiction is a brain disease. Leshner makes it clearer what he means by brain disease by presenting other so-called brain diseases, including schizophrenia, stroke and Alzheimer's disease. As we saw in Chapter 2, I like to distinguish illness from disease in general, and mental illness from brain disease in particular. Regardless of the fine-grained theoretical details, only the latter two are widely and naturally spoken of as brain diseases. As a result, these provide a very useful comparison class. To say that addiction is a brain disease is to say that it is, in certain relevant ways, like a stroke or Alzheimer's.

Suppose we wanted to criticise this, what could we do, logically speaking? There are two clear options. First, you can criticise the data itself: there aren't robust differences between the brains of addicts and non-addicts (or perhaps there are differences in *some* addicts, but not *all* addicts). Second, you can agree with (or simply bracket the validity of) the data on brain differences, but put pressure on the idea that brain differences *mean disease*. Since the former is simply contradicting Leshner's starting point, and the latter is more philosophically interesting anyway, we'll focus on the latter. So, we are granting that there are brain differences but dissecting how far this gets us argumentatively.

Let's cast our minds back to Chapter 2, where we looked at different accounts of illness and disease. At each extreme, you have the values-based theorists on the one hand, and the objectivists on the other, exemplified, most prominently, by Chris Boorse. You might think that the values-based theorist will be most strongly opposed to the brain disease hypothesis, but it is worth noting that not even Boorse's view justifies the inference from brain difference to brain disease. Far more needs to be said, especially when we reflect on the fact that the brain exhibits an impressive degree of plasticity. Life itself causes changes to the brain, and these changes can't all be diseases.

This relates to the criticism made by Levy (2013) in a paper entitled "Addiction is not a brain disease (and it matters)." In effect, Levy's criticism is what we might call a step-by-step "concede and dismantle" approach. Little by little, more and more is conceded to the brain disease hypothesis, and it *still* isn't deemed to be plausible.

What is needed to make the step from brain difference to brain disease? As we saw in Chapter 2, the objectivist needs at least some kind of biological malfunction. The notion of biological function is a thorny debate in philosophy of biology (played out mainly between selectionist (Millikan 1989) and systemic (Cummins 1975) accounts). Nevertheless, whatever our view of malfunction, it involves a biological mechanism or organ not doing what it is supposed to, what it has been designed to do by evolution and what it was selected for. The concern here is that there are quite plausible accounts of addiction according to which it involves the nervous system behaving in accordance with biological design or function. For example there are mismatch accounts (Durrant et al. 2009) which take addiction to involve well-functioning, but evolutionarily ancient, systems that are no longer suited to the contemporary ecology (most relevantly the ready availability of extremely rewarding substances and activities).

However, suppose mismatch accounts are wrong. Let's suppose that some other accounts (that we will examine in more detail in Section 11.3) are the correct ones. For example some reward prediction accounts, or incentive salience accounts, take there to be some kind of malfunction with the dopaminergic system. Even conceding this, it still doesn't mean that addiction is a brain disease, since no one thinks that malfunction is sufficient for disease. Many add that there needs to be a negative impact on the person or organism. But even if we grant that addiction can be harmful in the requisite way, it is only harmful in very specific contexts. Consider the comparison with the aforementioned "brain diseases," such as Alzheimer's or a stroke. These typically cause impairment in a great many life contexts. However, the urges that lead to suffering and impairment in addiction are profoundly cue-sensitive and hence only come into play in certain contexts. Certainly, some addicts have, through their addictions, impaired themselves more globally, but we need to keep those conceptually separate from the core phenomenon of addiction. To take an example, we must separate the chronic memory impairment of Korsakoff syndrome, caused by alcohol poisoning, from the addiction to alcohol that gave rise to it. The point is that people can be highly addicted to something and yet be entirely unimpaired in the vast majority of contexts in daily life.

As Levy himself notes, his criticism amounts to a two-stage approach to disease, more or less in line with Wakefield's harmful dysfunction account (Wakefield 1992, see Chapter 2 for more details). Furthermore, also like Wakefield, Levy takes disease and illness to be synonymous. But what if, along the lines that I and others (e.g. Boorse 1975; Cassell 1976) have suggested, we adopted a distinction between disease and illness? This might lead us, like Levy, to claim that addiction is not a brain *disease*. But this would still allow it to count as a mental *illness*.

We will examine this hypothesis in the next section. For now, however, it is worth noting that this move points to another problem for the brain disease view. We can grant that the brains of addicts exhibit robust differences, and we can even grant that these differences constitute not only malfunction, but also disease. But even if we do concede all this, it still doesn't follow that this captures the essence of addiction. For suppose we were presented with someone who bore all of the behavioural hallmarks of addiction, but who didn't have these anatomical differences, would we say that the person was not really addicted? I suggest that we would not. The person is still an addict, because an addict is someone whose life is impeded (willingly or reluctantly) in certain ways, regardless of what we discover about their brain. Again, we see the importance of the status/constitution distinction (Roberts, Krueger and Glackin 2019). It may just so happen that the only people who exhibit the behaviours that make them have the status of addicts are those with a particular brain disease, but even that highly speculative scenario doesn't make the status of addiction be about that disease. In a nearby possible world, there could be addicts who didn't have that disease.

From a different perspective, we might say the following. To say "Addiction = Brain disease" is just as much a category mistake as saying "Person = Brain." Ultimately, "addiction" is part of the conceptual scheme (the language game, if you will) of daily interactions, even forensic interactions, with entire human beings embedded in multifaceted lives with goals, plans, aspirations, value-systems, etc. "Brain disease" is emphatically not. It is part of a scientific and subpersonal language game and a potentially dubious one at that. With this in mind, let's now examine the idea that addiction might be a mental illness, rather than a brain disease.

11.2.3 Addiction as a mental illness

While it might seem progressive (and exculpatory, for reasons we have touched upon already and will unpack further) to think of addiction as a mental illness, it was added to the second edition of the DSM in 1968 (APA 1968), so the idea has a significant historical pedigree.

As mentioned, many don't distinguish disease from illness and won't distinguish the brain disease view from the mental illness view. This is because both involve seeing addiction as a pathology, and this is seen, at a relevant coarseness of grain, as where the argumentative action is. However, at a finer-grained level, there are nuanced positions that are strongly opposed to the brain disease view but are still open to a mental illness view. What does such a view look like? First of all, it is worth noticing that a unified account of pathology like Wakefield's, that doesn't distinguish illness from disease, will group brain disease and mental illness together. The question is simply: is there harmful dysfunction?

Boorse, on the other hand, distinguishes disease and illness, where the former is a scientific notion, and the latter is a practical one. Although they will downplay the importance or viability of the scientific notion of disease, values-based theorists will focus on what Boorse puts to one side as merely practical. On such a view, what could make addiction a mental illness (but not a brain disease, although some instances may involve brain disease) would be its negative impact upon human flourishing (see Chapter 2 for more details). Viewing addiction through this lens raises a very different question that is not (at least not directly) about brain differences but about whether the person's problems are best categorised and conceptualised as psychiatric, as mental illness or as a social problem, a problem with living. We saw in Chapter 2 how the big challenge for values-based theories is distinguishing psychiatric undesirability from other forms (e.g. poverty, being in an unhappy marriage, etc.). One way of doing this (and interestingly this fits with Boorse's insight) is practically, namely, by asking, what are the practical implications of a condition? What would a successful or at least beneficial intervention look like?

Certainly, addiction can be the result of societal issues, but, viewed in isolation, it is not the same kind of problem as those that cause it. In particular, if someone is living in contexts of poverty, or under threat of gang violence, there is nothing you can do *to them psychologically* that will rectify the situation. However, with addiction, a removal of the addictive urges, by whatever means, would be an alteration of the person that would in principle solve the issue (viewed in isolation). None of this is to deny the cue-driven and contextually embedded nature of addiction. The problem as it is found in real life is a problem with the mind and its environment, and although you could intervene on either to improve things, in principle, an intervention on the mind alone would improve matters (even though in practice, changes of context and nature are crucial). Other problems of living contrast with this. A pill won't fix your living conditions, but it could in principle make your craving for a drug go away.

In sum, there is a clear values-based route to thinking of addiction as a mental illness, which would eschew the need or inclination to talk about brain changes, let alone brain disease, but which would still mandate a compassionate, and treatment-based, response on the part of society.

11.2.4 Addiction and self-control

Having said all of this, one might worry that addiction doesn't seem like it is context-general enough to count as either a brain disease or a mental illness. Furthermore, we all have appetites for things, and some are stronger than others. I like coffee, and I drink it every day. I might find having to forego coffee unpleasant. Am I an addict? It seems like the sorts of behaviours that might tempt one to attribute addiction are potentially gradual rather than categorical. In light of this, we might think that looking for the nature of addiction among binary notions of illness or disease is to miss the point.

The strength of our appetites will come in degrees, and sometimes it will be strong enough that we will call these addictions. Some, but not all, of these addictions will ruin lives. What gives strong appetites the ability to cause harm and ruin lives? One answer to this is that they prevent the individual from exhibiting the degree of *self-control* that is often considered necessary for a happy, flourishing life.

What do we mean by self-control? Let's start with a straightforward answer that we will subsequently refine. Someone exhibits self-control when they behave in ways that are in accordance with motivational states that are somehow constitutive of their "self." A lack of self-control involves behaving in a way that is at odds with how you really, deep-down want to behave.

However, this admits of an ambiguity (that has often been overlooked). As Edmund Henden (2017) points out, we can distinguish between two kinds of motivational states that constitute how we want to behave "deep down." One of these is our all-things-considered better *judgment*. The other is what we truly *value*. Based on this distinction, Henden distinguishes judgment-based self-control and values-based self-control. Since this is a useful and important distinction, let's unpack it further.

Judgment-based self-control is what most theorists have traditionally meant by self-control. Like Mele (2002, 2010), it is natural to think of self-control as the opposite of the Aristotelian notion of *akrasia*, or weakness of will, "a trait of character exhibited in uncompelled, intentional behavior that goes against the agent's best or better judgment" (Mele 2002, p. 531, quoted in Henden 2017). An agent's "better judgment" is the outcome of an event of conscious deliberation according to which they judge that a given course of action is the best course of action (out of relevant alternatives). This is perhaps the most natural reading of a view we saw back in Chapter 4, the hierarchical view, according to which there are first- and second-order desires. You lack self-control when you give in and act in accordance with your first-order desires, but counter to your second-order desires. For example you give in to your desires to smoke, even though you don't want to want to smoke. Where does this second-order desire come from? Answer: your better judgment that you really ought to quit smoking.

There is nothing obviously wrong about this view, but it seems, on reflection, incomplete, for the following reason. Consider this example:

> An alcoholic wishing to stay abstinent may now judge that, all things considered, it would be better to stay home and watch television than go to the pub and have a drink with a friend. But as evening draws near, the temptation to drink increases, making him change his mind about what would be the better course of action – by giving too much weight to certain considerations that appear to provide reasons to meet with his friend, for example – and then revising his all things considered judgment accordingly. Is this a failure of self-control? Not if self-control is the contrary of akrasia. The alcoholic (unlike the *akratic*) does not act *against* his all things considered judgment when he chooses to meet up with his friend. Nevertheless, he plausibly exhibits a lack of self-control.
>
> (Henden 2017, p. 44)

Why is this? This is because the "self" in "self-control" is not only determined by what one judges to be best at a given point in time (as Mele puts it, as the result of conscious act of deliberation). It is also about a more stable, background state of what one values. This is especially important when considering addiction, since (as we can see in the example just given) one's deliberative judgements can be interfered with by addictive impulses and desires, by influencing patterns of reasoning.

Values can change over time, but in a way very different from the way in which judgments do, and they certainly exhibit more in the way of stability. Ultimately, the point is that we attribute certain values to an individual even when they are not thinking about them (indeed, you still have certain values when you are fast asleep).

This difference between judgment-based and values-based self-control is not merely a pedantic distinction. It has deep repercussions for how we are to think of failures in self-control. A failure in judgment-based self-control is a volitional/agentive failure. In these cases, we are driven to act in a way that is at odds with our better judgement. There is a synchronous conflict here. A failure in values-based self-control is a cognitive failure. We are, however momentarily, turning our back on our values, on who we take ourselves to be or want to be. To the extent that there is conflict here, it is asynchronous. The offending acts are reflected on later, retrospectively regretted but not committed in anguish as they are in failures of judgement-based self-control. If we were to identify an addict's failure of self-control at a given time, with one or with another, it would amount to making very different claims. These are undoubtedly very different psychological processes.

Which of these is more characteristic (if not necessarily definitive) of addiction? Let's examine each in turn. The first option, namely, viewing addiction as a volitional failure of judgment-based self-control, views addictive urges as so strong that they are (or can be) literally irresistible. In other words, the actions themselves are *compulsive*. Compulsion gets usefully operationalised as an insensitivity to variations in costs. On this view, the addict knows that failure to abstain comes at a high cost, but they can't help not abstaining. Raise the cost still further, and it still won't help matters.

This is problematic for the following reason. As Hanna Pickard (2017) aptly points out, there are several studies that demonstrate a significant degree of sensitivity to incentives on the part of addicts. For example Higgins et al. (1994) demonstrated that abstinence from cocaine use can be achieved with the use of relatively modest monetary incentives. Furthermore, results from the US National Comorbidity Survey (Warner et al. 1995) strongly suggest that many addicts "mature out" of their addictions, which seems to correlate with when the responsibilities of adult life start to emerge (especially parenthood). This could be interpreted in a number of ways, but one way of viewing this is through the lens of competing motives, interests and incentives. Addiction is therefore not compulsion. It leads to a high weighting in the rational calculus but doesn't bypass it altogether.

Perhaps, then, when we say that addicts lack self-control, we are talking about values-based self-control. In other words, it is not about urges and ensuing behaviour necessarily conflicting with current better judgments. Indeed, the addict can act in accordance with better judgment but still be deemed to lack self-control, or to have exhibited a failure of self-control. This is because, as Levy (2014, 2019) has persuasively argued, addiction has an impact on your better judgement at a given time. So perhaps a better characterisation (at least for much addictive behaviour) is to do with *oscillations*, rather than *conflicts*. This then might point to a values-based notion of self-control. The addict's behaviour is not (at least not necessarily) a compulsive and reluctant act in the face of judgements to the contrary: it is (or can be) an act in accordance with better judgement at the time that is at odds with diachronic values (namely, values held over an extended period of time). On this view, there is a conflict involved in this form of lack of self-control, but it's not a conflict with current mental states, but with overarching values, self-image etc.

This is a neat nuance and fits with a great deal of anecdotal evidence about what addiction is like. The regret often comes after rather than during the act, which itself is often freely, even joyfully, partaken in. So, have we reached a solution? Is addiction fundamentally about failures in values-based self-control, about failure to act in accordance with your genuine values, regardless of whether at the time of the act it aligns with your better judgment?

Sadly, there are reasons to think that the answer is No. In particular, what is problematic is the case of what might be called the "profoundly willing addict," or the "severe chronic addict," namely, someone who identifies with – and fully embraces – being an addict. One might dispute just how common these individuals are, although I don't think we need look very far. Legal addictive substances that don't impair daily functioning, like nicotine in tobacco smoking, are a rich source of this, especially among those for whom it is part of an identity (e.g. a "cool" identity). But the prevalence of these examples is not strictly relevant: the point is that we are still tempted both to think of them as addicts and to think of them as exhibiting failures of self-control. So how do we account for this?

So far, we have thought about self-control as an achievement or task and different kinds of self-control and different kinds of achievement or task. But we can think of self-control in a different way, more specifically as a psychological *capacity*. This might also be called "self-regulation," "executive control," and this is not about whether one behaves in accordance with certain aspects of the self or not, but it is rather about whether one has a certain *capacity* to control oneself, whether one actually wants to or not. In a sense, we began by reflecting on an ambiguity about the "self" in "self-control" (better judgment at a given time, versus self-identifying values over a longer period), whereas now we are turning our attention to an ambiguity in the "control" of "self-control."

One way in which this general "executive control" capacity gets tested is with performance on the Stroop task (Stroop 1935). The original task was reported as an "interference effect" to explain robust findings that there are delayed reaction times to incongruent stimuli as opposed to congruent stimuli. The basic task involves responding verbally as quickly as possible to the question "What is the colour of the word?" The congruent conditions involved being presented with colour words in their respective colours (GREEN in green, RED in red, etc.). Incongruent conditions involved being presented with colour words in incongruent colours (so, GREEN in red, RED in green, etc.)

Several studies, starting with Johnsen et al. 1997, modified the Stroop task to test attentional bias and executive control in addicts (in the first instance, tobacco smokers). The task involved colour-naming in three conditions: incongruent colour-words, smoking-related words and neutral, unrelated words. Unsurprisingly, among non-smokers, there was no significant difference in reaction times between the smoking words condition and the neutral words condition. For smokers, on the other hand, there was a significant reaction time delay with the smoking words condition, slowing it down almost to the level of the incongruous (Stroop) condition. Variants of this study have been conducted (e.g. for alcohol, Sharma, Albery and Cook 2001; gambling, Boyer and Dickerson 2003), with various accompanying interpretations (see Cox, Fadardi and Pothos 2006, for a theoretically sophisticated review of this literature). To summarise, the central idea behind these studies is that addicts have issues with controlling their responses to addiction-related stimuli. This interprets "control" as a very general capacity.

We may question this interpretation, but what it means for the deeply willing addict, the self-identifying addict, is that we still think of them as addicts, and as lacking self-control, because that is about their psychological *capacity* for control. And capacities are dispositions, they are only actualised *conditionally*. So, the willing smoker is addicted to smoking because their executive control, when it comes to smoking, is impaired. In other words, they need not actually exhibit any kind of conflict, but they are still addicts, because, *if* they were to try to cut down or stop smoking, they'd have a difficult time doing it, because of the inflexibility of their psychological constitution.

11.2.5 A socio-pragmatic account

Does *this* successfully define addiction? I'd be tempted to say that it comes close but is not quite there. The reason I say this is because there are all sorts of activities that we have dependencies on, both innate and acquired. We need to drink water, eat food, breathe air. We may also acquire moderate dependencies on exercise or sexual activity. But what do all canonical addictions have in common? The answer is that they are *bad* in some way. Note that dependencies on (strong appetites to engage in) sex, exercise, Internet use, computer games and so on only become addictions at the point where they become detrimental to the person's life. Cigarettes, alcohol and (in principle) illegal narcotics can be consumed in a controlled manner, and there are many individuals with dependencies on their consumption who are happy and "high functioning." Nevertheless, we are tempted to say they are addicted because these substances are seen as intrinsically bad for one. Furthermore, social stigma plays a role and varies cross-culturally. Habitual heavy drinking is not deemed problematic within UK pub culture, smoking is more stigmatised (and we have seen that stigma increase in recent decades), heroin is more stigmatised still. As a result, the threshold at which heroin use is deemed an addiction will be far lower in that cultural context than the threshold at which drinking would be.

More generally, the extent to which we are reluctant to describe a dependency as an addiction is a function of how we don't think it's a problem. This is seen with specific labels for specific additions. For example we may be reluctant to call someone with a clear dependency on alcohol an "alcoholic" (or indeed an addict) because we actually don't see their dependency as problematic. Alternatively, we might find it potentially problematic, if not to daily functioning and happiness, at least to their health in the long term, but still be reluctant to call them an alcoholic or addict, because of the weight that it carries: it might be rude or inappropriate, since it signals "this is a problem and you need help, or at least to do something about it." Notice how regular alcohol consumption has been normalised and is often seen as a necessary bringer of social benefits. It is certainly interesting to reflect on how tobacco smoking used to be very much the same but has changed as the health impacts became clearer, smoking became less normalised, etc. All this is to say nothing about which way things should go, but simply to flag the social function of calling someone an addict.

Notice, too, how this ties in with my view about "defining" delusion. Defining addiction on such a view is misguided because it is intrinsically bound up with negative evaluation and disapproval. One question is whether addiction may have a relatively cleanly definable descriptive core that is to do with self-control (we might call this "dependence"). On top of this core, there is a necessary layer of negative valence to get from dependence to addiction.

11.3 The science of addiction

Now that we have picked out the sort of thing that addiction is, we can focus on giving an account of the canonical phenomena once picked out. Crucially, then, these accounts do not seek to define addiction, but rather to explain instances of it. To critique them, you can't pick any case that you are tempted to call addiction and claim that their model fails to account for it. Rather, you need to point to the very phenomenon in the world that they take themselves to be accounting for and show how they are failing at accounting for *that*.

At this point, it is worth mentioning an interesting consequence of different scientific accounts of addiction. If one has objectivist leanings (see Chapter 2) and wants to tie illness or disease to notions of biological malfunction or dysfunction, then the model one takes to apply to addiction will potentially have direct consequences for whether it is to be seen as an illness or a disease. We'll start by looking at one hypothesis that construes addiction in terms of proper function.

11.3.1 Addiction as evolutionary mismatch

Durrant et al. (2009) view addiction from an evolutionary perspective. According to them:

> Addiction arises through interaction of these evolutionarily ancient systems, designed to promote the pursuit of natural rewards, and contemporary environments where purified and potent forms of drugs are readily available.
> (Durrant et al. 2009, p. 1049)

Roughly speaking, the idea is that in human (biological) evolutionary history, our nervous systems were designed to encourage the consumption of rewarding substances without self-control, since the environments in which we evolved were not so plentifully endowed with these substances. These "rewarding substances" or "natural rewards" encompass not only sugary and fatty foodstuffs, but also the consumption of mind-altering substances. Addiction has come on the scene, according to this hypothesis, not because of any change (or at least dysfunctional change) *in our brains and nervous systems*, but rather as a result of *environmental changes*. Indeed, they hypothesise, drug addiction has only been able to become a problem for humans relatively recently, and at least not until the development of agriculture and perhaps not before the growth of cities. Crucially, on this hypothesis, you get addiction as a result of the brain and body doing exactly what it was designed to do. There is merely a mismatch between the ancient context in which the evolutionary design took shape and the contemporary context in which we currently live.

One very general upshot of this view is not only that, taking objectivism as given for the time being, addiction is not a disease, but also that mental well-being should be seen as an environmentally embedded concern, rather than merely something to be considered within the bounds of skin and skull (see Chapters 3 and 12 for more on this).

11.3.2 Addiction as dysfunctional reward learning

Another scientific model agrees with significant parts of the evolutionary mismatch account, in particular, with the focus on rewards. However, a competing view insists that there is not enough focus on the neural machinery and not enough acceptance of the potency of narcotics, for example, in contrast to food stuffs. Also, they are not only potent, but also *neurobiologically misleading*. The psychological rewards we get from foodstuffs are somehow *real*: we need calories. In stark contrast, the rewards we get from drugs are illusory: we don't need drugs.

These models take drug addiction to involve dysfunctional reward signalling in the brain. This reward signalling is tied to the midbrain dopamine system. To simplify (in what I hope is a helpful manner), since the seminal work of Schultz (1992), the amount of dopamine in the brain was thought to correspond not so much to the strength of reward *per se*, as had been previously thought, but rather to whether the reward was better than expected. This was determined by measuring activity in the midbrain dopamine neurons of monkeys. In one of these experiments, monkeys learnt that if they pressed a lever following a cue, they would get a reward (in the form of fruit juice). During the learning phase, the midbrain dopamine neurons responded strongly to the delivery of the reward, but once the association was learnt, the neurons responded to the cue, but not the reward. In other words, the reward had become predictable. This, however, is not to say that it hasn't become rewarding: the monkeys clearly continue to be motivated by it. However, it is the cue that is surprising, not the reward, given the cue. This has led some theorists (Montague, Dayan and Sejnowski 1996) to hypothesise that midbrain dopamine is involved in signalling reward prediction, not the reward itself. This would mean that dopamine neurons would spike if the rewards were greater (or "better") that expected.

What does this have to do with addiction? Well, since drugs (some drugs, and certainly the most addictive ones) involve direct interference with the amount of dopamine in your brain, this alters your nervous system's appreciation of expected reward. Unlike a normal reward where the more you take it, the less surprising it is, and so dopamine levels will decrease, with drugs the dopamine response is kept artificially high. As Levy (2013) neatly puts it, as a result "the system treats the drug as of ever increasing value" (p. 3). This is clearly not how the reward system is *supposed* to work. It is supposed to allow the individual (the organism) to learn the reward value of something. In the case of repeated drug use, the reward system is dysfunctional, since the drug's value cannot be learnt.

11.3.3 Addiction as incentive salience

The evidence surrounding midbrain dopamine is open to different conceptualisations and interpretations, and not all theorists take it to be involved in reward learning. According to an influential alternative (Robinson and Berridge 1993; Berridge 2007), the role of dopamine is what they call "incentive sensitisation" or "incentive salience." One reason to prefer this interpretation is that learning about the relationship between a cue and a reward occurs without dopamine (e.g. in mice that have been genetically engineered to be unable to produce dopamine (dopamine deficient, or "DD") (Cannon and Palmiter 2003)).

The incentive salience hypothesis starts with the claim that there are three components to reward: *wanting*, *learning* and *liking*. (Note that these are terms of art and do not correspond to the subjective, personal-level phenomena we refer to by those terms). Once these three are distinguished, the claim is that midbrain dopamine is involved in *wanting*, but not in either learning (as claimed by the reward learning hypothesis discussed in the previous subsection) or liking (as initially widely accepted before the Schultz et al. experiments). More specifically, dopamine mediates wanting "by mediating the dynamic attribution of incentive salience to reward-related stimuli, causing them and their associated reward to become motivationally 'wanted'" (Berridge 2007, p. 408).

What is incentive salience? What it is *not* is liking or any sort of subjective pleasure. Liking, without incentive salience, would be a free-floating hedonic state without incentive target. Incentive salience attribution "makes a specific associated stimulus or action into an object of desire and can tag a specific behaviour as the rewarded response the individual is motivated to perform" (ibid.). An important point of clarification is that incentive salience is not the only kind of incentive motivational value carried by a reward. For example there are more "high-level," cognitive and cortically mediated forms of motivational incentive that rely on appreciations of act-outcome relationships (what we might think of as more "rational" forms of motivation).

I could go into much more detail (I refer the interested reader to Berridge 2007), but I am limited by space and want to present what is essential for the purposes of understanding addiction. Within classical conditioning paradigms (e.g. Pavlovian conditioning), it is common to distinguish the CS (conditioned stimulus) from the UCS (unconditioned stimulus). The CS is the stimulus that has no intrinsic reward value (e.g. the ring of a bell), but which, through conditioning, becomes predictive of the UCS, which is the reward. Viewed through the lens of the incentive salience hypothesis, what happens in conditioning is that incentive salience become attributed to the CS. The CS itself becomes "wanted" and motivates action. The Schultz (1992) dopamine data gets re-interpreted in light of this not in terms of reward prediction, but rather in terms of incentive salience. The cue (the CS) becomes *wanted*, and that is why you see a spike in dopamine at the time of the cue, but not the reward. The reward itself, however, is presumably very much liked.

What this has to do with drugs is that, by directly increasing dopamine levels, wanting becomes amplified to a dysfunctional extent and becomes disconnected from genuine biological rewards that are typically wanted when one is in specific physiological states (such as hunger, thirst, salt appetite). This also means that wanting can become disconnected from, and disproportionate to, liking. This fits with anecdotal evidence that addicts have to take their drug (or do their gambling, or whatever) in order to achieve a baseline of normality. They don't derive pleasure from doing it – only displeasure from not doing it.

11.3.3 Predictive processing accounts

We have encountered the predictive processing framework (PPF) at a number of junctures, and it may come as no surprise that it has also been used recently to make sense of addiction (Miller, Kiverstein and Rietveld (2020)). According to the PPF, the brain and nervous system is a hierarchically arranged prediction machine, whose central imperative is the minimisation of prediction error. The brain generates hypotheses (generative models) about what is going on in the world, based on how well they predict subsequent neural activity (see Chapter 7 for a more thorough presentation).

You may recall that there are a number of tweaks to this general, "prediction error minimization" picture. One, which is of least relevance to us here (although not entirely without relevance) is that these hypotheses are hierarchically arranged, with rapid-timescale predictions generally being closer to sensory surfaces, and longer timescale predictions being more abstract in nature and further away from sensory surfaces. The second and third added elements are of huge importance here, and they are related.

The second is that, not only does the brain have to predict input. In order to do this successfully, it needs second-order predictions as well, namely, predictions about how noisy and volatile the environment is, which amounts to how seriously it should take its predictions and hence prediction error. For example in good lighting, your appreciation that you are in an environment of high informational quality means that you'll tend to take bottom-up sensory input (prediction error) very seriously and rely less on top-down prediction. This amounts to high precision-weighting, which is, roughly, turning up the gain on prediction error. Conversely, in poor lighting, you are likely to turn down the gain on prediction error and are more likely to rely on top-down prediction. That is why you might experience a bush as a lurking person, only to quickly correct yourself as you examine it more closely. Crucially, this general mechanism of precision weighting or modulation is thought to be mediated by dopamine. It is also thought to underpin the phenomenon we (sometimes) call attention (Hohwy 2012).

The third, which is less important for psychosis, but especially important here, is that the PPF doesn't just account for perceptual experience, but also for action. Whereas the former involves "passive (or perceptual) inference," the latter involves "active inference." Active inference operates according to the very same principles as passive inference, namely, prediction error minimisation. However, whereas in passive inference you generate hypotheses to fit the world, in active inference, you generate a hypothesis, in particular a tactile and proprioceptive prediction, which you then make accurate via action. In other words, there are two ways of making your predictions accurate: one is passively, by making your models adhere to the world, and the other is actively, by making the world (most proximately the body) adhere to your models.

Given this, what does a predictive processing account of addiction look like? As with the other scientific models, dopamine plays a key role, but the key is to interpret the role of dopamine appropriately. (Interestingly, Miller, Kiverstein and Rietveld (2020) claim their proposal has the potential to unify learning and incentive salience hypotheses (see their fn.3).) Dopamine in predictive processing is associated with precision weighting. What this looks like for perceptual inference is an increased reliance on, or trust in, prediction error. What does this look like for active inference? According to Miller, Kiverstein and Rietveld (2020), the answer is: a high degree of agentive certainty, which drives action. Ultimately what their framework does is replace rewards as drivers of action, with highly precise *predictions*.

For Miller, Kiverstein and Rietveld (2020), human cognition in general does not cleanly separate passive and active inference, and their account of addiction reflects that. In particular, their general approach to predictive processing comes from the perspective of *embodied cognition* and, in particular, makes central use of the notion of affordances. Affordances are, in a sense, action-inviting perceptions: you don't just perceive the mug, you perceive it as graspable. It is within this approach that they put forward their proposal, and it is one that does away with the clean separation between action and perception. To sum up their approach to agentive motivation within the PPF, "agents act to minimize uncertainty in relation to their engagement with a field of multiple relevant affordances" (p. 3). In some cases, agents have high degrees of certainty (precision) in certain courses of action, and Miller et al. call these "action policies."

I inevitably have to skip over some important details, but, roughly, the central idea is that the midbrain dopamine system (following Friston 2012) functions to weigh confidence about what to do next (precision). Continued drug use leads to misleading "confidence in top-down predictions of future sensory states for policies of drug-seeking and drug-using behaviour" (Miller, Kiverstein and Rietveld 2020, p. 4). This, crucially, is best understood in deeply ecologically embedded terms:

> Thus, addiction is not due to substances of addiction hijacking midbrain dopaminergic systems. . . . Instead addiction should be understood in terms

> of how the whole person attunes to relevant affordances. What substance-induced dopamine does is *give the illusion of an agent attuning well to their environment when in fact in many cases the opposite is occurring.*
>
> (ibid., emphasis original)

This view has a number of interesting consequences. First, there is no reason to postulate a malfunction, Second, addiction is not only phenomenologically, but physiologically as well, like a hunger: the urges are a striving towards a homeostatic state just as they are with hunger. It is nice that there is this symmetry between the neuroscience and the phenomenology of addictive cravings. A third nice consequence is that this account fits with conditions under which people turn to drugs and become addicted. Indeed, contrary to many views, addiction is not an unfortunate side effect of engaging in a rewarding activity – it is what is often aimed for, namely, by providing certainty in a deeply volatile world (Lewis 2018). Finally, as well as it also fitting with addiction having the positive side-effect of conferring and constituting identity (Flanagan 2018), especially at troubled times where that might be being eroded, it also points towards an embedded/ecological perspective on mental health (as we will explore in the final chapter).

All of this doesn't only serve to down-moralise, but also to normalise and de-stigmatise addiction, and may also have positive practical and societal consequences.

11.4 Remaining philosophical issues

Aside from definitional issues arising from the concept of addiction, the sorts of phenomena that typically get called cases of addiction raise interesting philosophical issues.

11.4.1 Addiction and rationality

What is the relationship between addiction and rationality? Is the addict's behaviour rational? The answer to this clearly depends on what we mean by rationality, and addiction can be seen as a useful testing ground for reflections on rationality.

Whatever the finer details of what one thinks about rationality, we need to distinguish *epistemic rationality*, which one attributes to belief (and theoretical reasoning, reasoning about what to believe), and *practical rationality*, which one attributes to actions (and to practical reasoning, reasoning about what to do, or what one ought to do). If one thinks (like Levy 2014 does) that addiction is a problem with belief or judgment, namely, with believing that something has more value than it actually has, then we might be tempted to assess the relationship between addiction and rationality in terms of epistemic rationality. (This in turn might be accounted for in terms of biases of attention etc.) However, it has been far more typical to reflect on the practical rationality of addiction.

What is practical rationality? This is a hugely thorny issue, and several books have been entirely dedicated to this topic alone (e.g. Bratman 1987; Searle 2001; Price 2008). However, for our purposes, let's start with a simple claim that we will progressively modify or clarify: practical rationality is about aligning your actions with your desires. Here, again, we come up against the idea that there might be different kinds of desires. Bad, unhappy, reluctant cases of addiction are typically thought to involve some kind internal conflict between "urges" and more "reasoned desires" or "overarching values." Do these just straightforwardly get traded off against each other? Becker and Murphy (1988) famously put forward a "theory of rational addiction" along these lines. Their view is that drug-taking, even when the addict professes to the contrary, is a rational choice, which, as economists, they operationalise as attempts to maximise utility over time. It is an intriguing hypothesis but has come under various different kinds of criticism.

Perhaps the most prominent criticism (which we find in both Elster 1997; Rogeberg 2004) is that is it unfalsifiable. Becker and Murphy's claim is that addicts choose harmful drug use because they just value it more highly. But if the valuing is simply something that is "read off" the action (it's a "revealed preference" Samuelson 1948), then it cannot serve to explain it. At best, it is a trivial redescription. Furthermore, as Elster (1997, p. 760) points out, even if we grant that an addict is choosing to take drugs because the cost of the negative consequences is deemed a price worth paying, how would we know that the addict's appreciation of these consequences isn't profoundly impaired? And this ties in with a general concern with this view. The profound unhappiness and regret of addicts, the acceptance that their decisions are bad make the theory unappealing to many.

It does, however, raise two ambiguities about the notion of "utility maximization" or, in less econometric and more philosophical terminology, *doing what's best*. The first ambiguity is about doing what's best *for whom*? The second ambiguity is about whether this is an objective or subjective notion: is it about doing what you *think* is best or doing what is *actually* best? The latter notion, of doing what is actually best, can be used in a more externalist notion of practical rationality. According to this notion, even an addict who fully embraces, values and identifies with their addiction could still count as irrational, so long as they are deemed to be actually doing harm. Whether they think they are doing what is best is neither here nor there.

I have only scratched the surface, but reflections on addiction and rationality are a rich source of philosophical theorising.

11.4.2 Addiction, responsibility and blame

We talked earlier in the chapter about medical versus moral responses to addiction, but medical approaches aren't necessary for compassionate treatment of addicts: after all, we are compassionate and forgiving about a wealth of things that aren't illnesses or diseases. Furthermore, even if illness is involved, it doesn't necessarily fully exculpate (see Chapter 4). So, what is the relationship between addiction and questions of moral responsibility and blameworthiness?

A particularly interesting and influential contribution to this issue comes from Hanna Pickard (2017). In agreement with what we've just said, Pickard begins by pointing out that addiction is a heavily stigmatised condition (even our language reflects this, with words like "junkie" and expressions like "going clean"), and there tend to be two opposing ways of responding to it. One, which, like others (e.g. Levy 2013), she calls *the moral model*, is broken down into two parts: (1) drug use is a choice and (2) a critical moral stance towards that choice and the person who has made that choice. In response to this, *the disease model* seems to come as a welcome ideological corrective. However, it's not a good model *either*. First of all, it pathologises addiction (which, as we've seen in detail, is very much up for dispute), and, more importantly for current purposes, it removes the agency, autonomy and responsibility of addicts. According to the disease model, drug use is, for the addict, no longer a choice. It is an involuntary compulsion. But this seems neither accurate (for reasons we saw before about addicts responding to competing motives) nor helpful. As Pickard (2017) puts it: "self-conceiving as a helpless victim of a disease . . . risks placing addicts in a position whereby they view themselves as dependent on medical and associated professionals for a cure" (p. 171). Furthermore, evidence from Miller et al. (1996) strongly suggests that the disease model correlates with relapse. No, what we need to instil in addicts for therapeutic success is "a sense of agency and empowerment, alongside the fashioning and enacting of a life narrative that makes sense of the past while telling the story of a different future" (Pickard 2017, p. 171). Pickard suggests that this involves treading a middle ground between the moral model and the disease model. But how do we do this?

Drawing on her own clinical experience in a Therapeutic Community, Pickard suggests that the optimal stance to take towards addicts (and other forms of "disorders of agency," like borderline personality disorder, which is often characterised by self-harm, and eating disorders) is one of *responsibility without blame*. In effect, we need to accept the first part of the moral model (the choice part) but reject the second (the moral disapproval part).

Why does this apparently simple solution seem like a conundrum? (As Pickard herself admits: "I initially had no idea how this stance was conceptually possible, let alone achievable for myself within my own clinical practice" (p. 174).) It is because there is a deep-rooted but ultimately misguided tendency to tie responsibility and blame together. A particularly strong expression of this is to be found in the quote by Gary Watson (quoted in Pickard 2017, pp. 174–5): "to regard oneself or another as responsible just is the proneness to react to them in these kinds of ways." However, Pickard contends, the judgement that someone is responsible for something and "our practice of responding to [. . . that someone] with what is in effect an affective form of blame – a set of hostile feelings typically accompanied by equally hostile thoughts and actions" ought, both in principle and at times in practice, to come apart.

Addicts, in clinical contexts, are both most accurately and most usefully thought of as responsible (after all, "people will only change what they believe is in their power to change" (p. 175)), but the negative affective response of blame is both unhelpful and inappropriate. As Pickard very nicely puts it:

> In the clinic, the purpose of employing the concept of responsibility is therefore not fundamentally a form of backwards-looking moral evaluation, whereby a person is judged and potentially condemned for their past behaviour. Rather, the purpose of employing the concept of responsibility is fundamentally forwards-looking, serving to identify where there exists capacity for change thanks to the presence of choice and control, and, through clinical practices of holding responsible and to account, to motivate and encourage people to break the cycle – to develop, learn, and ultimately change what they choose to do and their sense of who they are and can be.
>
> (Pickard 2017, p. 175)

What is more, the separation of responsibility and blame gives rise to positive societal consequences. Or rather, the tying of responsibility and blame together forces us to adopt one of two extremes, either the moral model or the disease model.

> [And,] in placing blame squarely on addicts or their disease respectively, both models are united in enabling us to keep the focus of our attention away from ourselves and our society, avoiding the question of whether we, as a society, also collectively bear some responsibility for drug use and addiction and their consequent harms.
>
> (ibid., p. 177)

This, when we reflect on the sort of societal problems that give rise to addiction, is a very welcome corrective. After all, there is nothing intrinsically bad about taking drugs. As Müller and Schumann (2011) convincingly argue, drugs have been used as instruments throughout human history. In particular, they have been used as helpful means to understandable ends. As Pickard puts it, drugs "make us feel good, provide relief from suffering, and help us do various things we want to do better" (p. 177). It is only when drug use becomes harmful, often as caused by societal issues, that it becomes problematic. And this leads us on to a final question.

11.4.3 Addiction and pathology

Is addiction pathological? We've examined views that present addiction as a mental illness or a brain disease and hence as pathological. But what is the relationship between the concept of addiction and pathology? Ultimately, this requires us to carve up the conceptual space carefully: what is pathology and is addiction necessarily pathological? It is important to recognise that disputes about this matter may often only seem substantive. They will often turn out to be about terminological stipulation. In this spirit, it useful to think of it as a 2 × 2 matrix, as follows, with the four positions being represented by A, B, C and D.

	Values-based theorist	**Objectivist**
Addiction as pathological by definition	A	B
Addiction not pathological by definition	C	D

Position A takes pathology to involve some sort of impairment to flourishing (or similar), and addiction is seen as such an impairment by definition. Thus, someone might have a high-functioning dependence on something, but they are not an addict. Many people have a dependency on, say, alcohol (they drink every day and would really struggle to stop drinking for a long period), but are not "addicts" or "alcoholics." Their relationship with alcohol enhances, or at least doesn't diminish, their lives.

Position B takes pathology to be grounded in some form of biological dysfunction, for example some dysfunctional change to the brain. Addiction is only properly addiction if it involves this. There might be problematic relationships with certain substances or behaviours, but, unless they involve these objective dysfunctions, we are not talking about addiction, defined as pathology. We might, instead, be talking about plasticity at the neural level rather than pathology, or habit at the personal level, but not addiction.

Position C, like Position A, takes pathology to involve some sort of impairment to flourishing (or similar), but addiction is not seen as such an impairment by definition. On such a view, addiction might be tied to problems with self-control (e.g. they may have slow reactions in the addiction Stroop task), but these don't have to be problematic. The alcohol-dependent individual is an addict, but happily addicted (and hence is not an "alcoholic" if this is in contrast deemed to have an intrinsic negative valence), and would make their addiction more problematic if called upon to give up drinking, since they would presumably struggle to do so.

Position D, like Position B, takes pathology to be grounded in some form of biological dysfunction, but only some of the things that we rightly call addictions involve this. Some folk narratives seem to reflect this view in the (poorly specified) distinction between being "chemically addicted" and "psychologically addicted." Some drugs are presented as pleasurable, but not addictive, like cannabis. But people develop long-running cannabis-smoking habits that are hard to break. Some say that it is "psychologically addictive" but not "chemically addictive" in the way that, say, cocaine is.

The landscape becomes a bit more complicated if you introduce a nuanced distinction between brain disease and mental illness (as we have done earlier in this chapter and in Chapter 2). For example one might embellish Position D by saying that addiction is always a mental illness (indeed along values-based lines), but it only sometimes involves brain disease (where this is construed along objectivist lines).

Summary

- We examined four different views of addiction.
- First, we looked at the "lay view" (or "moral model"), according to which addiction is a moral failing. Although highly problematic, it introduces us to the complex relationship between psychiatric thinking and moral thinking.
- We then looked at views according to which addiction is a brain disease or a mental illness. These are importantly different propositions from one another.
- Finally, we examined views according to which addiction is a certain type of problem with self-control.
- There is an ambiguity about whether these views are getting at the concept of addiction, or the real-world phenomena that we call cases of addiction. Our practices of deeming things to be addictions (or people to be addicts) are sensitive in interesting and revealing ways to various socio-pragmatic factors.
- We then examined some scientific accounts of addiction, which differ in terms of whether they take addiction to be a biological malfunction or a mismatch between the environment and the nervous system.
- We ended by examining remaining philosophical issues, such as whether addiction is rational, whether an addict's behaviour is moral responsible or blameworthy and what exactly the conceptual relationship between addiction and pathology might be.

Questions

- In what ways, direct and indirect, can addiction harm?
- What sorts of things can you be addicted to? Is there anything that you can't be addicted to, and if so, why?
- Is there such a thing as a good addiction in a literal sense?
- In what sense does the addict value what they are addicted to?
- Do addicts always lack self-control?

Notes

1. Note, too, how well this fits with the fact that conscientious parents will follow the directive "keep your kids away from drugs!", but there is no similarly clear precautionary advice when it comes to preventing depression or psychosis.
2. Note how these completely come apart. You can think that something is inaccurate, but it is an ethically useful approach, or accurate, but ethically problematic.

Further reading

Berridge, K. 2017. Is Addiction a Brain Disease? *Neuroethics* 10 (1):29–33.

Glackin, S., Roberts, T. & Krueger, J. 2021. Out of Our Heads: Addiction and Psychiatric Externalism. *Behavioral Brain Research* 398:1–8.

Levy, N. 2013. Addiction Is Not a Brain Disease (And It Matters). *Frontiers in Psychiatry* 4 (24):1–7.

Pickard, H. 2017. Responsibility Without Blame for Addiction. *Neuroethics* 10 (1):169–80.

12 The future of philosophy of psychiatry

Overview

In this final chapter, I want to speculate about where I think the future of philosophy of psychiatry lies or, perhaps, could helpfully lie. First of all, there is a rather straightforward direction that philosophy of psychiatry can go in, and that's quite simply to examine some overlooked phenomena. There has been some work in all of these, but they have not received as much attention as the traditionally core psychiatric phenomena, and that is a shame, because they are at least as philosophically interesting. The next area where I see things heading in the future is the result of a fundamental change in our ecology: the rise and ubiquity of the Internet. What are the harmful and beneficial effects on our mental health and well-being? Third, an especially promising direction emerges from advancements in computational techniques. Although this is very exciting, it benefits from philosophical reflection and engagement. Fourth, there is an increased interest in, and development of, a family of views that I'm calling "Externalistic Psychiatry." Finally, and this is more of a hope than a bet, I present a shift in how many of the central questions in philosophy of psychiatry have been addressed. We should move from asking: "what is X?" and ask instead: "what does discourse about X do?" This then has consequences for how we can improve our discourse about, and by extension our societal responses to, mental illness.

DOI: 10.4324/9781315450124-14

12.1 Overlooked phenomena

Part two of this textbook looked at philosophical engagement with mental health conditions. As a result, there is an obvious sense in which you can produce new work by looking at mental health conditions that have not been looked at yet. However, there needs to be further motivation beyond the simple fact that philosophers haven't examined something yet. The phenomena in question need to be of philosophical interest.

Here, I present three phenomena that have received *some* attention, but arguably deserve more, because they have an especially high degree of philosophical interest. They are, first, psychological trauma, usually under the diagnostic label "post-traumatic stress disorder" (PTSD); second, eating disorders, and most specifically under the diagnostic label *anorexia nervosa* and third, the category of phenomena known as personality disorders.

12.1.1 Post-traumatic stress disorder

What is particularly interesting about psychological trauma in general, and PTSD in particular, is that it is an example of where cognition, where experiencing something, can leave its mark on the nervous system. To see why this is *so* interesting, compare psychological trauma to, say, traumatic brain injury: the latter involves a direct neurological impact, whereas the former involves the nervous system not being directly impacted, but somehow cognising, *appreciating* what is happening, its significance, and responding to that in a manner that has lasting consequences. It is, for example, the realisation that an experience is life-threatening that makes it psychologically traumatic, not its brute sensory intensity. By analogy, my eardrum could burst from a very loud shout, and this is a very different kind of lasting change from being emotionally disturbed by what someone tells me. Of course, this is all with the caveat that sensory intensity contributes to psychological trauma – it is the loudness of the bomb, as well as its life-threatening qualities, that makes it traumatic.

How are we to think of psychological trauma? Although there is little philosophical work on this (although see Ratcliffe et al. 2014), there is a significant amount of clinical theorising, especially structured around the diagnostic category of PTSD. The relationship between the loose, intuitive concept of psychological trauma and PTSD is relatively unsurprising. Psychological trauma simply refers to a lasting and negative impact from an experience or a series of experiences. PTSD is a diagnostic category characterised by a specific symptomatic presentation. Most prominent of these, at least in the public representation of PTSD, are so-called "flashbacks," the vivid re-living of (aspects of) the initial traumatic experience. These, in fact, are characterised as "Type 1" PTSD (Terr 1991). There is also "Type 2" PTSD (roughly synonymous with "complex PTSD") which lacks these flashbacks, but involves anxiety, hypervigilance and depersonalisation. The former is generally caused by one traumatic event of cataclysmic (often life-threatening) significance, which is what provides the theme and content of the flashback. The latter is usually caused by repeated trauma or extended periods of heightened threat (e.g. combat, especially ambush, scenarios). There is much more to say about this, but suffice it to say that there are many different theories about what is going on in both Type 1 and Type 2 PTSDs. I quickly summarise three different such theories.

Some of the earliest theories were "conditioning theories" (e.g. Keane, Zimering and Caddell 1985), which claimed that flashbacks were "over-consolidated memories." Shortly after these, "emotional processing theories" (Foa and Kozak 1986) argued that "complex fear structures" are stored in memory and produce cognitive, behavioural and physiological reactions when they become "activated." These structures are important in generating adaptive reactions to danger, but they can become maladaptive in certain contexts, giving rise in extreme circumstances, to PTSD, where a benign stimulus becomes associated with danger and activates these complex fear structures. This makes PTSD somewhat akin to a learnt phobia.

Third, there are cognitive theories (e.g. Ehlers and Clark (2000)) that present PTSD in terms of beliefs (or "appraisals"). Roughly speaking, PTSD sufferers develop excessively negative appraisals about both the external world and themselves: they view the world as a dangerous place, and they view themselves as incapable. Episodic memories and fear responses become biased by these negative appraisals. One consequence of these theories is that these appraisals need to be targeted by talking therapies (most prominently Cognitive Behavioural Therapy).

Wilkinson, Dodgson and Meares (2017) make three observations about these theories. The first is that they sometimes, but not always, seem to be talking about different phenomena. Second, even when they are talking about the same phenomena, they don't obviously seem to be in conflict with one another. Third (and in a related manner), they also seem to redescribe (re-conceptualise) what we already know is happening, rather than providing much in the way of a genuinely explanatory story. In response to this, Wilkinson et al. seek to subsume all of these theories within a predictive processing account, which also has the flexibility to account for Type 1 and Type 2 PTSD. Future directions would involve enriching this picture and also exploring consequences for successful treatment options.

12.1.2 Anorexia nervosa

Eating disorders are a relatively common mental health problem, with some statistics showing that anorexia nervosa (henceforth simply "anorexia") has the highest mortality rate of any psychiatric condition for adolescents. To that extent alone, they are a pressing societal concern. However, they are also of philosophical interest because they raise problems about how they are to be adequately characterised. This in turn is important for effective treatment.

Take anorexia, for example. It involves someone willingly depriving themselves of food to the (often severe) detriment of their physical health. Why do they do this? Within the standard belief-desire account (our old friend, the Humean account of motivation), when someone freely acts as they shouldn't, it is either explained by a belief that they shouldn't have, a desire that they shouldn't have, or a combination of both. In this particular case, we might think that people with anorexia wrongly believe that they are fat. This would in some respects give it affinities with delusion, with bad belief. On the other side of things, one might think that anorexia is about problematic desires: they have no appetite for food, or they have a strong desire for starvation. This would in some respects give it affinities with addiction.

Rather obviously, neither of these approaches seems right. People with anorexia do not seem adequately characterised as either delusional or addicted. Indeed, many find anorexia is not a typical "mental disorder" at all, precisely because it afflicts individuals who are not psychologically impaired or unusual. Anorexia is a particular source of anxiety for parents precisely because it might afflict their children with very little prior warning. Furthermore, whereas delusional and addicted individuals seem to have impaired agency and control, those with anorexia often talk about their extended periods of not eating as the establishment of control, the triumph of the will over the biological imperatives of nourishment.

These problems have led to approaches (somewhat similar to the moves made in philosophy of depression) that go beyond the belief-desire framework to more embodied and phenomenological approaches. Hannah Bowden (2012), for example, draws on the phenomenology of Merleau-Ponty and Sartre to resolve the apparent contradiction that individuals with anorexia seem to simultaneously believe that they are too fat and too thin. She suggests that people with anorexia experience a "pathological corporealisation," a sense of the body-as-object (and opaque, something I experience), rather than body-as-subject (and transparent, something I experience the world *through*), and as such they "feel fat" in this specialised phenomenological sense, while being perfectly capable of visually acknowledging (e.g. in a mirror) that they are too thin.

In another similar account, Frederik Svenaeus (2013) focuses on the phenomenology of the lived body and characterises anorexia as a striving for control "triggered by a state of bodily alienation in which the body is perceived to be foreign and horrifying to its bearer." He builds into his account the aggravating effect of the social objectification that young women are subjected to.

One direction that I would be interested in seeing is an embellishment of these rich, phenomenological insights with the predictive processing framework. To what extent, for example, might *control* be characterised in terms of predictive processing in a way that is illuminating for eating disorders?

12.1.3 Personality disorders

We touched upon personality disorders in Chapter 4 and noted their somewhat contradictory nature: to say that someone has a disorder, on one attractive view, is precisely to say that it is not part of someone's personality, that it is something outside of them, something that afflicts them and something that they want to be rid of. So how can something be both a disorder and part of their personality, as the label personality disorder suggests?

As we discussed in Chapter 4, whose emphasis was on moral responsibility, part of this is about seeing the problematic aspects of someone's personality as not their fault, for example, as the result of childhood adversity. Thus, personality disorder exculpates in a way that a personality does not. Indeed, personalities are rather canonical targets of moral approval or disapproval.

There is, however, a different angle from which this can be approached and one that doesn't focus on moral evaluation. It starts by asking: what is personality in general? In what senses might a personality be disordered? And what are these different categories of personality disorder? Are they legitimate? Are they useful? These are all deep and difficult questions, and I can't answer them satisfactorily here. However, I can make a start as a way of giving you a sense of where the future philosophical directions might lie.

A personality, at first estimation, is a pattern of behaviours and dispositions that we conceptualise as integral to who someone is. In keeping with this idea, comparing personality disorders with other mental disorders yields an important similarity and an important difference. Individuals with personality disorders and other mental disorders (e.g. depression, schizophrenia) will both be to some extent dysfunctional in the sense that the daily living of life will be in various ways (inter-personally, ergonomically) problematic. Where *personality* disorder differentiates itself from these other disorders is that these problems are seen as, first, the *kinds* of problems that can be chalked up to personality (compare "being irritable" or "being impulsive" with "hearing voices" – the former are the kinds of things that are building blocks of personality, whereas the latter is not) and second, as "ego-syntonic," namely, part of, or in keeping with, the individual's personality. As we've said, there is no different, underlying person afflicted by an external condition. The problematic behaviours, deemed disordered, pathological, are part of who this person is. Hence, we talk of a personality disorder, namely, a disordered personality.

This yields a very difficult problem: how do we draw the line between a disordered personality and a personality that is problematic but not disordered? There are surely people who have problematic personalities; for example they may be very hard to get along with, but we wouldn't want to pathologise them or, indeed (at least not intuitively, although one may argue for a revision of such intuitions), let them off the moral hook.

One way to go, which we have looked at already, is aetiology. Is it the result of childhood abuse or biological harm (see discussion of FASD from Chapter 4)? Another way is as follows. Problematic or otherwise, there are certain things that we expect of a properly functioning, non-pathological personality. For example we might think that a degree of stability is part of it. Somebody might be overly irritable, unfriendly, stubborn and so on in ways that make very few people like them and want to spend time with them, but if they are stably and predictably so, then (we are tempted to say) they are just an unpleasant person, not a disordered person. Among the instability of disordered personalities, we might distinguish between uncontrolled instability and strategic (manipulative) instability. Some people change their personalities for different contexts, that is they *weaponise* their personalities. This is another sort of things that might be deemed a disordered personality but in a very different way. It flies in the face of what we expect from personalities: that they be stable.

Of course, this gives us clues about how personality disorder might emerge in a traumatic, abusive childhood. Your primary caregiver might beat you aggressively, but then hug you and offer affection: what sort of model is that for a well-functioning personality? Alternatively, a child may find themselves having to change their own personalities as a survival strategy. This would lead to strategic (manipulative) shifts in personality.

Now let's move on to the subtypes. Needless to say, just as there are many dimensions of personality, there are many different taxonomies of personality disorder. In the DSM-5, they are broadly divided into three clusters, A, B and C (and a fourth "not otherwise specified" category for that which fails to fit the first three). Cluster A (sometimes called the "odd cluster" for obvious reasons) includes "paranoid," "schizoid" and "schizotypal" personality disorders. These are bordering on psychosis, on a detachment from reality (hence "odd"), but are deemed more akin to a personality trait: a way of being, thinking, experiencing the world. Cluster B (sometimes called the "dramatic cluster") include what might be thought of as the canonical personality disorders: "antisocial", "borderline", "histrionic" and "narcissistic" personality disorders. Cluster C (sometimes called the "anxious cluster") includes "avoidant", "dependent" and "obsessive-compulsive" personality disorders. All of these labels are relatively self-explanatory except for "borderline personality disorder" (BPD), which is called that for the historic reason that psychiatrists thought it was on the "border" between psychosis and neurosis. A more helpful label (which is sometimes used) is "emotionally unstable personality disorder."

What are we to make of this taxonomy? On the surface, it seems like quite a helpful descriptive carving up of the territory. But (applying some of the reasoning from Chapter 3) is it picking out real conditions in the world? One way of reflecting on this is by asking: is it conceivable that we might *discover* a new personality disorder? I intend this to be a question about *conceptual* possibility, rather than actual scientific possibility. To answer this in the affirmative would suggest that we are talking about things that objectively exist in the world. But to do so, would, in my opinion, reveal an important misconception. Though helpful, this taxonomy of personality disorders is not like a taxonomy of, say, birds. It is rather a description of a wide array of ways in which certain extremes of personality can be problematic for individuals who have them and those around them. A new category of personality disorder would instead involve a new way of carving up old territory.

Also, as a result of this, people who get the same personality disorder diagnoses are simply problematic in similar ways. They don't have the same "underlying conditions" (or at least it would be a huge assumption to claim that they did). This also fits with how very often personality disorder diagnoses are "co-diagnosed" with other more traditional psychiatric disorders (e.g. schizoaffective disorder and borderline personality disorder).

12.2 Mental well-being in a digital world

Technological advancements in general, and the Internet in particular, have radically changed almost every aspect of our lives. The Internet is not only a tremendous source of information, as well as a way of browsing and purchasing products and services, it also serves to connect people through various forms of what we might call *online social networking*. Narrowly construed, online social networking, often called "social media," refers to platforms like Facebook and Twitter. However, the broader notion of social networking I have in mind includes slower timescale interactions like emails, blogs, discussion forums, as well as rapid timescale video-conferencing platforms and software (like Zoom, Teams and Skype). Ultimately, these social uses of the Internet have radically changed the way in which we can form and maintain social interactions. Aside from being philosophically interesting in its own right (How does it change humanity or threaten democracy, liberty, etc.?), this also has consequences for mental health and well-being, and it is timely for philosophers to reflect on this.

12.2.1 The self and self-esteem online

One area where the Internet has philosophically interesting consequences for mental health and well-being is with regard to the self and self-esteem. Canonical forms of social media involve the careful curation and presentation of content about oneself, usually via some form of profile. It also involves a huge amount of exposure to the lives of others and to various unrealistic ideals of body, beauty, wealth and success. These two features relate to two quite different mental health domains. The first is about how the flexibility and inauthenticity of self-presentation online erode the sense of self (see Dreyfus's (2001) reflections on the Internet through the lens of Kierkegaard's philosophy of self and authenticity) in a way that could yield mental health issues (see also our discussion of personality disorders and instability of personality, from Section 12.1.3). The second is about how the presentation of idealised versions of others can lead to low self-esteem, anxiety and depression, especially, but not only, when coupled with online abuse and trolling (think about the concept of "body shaming" for example). The Internet is a minefield of potentially psychologically damaging content and interactions, especially for children and adolescents.

12.2.2 Loneliness and togetherness online

But the contribution of the Internet to our social lives isn't all bad. We are social animals, and the increased scope for social contact can be aided by the powers of the Internet. There are two ways in which this can be seen, one indirect and the other more direct.

First of all, the Internet allows people to cast their social net more widely. This in particular can be very helpful when thinking about mental health support groups. Although this applies in a variety of contexts, consider the following example. Imagine you are someone who lives in an isolated village and who starts hearing voices. In addition to the first-order negativity of hearing voices (and indeed, as we saw in Chapter 7, they needn't even be negative), there is the second-order negativity of thinking, or worrying, that you are weird, different, that something's wrong with you, that all manner of bad things will happen to you if people find out you hear voices and so on. These fears could feed back into your experiences and make them worse, more negative, more harmful. But now imagine that you can find out, thanks to the Internet, that there are a great many people out there just like you, who are learning to live with and make sense of their voices and are capable of flourishing in life. Suddenly, you have all of the emotional support that comes from feeling part of a community, you see that there are people like you and you also have access to their advice. This is a tremendously powerful tool. Of course, as is so often the case, there is the flipside to this, and these powers can be used for ill as well as good. In other words, there are online communities that don't support and promote psychological health and well-being but rather encourage harmful behaviours (e.g. pro-self-harm communities, pro-anorexia communities (Giles 2006)).

Looking more directly at the impact of the Internet on social bonding, Lucy Osler (2020) has explored feelings of togetherness and "we-experiences" online, drawing on Walther's (1923) phenomenology of communality and applying it to the online realm. We live in a world where some have claimed we are in the midst of a loneliness epidemic. And although loneliness itself is not a mental illness, it surely aggravates mental health problems and is often seen as a strong predictor of various serious mental health conditions: especially psychosis (delusionality, paranoia), anxiety and depression. Reflecting on ways in which the Internet can foster togetherness could help us combat loneliness more effectively.

12.2.3 Internet addiction

Another respect in which the Internet affects us is not directly through what it enables as a technology, but rather how our appetite to engage with it might interfere with other aspects of our lives. In other words, our relationship to the Internet, and especially to smartphones, can be seen as highly problematic. Although not recognised by either the ICD or the DSM, some have suggested that there is something called "Internet Addiction Disorder." This in truth is not one addiction, but a cluster of them, since the Internet affords many different activities. So, individuals can have problematically strong appetites for: online gaming, social media (especially Facebook, Twitter, Instagram), online pornography, online shopping (especially auction sites) and so on.

Of course, one might question whether these could ever really count as addictions? These are not, of course, drugs. However, gambling addiction is a canonical addiction that doesn't involve the ingestion of anything. And various other rewarding activities like sex are deemed by many to be viable objects of addiction. As we saw in Chapter 11, one plausible way of thinking of "addiction" as a concept is not as a brain disease, but as an appetite that is problematic in some way that interferes with human flourishing. So, the adolescent, who sees their grades slip, or who sees their personal hygiene disintegrate, who loses their friends, their job, etc., because they are always at their computer, is problematic in the relevant ways. Of course, this comes in degrees, but that does not make it any less serious. Of course, the materiality of sitting at home on your computer constrains what that engagement looks like. Smartphones are portable, and that changes the nature of the addiction: you are no longer a recluse, you move around, but that is a double-edged sword. You can drive while your phone is in easy reach, and a strong appetite for it might lead you to crash the car. You can be looking after young children while scrolling obsessively through Instagram and lack the attention required to prevent the toddler from falling down the stairs. This is not drug addiction but the negative consequences can be just as bad.

To combat this, our conversations need to be open and honest. And because so many of us, many of whom are affluent and empowered, are guilty to some extent of this, and because it is in the interest of the most powerful forces in society for us to remain so, it is hardly surprising that "Internet addiction" is as under-stigmatised as "drug addiction" is over-stigmatised.

12.3 The rise of computational psychiatry

Computing and AI have made very significant contributions to both clinical and theoretical cognitive science. When computational techniques are applied to psychiatric phenomena, we call this "computational psychiatry." There is a broader and narrower understanding of computational psychiatry. According to the narrower sense, it refers specifically to computational approaches to cognition, where the cognition in question is of psychiatric interest. The broader sense incorporates this but includes computational approaches to psychiatry more broadly, to include, for example, epidemiology and diagnostic procedure. In other words, it is not only cognition that benefits from computational treatments but other phenomena relevant to psychiatry, e.g. diagnostic categories, environmental patterns, behavioural patterns as well. Let's start, however, with the narrow sense, which also falls under the broader concept anyway.

12.3.1 Clinical computational neuroscience

As we've seen at numerous junctures, the predictive processing framework proves itself to be rather useful when it comes to thinking about various psychiatric phenomena. We have, however, mainly considered it with a view to solving quite broad theoretical issues (e.g. what voice-hearing experiences are in general, within the predictive processing framework). However, predictive processing and, more broadly, Bayesian approaches to cognition allow also for far more fine-grained and applied work. In a sense, we have been so far talking about where you would plug in some numerical values, but we haven't plugged them in yet.

Adams, Huys and Roiser (2016) review computational work on depression and psychosis with a view to promoting computational psychiatry. As they put it:

> Computational Psychiatry aims to describe the relationship between the brain's neurobiology, its environment and mental symptoms in computational terms. In so doing, it may improve psychiatric classification and the diagnosis and treatment of mental illness. It can unite many levels of description in a mechanistic and rigorous fashion, while avoiding biological reductionism and artificial categorisation.
>
> (p. 53)

One exciting feature of computational psychiatry is that it allows us to see significance to cognition (both normal and abnormal) of many different kinds of empirical observation, whether that be anatomical or neurochemical (e.g. at the level of neurotransmission). As we saw in Chapter 6, it is not enough to see a correlation between something (e.g. dopamine levels) and a symptom or symptom severity. You need to fill in the gaps between one and the other: what does the former contribute to the overall picture such that the latter can be expected? Adams, Huys and Roiser (2016) make a very convincing case that computational approaches have the power to do this bridging work, or at least a significant part of it.

12.3.2 Computational psychiatric epidemiology

To the extent that psychiatric epidemiology is a branch of psychiatry, computational psychiatric epidemiology, which is simply a subset of computational epidemiology more generally, counts as computational psychiatry broadly construed. It involves zooming right out and looking at patterns of mental ill health across society as a whole.

Of course, epidemiology is usually associated with the spread of infectious diseases, but this is an overly narrow understanding of the discipline: it seeks to understand and generate hypotheses about the prevalence of certain conditions and what they correlate with and what may cause them (we all know that one cannot infer the latter from the former). In short, it is all about discernible statistical patterns. If the data sets are big, and the patterns are hard to ascertain, that is where computational techniques become extremely valuable.

Certainly, one individual cannot catch, say, psychosis off another (although delusional ideas can spread in certain contexts), but, if you take a set of people who have psychosis, you can look at what it is about their lives that they may all have in common and which might be postulated as the "hidden cause." Of course, one thing that computational techniques, especially deep neural networks ("deep learning"), are especially good at is finding these "hidden causes."

12.3.3 Deep learning in diagnostics and early intervention

Durstewitz, Koppe and Meyer-Lindenberg (2019) provide a thorough review of the applications of deep learning in psychiatry. As we have touched upon, deep neural networks are particularly good at learning abstract representations from large quantities of raw data. This means that, once trained, they are very good at "making sense" of varied inputs. So, for example they are good at image recognition and categorisation. More recently, companies like Google's DeepMind Health have been training deep neural networks to achieve superhuman levels of image analysis for tumour detection.

The very same principles have been applied to psychiatric contexts. Of course, since psychiatric phenomena are rather more conceptually complex, and ontologically distributed, than brain tumours, the sources of input that are used by the neural networks vary to a striking degree.

For example some work attempts to use these networks for diagnosis and prognosis based on neuroimaging inputs. Although these seem like promising sources of data (not least because the relevance of the brain to mental illness is clear), the sample sizes are relatively small. Much larger amounts of data are to be found from mobile phones and other wearables. Here, the neural networks are trained on a variety of extended and environmental phenomena (i.e. beyond the brain) including movement patterns, physical activity, quantity of social interaction and even specific patterns in linguistic output in text messages or Twitter feeds. Indeed, this might be used to detect people who are "at risk" long before they would typically come to the attention of mental health services and to allow people to intervene early and take preventative measures.

These techniques are in their infancy, but they are producing promising results. Having said that, they raise interesting philosophical issues. First of all, a highly trained neural network that performs at superhuman levels functions like a "black box": it gives us results, but we don't fully understand how it arrives at such results. Second, in diagnosis, we are constrained by our existing coarse-grained categories. As Durstewitz et al. put it, the "expert-provided diagnostic labels used for training defines an upper bound on the system's performance" (p. 1591). In short, these networks are so advanced that there are problems raised by the somewhat paradoxical fact that they outstrip us, and we constrain them.

12.4 Externalistic psychiatry

In a paper we already mentioned in Chapter 2, Will Davies (2016) notes that some theorists (Drayson 2009; Sprevak 2011) are putting forward a programmatic vision for externalistic psychiatry. He then proceeds to show various ways in which much of psychiatry is *already* implicitly externalist. This is a valuable observation, which gets further extended (and hence somewhat deflated, perhaps) by Glackin (2017), who says that this also applies to somatic medicine, and not just to psychiatry. What I am interested in here is a return to the more programmatic, revolutionary kind of externalism, rather than a description of what is already implicitly going on. One reason for my interest in this is that what Davies and Glackin are talking about are the concepts and categories of psychiatry and medicine. I am interested in what the future holds for externalistic approaches to the phenomena themselves, not simply our way of categorising them. A mosquito bite is only a mosquito bite if it was caused by the bite of a mosquito. That much is hard to dispute. However, that very thing, which we categorize as a mosquito bite and which we can imagine an identical replica of that was not caused by the bite of a mosquito, to say that *that very thing* is constituted by things outside of the organism is a far more contentious issue, and, in this particular case is potentially preposterous. But there are other cases where it might not be. In particular, claims about the phenomena themselves being reliant on "external factors" might be precisely less preposterous because we are talking about minds, not bodies.

In philosophy of cognitive science, there is a family of views that is referred to as "4E cognition." These are views about the nature of the mind and cognition. If they are taken seriously, then they ought to apply to mental illness, too, as well as to the mind in general. These 4 "Es" are: Embodied, Embedded, Extended, Enactive. I now explore what each of these may look like. There is more work under some of the "E-banners" than others, especially the last of these.

12.4.1 Embodied psychiatry

Embodied approaches to cognition, and pathologies of cognition, have a long pedigree, especially in the phenomenological tradition (Merleau-Ponty being the archetypal phenomenologist of embodiment, who also engaged with both psychiatric and neurological case studies).

At various junctures in this book, we have seen embodied leanings. With depression, addiction and even psychosis, we have encountered theorists (usually either phenomenologists like Ratcliffe, Fuchs and Carel or other philosophers with phenomenological sympathies like Smith) who reject the idea of the mind as a free-floating centre of pure consciousness: it is embodied and constituted by a way of finding oneself as a body in a world. Thus, in psychosis, we might feel our body, usually so light and transparent, and, our viscera, usually imperceptible, in a heightened, even excruciatingly painful way. In depression, we might feel heavy, unable to move, because the world is devoid of embodied possibilities for action. In addition, we are physically moved, drawn this way and that. Embodiment is not the future of philosophy of psychiatry but a particularly interesting subset of its past and present, extending, no doubt, promisingly into the future.

12.4.2 Embedded psychiatry

"Embedded" usually refers to brute, environmental embedding but can also refer to social embedding. Environmentally embedded psychiatry would look at how different spaces, both natural or architectural, might promote mental well-being. Socially embedded psychiatry will take very seriously just how deeply social we are. Indeed, a mind that is not social is not recognisably human. Again, this might examine the ways in which social cohesion promotes, or social isolation and alienation impair, mental health and well-being.

12.4.3 *Extended psychiatry*

In discussion of 4E Cognition (e.g. Newen, Gallagher and de Bruin 2018), "Extended" is usually presented as the last E, and it does, to some extent, stand out on a limb. The other three Es are very much compatible and conducive to one another. Extended Cognition in contrast seems at odds with some of these. It is at once more conservative than the other three and more radical. It is conservative in that the Extended Mind Hypothesis (EMH) is based on quite a traditional form of functionalism (and by extension representationalism). It is more radical in that it literally takes the mind to extend into the environment.

I haven't the space to present EMH here, nor the arguments against it, but what it would mean for mental illness is that the pathological mind might be construed as literally extending out into the environment. We would be able to talk quite literally about pathologies of mind in terms of pathologies of the environment around the organism.

12.4.4 *Enactive psychiatry*

This is where a significant amount of recent work has taken place, especially the work of de Haan (2020) and Nielsen (2021). There are various different forms of enactivism. The most relevant for our purposes is autopoetic enactivism, according to which the meaning that we experience as beings in the world is "enacted," namely, emerges through our striving to "self-maintain" as organisms in a world. A central and related concept is "sense-making," which is the fundamental component of cognition: it involves making sense of the world in a primitive and fundamental way, in accordance with the organism's purpose. In de Haan (2020), mental disorder is defined as a stable pattern of sense-making going astray. This framework seems to me like an immensely fruitful direction and will no doubt flourish in the near future.

12.4.5 *Clinical consequences of externalistic psychiatry*

One of the things that I find most exciting about these views is how they reconceptualise not just mental illness but its treatment as well. It means that various external interventions, from prescribing dance, exercise, or cold-water swimming, or promoting befriending schemes, woodland walks and so on, aren't just ways of improving people's lives by skirting around the edges of their mental disorder in the hope that they may make it bearable or even feedback in and help to improve it. No, these interventions would get at the very heart of the disorder, because, from a perfectly legitimate, theoretically and scientifically grounded perspective, that disorder "just ain't in the head."

12.5 Understanding psychiatric discourse

As many philosophers have noticed, we are often misled by language in a number of ways. Most fundamentally, we can find ourselves thinking that certain things concretely and objectively exist, just because we talk about them in certain ways that seem to suggest that they do. In a very different respect, and at a practical level, the way we talk about certain things can have consequences that far outstrip our intentions or even our awareness of such consequences. All of this applies to psychiatry as much as to anything else, and, as such, I hope that a fruitful line of future research will involve a careful unearthing of the nature and consequences of psychiatric discourse.

I start by presenting some metaphysical (or perhaps anti-metaphysical) positions that I think usefully emerge from a careful reflection on how we use language. Then, I present what I take to be some important practical consequences, indeed obligations, that emerge from these somewhat distant and arcane theoretical positions.

12.5.1 Beyond realism and anti-realism: expressivism, fictionalism, constructivism

Does the claim that mental illness is not a concrete, objective thing mean that it is not real? It depends what you mean by "real," and many theorists don't question the objective reality of say, mental illness, in order to belittle it. On the contrary, it retains or even accentuates its importance. "Objective reality," to the extent that that idea even makes sense, is brute and relatively uninteresting. Social reality, on the other hand, is laden with nuance, meaning, importance and consequence.

There are three different ways of adopting this form of "quasi-realism" (to borrow Simon Blackburn's terminology), and we touched on these in Chapter 2. They are: expressivism, fictionalism and constructivism. Roughly speaking, expressivism is the view that certain forms of discourse that look fact-stating are, at least in part, doing something else. This is often put in terms of assertions in that domain not being the *expression* of beliefs, but of other mental states like emotions, desires, "pro-attitudes" and so on (so, "Murder is wrong" is analysed as "Murder, yuk!" or "Murder, I disapprove" etc.). The fictionalist and the constructivist are very similar in that they take the assertions in the discourse to be fact-stating alright, but the facts in question are either fictional (for the fictionalist) or constructed (for the constructivist).

Depending on the finer details of the form of fictionalism or constructivism, they can look extremely similar. I haven't the space to go into all of the details, but here is a very simplified presentation of a distinction, to the extent that one can be made at all. The fictionalist will tend to take the facts (the discourse) in question to be circumscribed to a particular domain (e.g. mathematics, morality, psychiatry), where certain institutional structures and practices will be responsible, if not consciously or deliberately, at least explicitly, in the generation of fictional (or fiction-like) discourse. In contrast, the constructivist will tend to take construction to be more pervasive and just is a greater or lesser part of many (perhaps all) areas of discourse. It is shaped and regimented, but not brought about, by institutional structures.

Another way of putting this is in terms of what is the relevant contrast, rather than in terms of differences in the positive accounts. A fictionalist about something is partly talking about a domain as fictional (or fiction-like) as *distinct from the majority of other domains* which are not fictional. In other words, fictional discourse is not the status quo. Constructivists are, in contrast, less likely to talk about constructed domains as opposed to non-constructed domains: they will assume that some processes of construction are at work in everything: the key is to look at the details. That is also why a fictionalist will tend to be a fictionalist about a broad domain of discourse (mathematics, morality), whereas a constructivist will look, for example, at specific concepts, some very broad (race, gender (Haslanger 2012; Sveinsdóttir 2013)), others very specific (dissociative identity disorder (Hacking 1995)), and examine the details of how they are constructed.

12.5.2 The ameliorativist agenda: optimising discourse for scientific progress and social justice

In Chapter 2, we touched upon constructivism about mental illness with a view to answering from an ontological perspective: "what is mental illness?" But the ontological position of constructivism has important – indeed, powerful – practical consequences. Sally Haslanger puts it beautifully when she writes:

> If the point of claiming that a phenomenon is socially constructed is not to take an anti-realist, anti-essentialist, and anti-naturalist stance towards it, what could the point possibly be? I suggest that the . . . goal is to show how the phenomenon in question depends, *in ordinary ways*, on us, particularly on our attitudes and our interactions. . . . if we know both that, and how, a phenomenon depends on us, we are in a better position to intervene to change it. . . . my own view is that if we are going to change the world, we need to know how the problematic parts are created and maintained.
>
> (Haslanger 2012, pp. 214–15)

In her book, Haslanger focuses on race and gender with a specific view to promoting social justice. But how might this apply to mental illness in general and specific diagnostic categories in particular? I can only gesture towards possible answers. We need to look closely at, among other things, public perceptions of psychiatric phenomena; institutional responses and connections between different institutional structures such as psychiatric diagnosis, the welfare state, law and criminal justice. We also need to look at various looping effects and self-fulfilling prophecies, especially in the context of self-stigma and its relationship with low well-being and pessimism of both outlook and outcome. Ultimately what we are after is a discourse surrounding mental health and psychiatry that both reflects and promotes a society that is humane, compassionate, optimistic, fair and so on.

Summary

- New advancements can be made in philosophy of psychiatry simply by examining currently under-examined but deeply philosophically interesting phenomena.
- These include psychological trauma, anorexia nervosa and personality disorders.
- Another area of advancement involves reflecting on the impact of the Internet in general and social media in particular on our mental health and well-being.
- A cutting-edge application of philosophy of psychiatry is a critical engagement of areas where scientific advancements are most rapid and exciting. One such area is in computational psychiatry, where this includes computational models of psychiatric cognition and computational and machine learning approaches to psychiatric epidemiology, as well as diagnosis and prognosis.
- Finally, there is more work to be done in understanding the nature and power of psychiatric (and, more generally, mental health) discourse, with a view to improving both its scientific utility and social impact.

Questions

- Can you think of any other mental health issues that have not been touched upon? What philosophical issues might be raised by them?
- Can you think of any other risks and benefits of the "information age"?
- In what ways might psychiatric conditions literally be distributed across the environment?
- How many of the ways in which mental health services are suboptimal can be helped by an improvement in how we talk and think about mental health?

Further reading

Charland, L.C., Hope, T., Stewart, A. & Tan, J. 2013. Anorexia Nervosa as a Passion. *Philosophy, Psychiatry, and Psychology* 20 (4):353–65.

Haslanger, S. 2012. *Resisting Reality: Social Construction and Social Critique*, Oxford: Oxford University Press.

Montague, P.R., Dolan, R.J., Friston, K.J. & Dayan, P. 2012. Computational Psychiatry. *Trends in Cognitive Sciences* 16 (1):72–80.

Osler, L. 2020. Feeling Togetherness Online: A Phenomenological Sketch of Online Communal Experiences. *Phenomenology and the Cognitive Sciences* 19 (3):569–88.

Wilkinson, S., Dodgson, G. & Meares, K. 2017. Predictive Processing and the Varieties of Psychological Trauma. *Frontiers in Psychology* 8.

References

Achinstein, P. 1983. *The Nature of Explanation*, Oxford: Oxford University Press.

Adams, F. & Aizawa, K. 2001. The Bounds of Cognition. *Philosophical Psychology* 14 (1):43–64. https://doi.org/10.1080/09515080120033571

Adams, R.A., Huys, Q.J. & Roiser, J.P. 2016. Computational Psychiatry: Towards a Mathematically Informed Understanding of Mental Illness. *Journal of Neurology, Neurosurgery, and Psychiatry* 87 (1):53–63. https://doi.org/10.1136/jnnp-2015-310737

Adams, R.A., Shipp, S. & Friston, K.J. 2013. Predictions Not Commands: Active Inference in the Motor System. *Brain Structure and Function* 218 (3):61143.

Alderson-Day, B. & Fernyhough, C. 2015. Inner Speech: Development, Cognitive Functions, Phenomenology, and Neurobiology. *Psychological Bulletin* 141 (5):931.

Alderson-Day, B., Moseley, P., Mitrenga, K., Moffatt, J., Lee, R., Foxwell, J., Hayes, J., Smailes, D. & Fernyhough, C. 2022. Varieties of Felt Presence? Three Surveys of Presence Phenomena and Their Relations to Psychopathology. *Psychological Medicine* 1–9.

Alexander, M.P., Stuss, D.T. & Benson, D.F. 1979. Capgras Syndrome: A Reduplicative Phenomenon. *Neurology* 29 (3):334–9. https://doi.org/10.1212/wnl.29.3.334

Alloy, L.B. & Abramson, L.Y. 1979. Judgment of Contingency in Depressed and Nondepressed Students: Sadder but Wiser? *Journal of Experimental Psychology: General* 108 (4):441–85. https://doi.org/10.1037/0096-3445.108.4.441

Altshuler, K.Z. 1971. Studies of the Deaf: Relevance to Psychiatric Theory. *The American Journal of Psychiatry* 127 (11):1521–6. https://doi.org/10.1176/ajp.127.11.1521

Ames, D. 1984. Self Shooting of a Phantom Head. *The British Journal of Psychiatry: The Journal of Mental Science* 145:193–4. https://doi.org/10.1192/bjp.145.2.193

Amundson, R. 2000. Against Normal Function. *Studies in History, Philosophy, Biology and Biomedical Science* 31 (1):33–53.

Andreasen, N.C. 1997. Linking Mind and Brain in the Study of Mental Illnesses: A Project for a Scientific Psychopathology. *Science* 275 (5306):1586–93. https://doi.org/10.1126/science.275.5306.1586

Andreasen, N.C. 2001. *Brave New Brain: Conquering Mental Illness in the Era of the Genome*, Oxford University Press.

Anscombe, G.E.M. 1957. *Intention*, Cambridge: Harvard University Press.

APA (American Psychiatric Association). 1968. *Diagnostic and Statistical Manual of Mental Disorders*, second edition, Washington, DC: American Psychiatric Press.

APA (American Psychiatric Association). 1980. *Diagnostic and Statistical Manual of Mental Disorders*, third edition, Washington, DC: Psychiatric Press.

APA (American Psychiatric Association). 2000. *Diagnostic Statistical Manual of Mental Disorders*, fourth edition, Text Revision (DSM-IV-TR), Washington, DC: Psychiatric Press.

APA (American Psychiatric Association). 2013. *Diagnostic and Statistical Manual of Mental Disorders*, fifth edition, Washington, DC: Psychiatric Press.

Armstrong, D.M. 1968. *A Materialist Theory of the Mind*, New York: Routledge & Kegan Paul.

Artaud, A. 1965. *Artaud Anthology*, City Lights Books.

Atkinson, J.R. 2006. The Perceptual Characteristics of Voice-Hallucinations in Deaf People: Insights into the Nature of Subvocal Thought and Sensory Feedback Loops. *Schizophrenia Bulletin* 32 (4):701–8. https://doi.org/10.1093/schbul/sbj063

Atkinson, J.R., Gleeson, K., Cromwell, J. & O'Rourke, S. 2007. Exploring the Perceptual Characteristics of Voice-Hallucinations in Deaf People. *Cognitive Neuropsychiatry* 12 (4):339–61. https://doi.org/10.1080/13546800701238229

Atran, S. 2010. *Talking to the Enemy: Violent Extremism, Sacred Values, and What It Means to Be Human*, London: Penguin.

Austin, J.L. 1962. *How to Do Things with Words*, Oxford: Clarendon Press.

Ayer, A.J. 1952. *Language, Truth and Logic*, 1st Dover edition, New York: Dover Publications.

Badcock, P.B., Davey, C.G., Whittle, S., Allen, N.B. & Friston, K.J. 2017. The Depressed Brain: An Evolutionary Systems Theory. *Trends in Cognitive Sciences* 21 (3):182–94. https://doi.org/10.1016/j.tics.2017.01.005

Balzan, R., Delfabbro, P., Galletly, C. & Woodward, T. 2013. Confirmation Biases Across the Psychosis Continuum: The Contribution of Hypersalient Evidence-Hypothesis Matches. *British Journal of Clinical Psychology* 52 (1):53–69.

Barrett, J.L. 2007. Cognitive Science of Religion: What Is It and Why Is It?. *Religion Compass* 1 (6):768–86.

Barrett, L.F., Quigley, K.S. & Hamilton, P. 2016. An Active Inference Theory of Allostasis and Interoception in Depression. *Philosophical Transactions of the Royal Society of London. Series B, Biological Sciences* 371 (1708):20160011. https://doi.org/10.1098/rstb.2016.0011

Batson, C.D. & Ventis, W.L. 1982. *The Religious Experience*, New York: Oxford University Press.

Bauer, R.M. 1984. Autonomic Recognition of Names and Faces in Prosopagnosia: A Neuropsychological Application of the Guilty Knowledge Test. *Neuropsychologia* 22 (4):457–69.

Bayne, T. & Fernandez, J. (eds.). 2008. *Delusion and Self-Deception: Affective and Motivational Influences on Belief Formation (Macquarie Monographs in Cognitive Science)*, Psychology Press.

Bayne, T. & Pacherie, E. 2004. Bottom-Up or Top-Down?: Campbell's Rationalist Account of Monothematic Delusions. *Philosophy, Psychiatry & Psychology* 11 (1):1–11.

Bayne, T. & Pacherie, E. 2005. In Defence of the Doxastic Conception of Delusions. *Mind & Language* 20:163–88.

Becker, G.S. & Murphy, K.M. 1988. A Theory of Rational Addiction. *Journal of Political Economy* 96 (4):675–700.

Bell, V. 2013. A Community of One: Social Cognition and Auditory Verbal Hallucinations. *PLoS Biology* 11 (12):p.e1001723.

Bentall, R.P. 2003. *Madness Explained*, London: Penguin.

Bentall, R.P. & Slade, P.D. 1985. Reality Testing and Auditory Hallucinations: A Signal Detection Analysis. *British Journal of Clinical Psychology* 24 (3):159–69.

Bentall, R. & Varese, F. 2013. Psychotic Hallucinations. In *Hallucination: Philosophy and Psychology*, p. 65.

Berridge, K.C. 2007. The Debate Over Dopamine's Role in Reward: The Case for Incentive Salience. *Psychopharmacology* 191 (3):391–431. https://doi.org/10.1007/s00213-006-0578-x

Berrios, G.E. 1991. Delusions as 'Wrong Beliefs': A Conceptual History. *British Journal of Psychiatry* 159:6–13.

Berrios, G.E. & Brook, P. 1982. The Charles Bonnet Syndrome and the Problem of Visual Perceptual Disorders in the Elderly. *Age and Ageing* 11 (1):17–23. https://doi.org/10.1093/ageing/11.1.17

Berrios, G.E. & Marková, I.S. 2015. Towards a New Epistemology of Psychiatry. In *Re-visioning Psychiatry: Cultural Phenomenology, Critical Neuroscience, and Global Mental Health*, pp. 41–64.

Bick, P. & Kinsbourne, M. 1987. Auditory Hallucinations and Subvocalizations in schizophrenics. *American Journal of Psychiatry* 14:222–5.

Blackburn, S. 1993. *Essays in Quasi-Realism*, New York: Oxford University Press.

Blakemore, S.J., Smith, J., Steel, R., Johnstone, E.C. & Frith, C.D. 2000. The Perception of Self-Produced Sensory Stimuli in Patients with Auditory Hallucinations and Passivity Experiences: Evidence for a Breakdown in Self-Monitoring. *Psychological Medicine* 30 (5):1131–9.

Blakemore, S.J., Wolpert, D.M. & Frith, C.D. 1999. The Cerebellum Contributes to Somatosensory Cortical Activity During Self-Produced Tactile Stimulation. *Neuroimage* 10 (4):448–59.

Blankenburg, W. 1971. *Der Verlust der natürlichen selbstverständlichkeit: ein Beitrag zur Psychopathologie symptomarmer Schizophrenien*. Stuttgart: Enke.

Bloom, P. 2007. Religion Is Natural. *Developmental Science* 10 (1):147–51. https://doi.org/10.1111/j.1467-7687.2007.00577.x

Bolton, D. 2008. *What Is Mental Disorder?: An Essay in Philosophy, Science, and Values*, Oxford: Oxford University Press. doi:10.1093/med/9780198565925.001.0001

Boorse, C. 1975. On The Distinction Between Disease and Illness. *Philosophy and Public Affairs* 5:49–68.

Boorse, C. 1976. What a Theory of Mental Health Should Be. *Journal for the Theory of Social Behavior* 6:61–84.

Boorse, C. 1977. Health as a Theoretical Concept. *Philosophy of Science* 44:542–73.

Boorse, C. 1997. A Rebuttal on Health, in J. M. Humber and R. F. Almeder (eds.), *What is disease?,* Totowa, NJ: Humana Press, 3–143.

Borsboom, D. 2017. A Network Theory of Mental Disorders. *World Psychiatry* 16:5–13. https://doi.org/10.1002/wps.20375

Bortolotti, L. 2009. *Delusions and Other Irrational Beliefs*, Oxford: Oxford University Press.

Bortolotti, L. 2020. Doctors Without 'Disorders'. *Aristotelian Society Supplementary Volume* 94 (1):163–84.

Bowden, H. 2012. A Phenomenological Study of Anorexia Nervosa. *Philosophy, Psychiatry, and Psychology* 19 (3):227–41.

Boyer, M. & Dickerson, M. 2003. Attentional Bias and Addictive Behaviour: Automaticity in a Gambling-Specific Modified Stroop Task. *Addiction (Abingdon, England)* 98 (1):61–70. https://doi.org/10.1046/j.1360-0443.2003.00219.x

Boyle, M. 1990. *Schizophrenia: A Scientific Delusion?*, 1st edition, Routledge.

Boyle, M. 2002. *Schizophrenia: A Scientific Delusion?*, 2nd edition, Routledge.

Bratman, M. 1987. *Intention, Plans, and Practical Reason*, Cambridge, MA: Harvard University Press.

Brentano, F. 1874. *Psychologie vom empirischen Standpunkte*, Duncker Und Humblot.

Broadbent, A. 2019. *Philosophy of Medicine*, Oxford: Oxford University Press.

Brookwell, M.L., Bentall, R.P. & Varese, F. 2013. Externalizing Biases and Hallucinations in Source-Monitoring, Self-Monitoring and Signal Detection Studies: A Meta-Analytic Review. *Psychological Medicine* 43 (12):2465–75.

Campbell, J. 2001. Rationality, Meaning, and the Analysis of Delusion. *Philosophy, Psychiatry, and Psychology* 8 (2–3):89–100.

Campbell, J. 2008. Interventionism, Control Variables and Causation in the Qualitative World. *Philosophical Issues* 18 (1):426–45.

Cannon, C.M. & Palmiter, R.D. 2003. Reward Without Dopamine. *The Journal of Neuroscience: The Official Journal of the Society for Neuroscience* 23 (34):10827–31. https://doi.org/10.1523/JNEUROSCI.23-34-10827.2003

Canstatt, C.F. 1841. Handbuch der medicinischen Klinik: Erster Band. Die Specielle. Pathologie u. Therapie, vom klinischen Standpuncte aus bearbeitet.

Capgras, J. & Reboul-Lachaux, J. 1923. L'illusion des 'sosies' dans un delire systematise chronique. *Bulletin de la société clinique de médecine mentale* 11: 6–16.

Carel, H. 2007. Can I Be Ill and Happy? *Philosophia* 35 (2):95–110.

Carel, H. 2016. II—Virtue Without Excellence, Excellence Without Health. *Aristotelian Society Supplementary Volume* 90 (1):237–53.

Cartwright, W. 2006. Reasons and Selves: Two Accounts of Responsibility in Theory and Practice. *Philosophy, Psychiatry, and Psychology* 13 (2):143–55.

Cassell, E.J. 1976. Illness and Disease. *Hastings Center Report* 6 (2):27–37.

Casu, G., Cascella, N. & Maggini, C. 1994. Homicide in Capgras' Syndrome. *Psychopathology* 27 (6):281–4. https://doi.org/10.1159/000284884

Champlin, T.S. 1981. The Reality of Mental Illness. *Philosophy* 56 (218):467–87.

Charland, L.C., Hope, T., Stewart, A. & Tan, J. 2013. Anorexia Nervosa as a Passion. *Philosophy, Psychiatry, and Psychology* 20 (4):353–65.

Chekroud, A.M. 2015. Unifying Treatments for Depression: An Application of the Free Energy Principle. *Frontiers in Psychology* 6:153. https://doi.org/10.3389/fpsyg.2015.00153

Chentsova-Dutton, Y.E., Ryder, A.G. & Tsai, J.L. 2014. Understanding Depression Across Cultural Contexts. In I. Gotlib and C. Hammen (eds.), *Handbook of Depression*, 3rd edition, New York, NY: Guilford Press, pp. 337–52.

Cho, R. & Wu, W. 2013. Mechanisms of Auditory Verbal Hallucination in Schizophrenia. *Frontiers in Schizophrenia* 4:155. https://doi.org/10.3389/fpsyt.2013.00155

Choi, H.D., Kim, K.K. & Koo, B.H. 2011. A Case of Catatonia and Neuroleptic Malignant Syndrome Probably Associated with Antipsychotic in Korea. *Psychiatry Investigation* 8 (2):174–7. https://doi.org/10.4306/pi.2011.8.2.174

Chomsky, N. 1965. *Aspects of the Theory of Syntax*, Cambridge, MA: MIT Press.

Chrisman, M. 2017. *What Is This Thing Called Metaethics?*, Routledge.

Chung, M.C., Fulford, K.W.M., Fulford, B. & Graham, G. (eds.). 2007. *Reconceiving Schizophrenia*, Oxford University Press.

Clark, A. 2013. Whatever Next? Predictive Brains, Situated Agents, and the Future of Cognitive Science. *Behavioral and Brain Sciences* 36 (3):181–204.

Clark, A. & Chalmers, D. 1998. The Extended Mind. *Analysis* 58 (1):7–19.

Cohen, J.D. & Servan-Schreiber, D. 1992. Context, Cortex, and Dopamine: A Connectionist Approach to Behavior and Biology in Schizophrenia. *Psychological Review* 99 (1):45.

Coltheart, M. 2007. Cognitive Neuropsychiatry and Delusional Belief. *Quarterly Journal of Experimental Psychology (2006)* 60 (8):1041–62. https://doi.org/10.1080/17470210701338071

Conrad, P. 1992. Medicalization and Social Control. *Annual Review of Sociology*, 18, 209–232.

Conrad, P. & Schneider, J. 1980. *Deviance and Medicalization: From Badness to Sickness,* St. Louis, MO: CV Mosley.

Cooper, R. 2002. Disease. *Studies in History and Philosophy of Science Part C: Studies in History and Philosophy of Biological and Biomedical Sciences* 33 (2):263–82.

Cooper, R. 2004. What Is Wrong with the DSM? *History of Psychiatry* 15 (1):5–25.

Cooper, R. 2007. *Psychiatry and Philosophy of Science*, Routledge.

Cooper, R. 2017. Where's the Problem? Considering Laing and Esterson's Account of Schizophrenia, Social Models of Disability, and Extended Mental Disorder. *Theoretical Medicine and Bioethics* 38 (4):295–305.

Cooper, R. 2018. Understanding the DSM-5: Stasis and Change. *History of Psychiatry* 29 (1):49–65. https://doi.org/10.1177/0957154X17741783

Corlett, P.R., Taylor, J.R., Wang, X.J., Fletcher, P.C. & Krystal, J.H. 2010. Toward a Neurobiology of Delusions. *Progress in Neurobiology* 92 (3):345–69.

Courbon, P. & Fail, G. 1927. Syndrome "d'illusion de Frégoli" et schizophrénie [Syndrome of the "illusion of Fregoli" and schizophrenia]. *Bulletin de la Société Clinique de Médecine Mentale* 20:121–5.

Cox, W.M., Fadardi, J.S. & Pothos, E.M. 2006. The Addiction-Stroop Test: Theoretical Considerations and Procedural Recommendations. *Psychological Bulletin* 132 (3):443–76. https://doi.org/10.1037/0033-2909.132.3.443

Cramer, A.O., Waldorp, L.J., van der Maas, H.L. & Borsboom, D. 2010. Comorbidity: A Network Perspective. *The Behavioral and Brain Sciences* 33 (2–3):137–93. https://doi.org/10.1017/S0140525X09991567

Craver, C. 2006. When mechanistic models explain. *Synthese* 153 (3):355–376.

Critchley, E.M.R., Denmark, J.C., Warren, F. & Wilson, K.A. 1981. Hallucinatory Experiences of Prelingually Profoundly Deaf Schizophrenics. *The British Journal of Psychiatry* 138 (1):30–2.

Cummins, R. 1975. Functional Analysis. *Journal of Philosophy* 72 (November):741–64.

Currie, G. 2000. Imagination, Delusion and Hallucinations. *Mind and Language* 15 (1):168–83.

Currie, G. & Jureidini, J. 2001. Delusion, Rationality, Empathy. *Philosophy, Psychiatry, and Psychology* 8 (2–3):159–62.

Currie, G. & Ravenscroft, I. 2002. *Recreative Minds: Imagination in Philosophy and Psychology*, Oxford University Press.

Cutting, J. & Dunne, F. 1989. Subjective Experience of Schizophrenia. *Schizophrenia Bulletin*, 15 (2):217–31.

Da Fonseca, D. & Fourneret, P. 2018. Schizophrénie à début très précoce [Very Early Onset Schizophrenia]. *L'Encephale* 44 (6S):S8–11. https://doi.org/10.1016/S0013-7006(19)30071-5

David, A.S. 1999. On the Impossibility of Defining Delusions. *Philosophy, Psychiatry, and Psychology* 6 (1):17–20.

Davidson, D. 1963. Actions, Reasons, and Causes. *Journal of Philosophy* 60 (23):685.

Davidson, D. 1970. *Mental Events*. In L. Foster & J.W. Swanson (eds.), Oxford: Clarendon Press, pp. 207–24.

Davies, M. & Coltheart, M. 2000. Pathologies of Belief. *Mind and Language* 15 (1):1–46.

Davies, M., Coltheart, M., Langdon, R. & Breen, N. 2001. Monothematic Delusions: Towards a Two-Factor Account. *Philosophy, Psychiatry & Psychology* 8 (2–3): 133–58.

Davies, W. 2016. Externalist Psychiatry. *Analysis* 76 (3):290–6.

Dawkins, R. 1976. *The Selfish Gene*, New York: Oxford University Press.

Dawkins, R. 2006. *The God Delusion*, Boston: Houghton Mifflin.

de Boer, N.S., de Bruin, L.C., Geurts, J. & Glas, G. 2021. The Network Theory of Psychiatric Disorders: A Critical Assessment of the Inclusion of Environmental Factors. *Frontiers in Psychology* 12:623970. https://doi.org/10.3389/fpsyg.2021.623970

Deamer, F. 2021. Why Do We Talk To Ourselves? *Review of Philosophy and Psychology* 12 (2):425–433.

Deamer, F., Palmer, E., Vuong, Q.C., Ferrier, N., Finkelmeyer, A., Hinzen, W. & Watson, S. 2019. Non-Literal Understanding and Psychosis: Metaphor Comprehension in Individuals with a Diagnosis of Schizophrenia. *Schizophrenia Research: Cognition* 18:100159.

Deamer, F. & Wilkinson, S. 2015. The Speaker Behind the Voice: Therapeutic Practice from the Perspective of Pragmatic Theory. *Frontiers in Psychology* 6:817. https://doi.org/10.3389/fpsyg.2015.00817

Deamer, F. & Wilkinson, S. 2021. Metaphorical Thinking and Delusions in Psychosis. In M. Amblard, M. Musiol & M. Rebuschi (eds.), *Coherence of Discourse: Formal and Conceptual Issues of Language*, Cham: Springer Verlag, pp. 119–30.

Deb, S., Lyons, I., Koutzoukis, C., Ali, I. & McCarthy, G. 1999. Rate of Psychiatric Illness 1 Year after Traumatic Brain Injury. *The American Journal of Psychiatry* 156 (3):374–8. https://doi.org/10.1176/ajp.156.3.374

de Haan, S. 2020. An Enactive Approach to Psychiatry. *Philosophy, Psychiatry, and Psychology* 27 (1):3–25.

Delay, J. 1952. Utilisation therapeutique psychiatrique d'une phenothiazine d'action centrale lective (456 ORP). *Ann Med Psychol* 110:112–5.

Delespaul, P., deVries, M. & van Os, J. 2002. Determinants of Occurrence and Recovery from Hallucinations in Daily Life. *Social Psychiatry and Psychiatric Epidemiology* 37 (3):97–104. https://doi.org/10.1007/s001270200000

Dennett, D.C. 1969. *Content and Consciousness*, London: Routledge.

Dennett, D.C. 1991. *Consciousness Explained*, London: Penguin Books.

Docherty, J.P., Van Kammen, D.P., Siris, S.G. & Marder, S.R. 1978. Stages of Onset of Schizophrenic Psychosis. *The American Journal of Psychiatry* 135 (4):420–6. https://doi.org/10.1176/ajp.135.4.420

Dodgson, G. & Gordon, S. 2009. Avoiding False Negatives: Are Some Auditory Hallucinations an Evolved Design Flaw?. *Behavioural and Cognitive Psychotherapy* 37 (3):325–34. https://doi.org/10.1017/S1352465809005244

Donzelot, J. 1997. *The Policing of Families*. Translated by Robert Hurley, Baltimore: Johns Hopkins University Press.

Drayson, Z. 2009. Embodied Cognitive Science and its Implications for Psychopathology. *Philosophy, Psychiatry, and* Psychology 16 (4):329–40.

Dreyfus, H.L. 2001. *On the Internet*, London: Routledge.

du Feu, M. & McKenna, P.J. 1999. Prelingually Profoundly Deaf Schizophrenic Patients who Hear Voices: A Phenomenological Analysis. *Acta Psychiatrica Scandinavica* 99 (6):453–9. https://doi.org/10.1111/j.1600-0447.1999.tb00992.x

Durrant, R., Adamson, S., Todd, F. & Sellman, D. 2009. Drug Use and Addiction: Evolutionary Perspective. *The Australian and New Zealand Journal of Psychiatry* 43 (11):1049–56. https://doi.org/10.3109/00048670903270449

Durstewitz, D., Koppe, G. & Meyer-Lindenberg, A. 2019. Deep Neural Networks in Psychiatry. *Molecular Psychiatry* 24 (11):1583–98.

Edwards, C. 2009. Ethical Decisions in the Classification of Mental Conditions as Mental Illness. *Philosophy, Psychiatry, and Psychology* 16 (1):73–90.

Egan, A. 2008. Imagination, Delusion, and Self-Deception. In T. Bayne & J. Fernandez (eds.), *Delusion and Self-Deception: Affective and Motivational Influences on Belief Formation (Macquarie Monographs in Cognitive Science)*, Psychology Press.

Ehlers, A. & Clark, D.M. 2000. A Cognitive Model of Posttraumatic Stress Disorder. *Behaviour Research and Therapy* 38:319–45.

Eilan, N. 2000. On Understanding Schizophrenia. In D. Zahavi (ed.), *Exploring the Self*, Amsterdam: John Benjamins, pp. 97–113.

Ellis, H.D., Quayle, A.H., Young, A.W. & de Pauw, K.W. 1997. Response from Ellis, Young, Quayle and de Pauw. *Trends in Cognitive Sciences* 1 (5):158.

Ellis, H.D. & Young, A.W. 1990. Accounting for Delusional Misidentifications. *British Journal of Psychiatry* 157:239–48.

Elster, J. 1997. More than Enough (Reviewing *Accounting for Tastes* by Gary S. Becker). *University of Chicago Law Review* 64 (2):Article 8.

Engelhardt, H.T. 1975. The Concepts of Health and Disease. In H. Tristram Engelhardt & Stuart F. Spicker (eds.), *Evaluation and Explanation in the Biomedical Sciences*, Dordrecht: Reidel, pp. 125–41.

Eshel, N. & Roiser, J.P. 2010. Reward and Punishment Processing in Depression. *Biological Psychiatry* 68 (2):118–24. https://doi.org/10.1016/j.biopsych.2010.01.027

Fabry, R.E. 2020. Into the Dark Room: A Predictive Processing Account of Major Depressive Disorder. *Phenomenology and the Cognitive Sciences* 19 (4):685–704.

Faye, J. 2007. The Pragmatic-Rhetorical Theory of Explanation. In J. Persson & P. Ylikoski (eds.), *Rethinking Explanation. Series: Boston Studies in the Philosophy of Science*, vol. 252, Dordrecht: Springer Verlag, pp. 43–68.

Fearon, P., Kirkbride, J.B., Morgan, C., Dazzan, P., Morgan, K., Lloyd, T., Hutchinson, G., Tarrant, J., Fung, W.L.A., Holloway, J. & Mallett, R. 2006. Incidence of Schizophrenia and Other Psychoses in Ethnic Minority Groups: Results from the MRC AESOP Study. *Psychological Medicine* 36 (11):1541–50.

Feinberg, I. 1978. Efference Copy and Corollary Discharge: Implications for Thinking and Its Disorders. *Schizophrenia Bulletin* 4 (4):636.

Fernyhough, C. 1996. The Dialogic Mind: A Dialogic Approach to the Higher Mental Functions. *New Ideas in Psychology* 14 (1):47–62.

Fernyhough, C. 2004. Alien Voices and Inner Dialogue: Towards a Developmental Account of Auditory Verbal Hallucinations. *New Ideas in Psychology* 22 (1):49–68.

Ferrari, A.J., Charlson, F.J., Norman, R.E., Patten, S.B., Freedman, G., Murray, C.J., Vos, T. & Whiteford, H.A. 2013. Burden of Depressive Disorders by Country, Sex, Age, and Year: Findings from the Global Burden of Disease Study 2010. *PLoS Medicine* 10 (11):e1001547. https://doi.org/10.1371/journal.pmed.1001547

Fierro, M., Hernández, A.M. & Malcolm, J. 2018. The Psychopathology of Delusion of Control According to Subjective Experience. *Revista Colombiana de Psiquiatría (English ed.)* 47 (4):221–8.

Fine, C., Craigie, J. & Gold, I. 2005. Damned if You Do; Damned if You Don't: The Impasse in Cognitive Accounts of the Capgras Delusion. *Philosophy, Psychiatry, & Psychology* 12 (2):143–51.

Flanagan, O. 2018. Identity and Addiction. In *The Routledge Handbook of Philosophy and Science of Addiction*, London: Routledge, pp. 77–89.

Fletcher, P.C. & Frith, C.D. 2009. Perceiving Is Believing: A Bayesian Approach to Explaining the Positive Symptoms of Schizophrenia. *Nature Reviews. Neuroscience* 10 (1):48–58. https://doi.org/10.1038/nrn2536

Foa, E.B. & Kozak, M.J. 1986. Emotional Processing of Fear: Exposure to Corrective Information. *Psychological Bulletin* 99:20–35.

Foddy, B. & Savulescu, J. 2010. A Liberal Account of Addiction. *Philosophy, Psychiatry, and Psychology* 17 (1):1–22.

Ford, J.M. & Mathalon, D.H. 2004. Electrophysiological Evidence of Corollary Discharge Dysfunction in Schizophrenia during Talking and Thinking. *Journal of Psychiatric Research* 38 (1):37–46.

Foucault, M. 1961. *Folie et déraison: histoire de la folie à l'âge classique* (Vol. 169), Paris: Plon.

Frankfurt, H.G. 1971. Freedom of the Will and the Concept of a Person. *Journal of Philosophy* 68 (1):5–20.

Freeman, D., Garety, P.A. & Phillips, M.L. 2000. The Examination of Hypervigilance for External Threat in Individuals with Generalized Anxiety Disorder and Individuals with Persecutory Delusions Using Visual Scan Paths. *The Quarterly Journal of Experimental Psychology A: Human Experimental Psychology* 53A (2):549–67. https://doi.org/10.1080/027249800390600

Fricker, E. 1987. The Epistemology of Testimony. *Aristotelian Society Supplementary Volume* 61:57–83. doi:10.1093/aristoteliansupp/61.1.57

Friston, K. 2005. A Theory of Cortical Responses. *Philosophical Transactions of the Royal Society of London. Series B, Biological Sciences* 360 (1456):815–36. https://doi.org/10.1098/rstb.2005.1622

Friston, K. 2012. Predictive Coding, Precision and Synchrony. *Cognitive Neuroscience* 3 (3–4):238–9. https://doi.org/10.1080/17588928.2012.691277

Frith, C. 1992. *The Cognitive Neuropsychology of Schizophrenia*, Psychology Press.

Frith, C., Blakemore, S.-J. & Wolpert, D.M. 2000. Explaining the Symptoms of Schizophrenia: Abnormalities in the Awareness of Action. *Brain Research Reviews* 31 (2–3):357–63.

Frith, C.D. 1979. Consciousness, Information Processing and Schizophrenia. *The British Journal of Psychiatry* 134 (3):225–35.

Frith, C.D. & Done, D.J. 1989. Experiences of Alien Control in Schizophrenia Reflect a Disorder in the Central Monitoring of Action. *Psychological Medicine* 19 (2):359–63.

Fuchs, T. 2013. Depression, Intercorporeality, and Interaffectivity. *Journal of Consciousness Studies* 20 (7–8):7–8.

Fulford, K.W.M. 1989. *Moral Theory and Medical Practice*, Cambridge: Cambridge University Press.

Fulford, K.W.M. & Jackson, M. 1997. Spiritual Experience and Psychopathology. *Philosophy, Psychiatry, and Psychology* 4 (1):41–65.

Gaebel, W. & Kerst, A. 2019. The Debate about Renaming Schizophrenia: A New Name Would Not Resolve the Stigma. *Epidemiology and Psychiatric Sciences* 28 (3):258–61.

Gallagher, S. 2004. Neurocognitive Models of Schizophrenia: A Neurophenomenological Critique. *Psychopathology* 37 (1):8–19.

Gallagher, S. 2018. The Extended Mind: State of the Question. *Southern Journal of Philosophy* 56 (4):421–47.

Garwood, L., Dodgson, G., Bruce, V. & McCarthy-Jones, S. 2013. A Preliminary Investigation into the Existence of a Hypervigilance Subtype of Auditory Hallucination in People with Psychosis. *Behavioural and Cognitive Psychotherapy* 1–11.

George, B. & Klijn, A. 2013. A Modern Name for Schizophrenia (PSS) Would Diminish Self-Stigma. *Psychological Medicine* 43 (7):1555–7.

Gigerenzer, G. & Selten, R. (eds.). 2002. *Bounded Rationality: The Adaptive Toolbox*, MIT press.

Giles, D. 2006. Constructing Identities in Cyberspace: The Case of Eating Disorders. *British Journal of Social Psychology* 45:463–77.

Glackin, S.N. 2017. Individualism and the Medical: What about Somatic Externalism? *Analysis* 77 (2):287–93.

Glackin, S.N. 2019. Grounded Disease: Constructing the Social from the Biological in Medicine. *Philosophical Quarterly* 69 (275):258–76.

Goldman, A. & Whitcomb, D. (eds.). 2010. *Social Epistemology: Essential Readings*, Oxford University Press.

Gould, L.N. 1948. Verbal Hallucinations and Activation of Vocal Musculature. *American Journal of Psychiatry* 105:367–72.

Gould, L.N. 1950. Verbal Hallucinations as Automatic Speech – The Reactivation of Dormant Speech Habit. *American Journal of Psychiatry* 107 (2):110–19.

Gould, S.J. & Lewontin, R.C. 1979. The Spandrels of San Marco and the Panglossian Paradigm: A Critique of the Adaptationist Programme. *Proceedings of the Royal Society of London. Series B, Biological Sciences* 205 (1161):581–98. http://www.jstor.org/stable/77447

Graham, G. 2015. Words, Worlds, and Addictions. *Philosophy, Psychiatry, and Psychology* 22 (1):45–7.

Griffiths, T.L., Kemp, C. & Tenenbaum, J.B. 2008. *Bayesian Models of Cognition*.

Gupta, R., Assalman, I. & Bottlender, R. 2012. Menopause and Schizophrenia. *Menopause International* 18 (1):10–14. https://doi.org/10.1258/mi.2012.011116

Guthrie, S., Agassi, J., Andriolo, K.R., Buchdahl, D., Earhart, H.B., Greenberg, M., Jarvie, I., Saler, B., Saliba, J., Sharpe, K.J. & Tissot, G. 1980. A Cognitive Theory of Religion [and Comments and Reply]. *Current Anthropology* 21 (2):181–203.

Guze, S.B. 1992. *Why Psychiatry Is a Branch of Medicine*, New York: Oxford University Press.

Hacking, I. 1995. The Looping Effects of Human Kinds. In D. Sperber, D. Premack & A.J. Premack (eds.), *Causal Cognition: A Multidisciplinary Debate*, Oxford: Clarendon Press, Oxford University Press, pp. 351–94.

Hacking, I. 1999. *The Social Construction of What?*, Harvard University Press.

Hare, R.M. 1952. *The Language of Morals*, Oxford: Clarendon.

Hare, R.M. 1986. Health. *Journal of Medical Ethics* 12 (4):174–81.

Haslam, N. 2002. Kinds of kinds: A conceptual taxonomy of psychiatric categories. *Philosophy, Psychiatry, and Psychology* 9 (3):203–217.

Haslanger, S. 2012. *Resisting Reality: Social Construction and Social Critique*, Oxford: Oxford University Press.

Hayward, P. & Bright, J.A. 1997. Stigma and Mental Illness: A Review and Critique. *Journal of Mental Health* 6 (4):345–54. https://doi.org/10.1080/09638239718671

Health Professions Council. 2009. *Standards of Proficiency: Practitioner Psychologists*, London: Health Professions Council.

Heather, N. 1992. Why Alcoholism is not a Disease. *The Medical Journal of Australia* 156 (3):212–5. https://doi.org/10.5694/j.1326-5377.1992.tb139711.x

Helmholtz, H. von. 1962. *Treatise on Physiological Optics*, J.P.C. Southall (trans.), vol. II, New York, NY: Dover (Original work published 1866).

Hempel, C.G. & Oppenheim, P. 1948. Studies in the Logic of Explanation. *Philosophy of Science* 15 (2):135–75.

Henden, E. 2017. Addiction, Compulsion, and Weakness of the Will: A Dual Process Perspective. In N.H. Gabriel Segal (ed.), *Addiction and Choice: Rethinking the Relationship*, pp. 116–32.

Higgins, S.T., Budney, A.J., Bickel, W.K., Foerg, F.E., Donham, R. & Badger, G.J. 1994. Incentives Improve Outcome in Outpatient Behavioral Treatment of Cocaine Dependence. *Archives of General Psychiatry* 51 (7):568–76. https://doi.org/10.1001/archpsyc.1994.03950070060011

Hinton, D.E. & Kirmayer, L.J. 2017. The Flexibility Hypothesis of Healing. *Culture, Medicine and Psychiatry* 41 (1):3–34. https://doi.org/10.1007/s11013-016-9493-8

Hirstein, W. & Ramachandran, V.S. 1997. Capgras Syndrome: A Novel Probe for Understanding the Neural Representation and Familiarity of Persons. *Proceedings of the Royal Society of London* 264:437–44.

Hoffman, R.E. 2007. A Social Deafferentation Hypothesis for Induction of Active Schizophrenia. *Schizophrenia Bulletin* 33 (5):1066–70. https://doi.org/10.1093/schbul/sbm079

Hofmann, B. 2002. On the Triad Disease, Illness and Sickness. *Journal of Medicine and Philosophy* 27 (6):651–73.

Hohwy, J. 2004. Top-Down and Bottom-Up in Delusion Formation. *Philosophy Psychiatry and Psychology* 11 (1):65–70.

Hohwy, J. 2012. Attention and Conscious Perception in the Hypothesis Testing Brain. *Frontiers in Psychology* 3:96. https://doi.org/10.3389/fpsyg.2012.00096

Hohwy, J. 2013. *The Predictive Mind*, OUP Oxford.

Horwitz, A.V. & Wakefield, J.C. 2007. *The Loss of Sadness: How Psychiatry Transformed Normal Sorrow Into Depressive Disorder*, Oxford: Oxford University Press.

Howes, O.D. & Kapur, S. 2009. The Dopamine Hypothesis of Schizophrenia: Version III—The Final Common Pathway. *Schizophrenia Bulletin* 35 (3):549–62.

Hume, D. 1739. *A Treatise of Human Nature*, Oxford: Oxford University Press.

Jablensky, A. 2010. Psychiatry in Crisis? Back to Fundamentals. *World Psychiatry* 9 (1):29.

Jackson, M. 1991. The relationship between psychotic and spiritual experience, D. Phil Thesis, Oxford University, Oxford.

Jackson, M. 1997. Benign Schizotypy? The Case of Spiritual Experience. In G.S. Claridge (ed.), *Schizotypy. Implications for Illness and Health*, Oxford: Oxford University Press.

Jackson, M. 2001. Divine Madness? The Overlap between Psychotic and Spiritual Experience and Its Implications. In I. Clarke (ed.), *Psychosis and Spirituality: Exploring the New Frontier*, London: Whurr.

Jackson, M. & Fulford, K.W.M. 2002. Psychosis Good and Bad: Values-Based Practice and the Distinction between Pathological and Nonpathological Forms of Psychotic Experience. *Philosophy, Psychiatry, and Psychology* 9 (4):387–94.

Jacobsen, E. 1931. Electrical Measurements of Neuromuscular States during Mental Activities. VII: Imagination, Recollection, and Abstract Thinking Involving the Speech Musculature. *American Journal of Physiology* 97:200–9.

Jaspers, K. 1913/1963. *General Psychopathology*, Johns Hopkins University Press.

Jeannerod, M. 1995. Mental Imagery in the Motor Context. *Neuropsychologia* 33 (11):1419–32.

Johns, L. & McGuire, P.K. 1999. Verbal Self-Monitoring and Auditory Hallucinations in Schizophrenia. *The Lancet* 353:469–70.

Johns, L.C., Rossell, S., Frith, C., Ahmad, F., Hemsley, D., Kuipers, E. & McGuire, P.K. 2001. Verbal Self-Monitoring and Auditory Verbal Hallucinations in Patients with Schizophrenia. *Psychological Medicine* 31 (4):705–15. https://doi.org/10.1017/s0033291701003774

Johnsen, B.H., Thayer, J.F., Laberg, J.C. & Asbjornsen, A.E. 1997. Attentional Bias in Active Smokers, Abstinent Smokers, and Nonsmokers. *Addictive Behaviors* 22 (6):813–17. https://doi.org/10.1016/s0306-4603(97)00010-5

Johnson, M.K., Hashtroudi, S. & Lindsay, D.S. 1993. Source Monitoring. *Psychological Bulletin* 114 (1):3.

Johnstone, L. 2018. Psychological Formulation as an Alternative to Psychiatric Diagnosis. *Journal of Humanistic Psychology* 58 (1):30–46. doi:10.1177/0022167817722230.

Jones, M. & Wilkinson, S. 2020. From Prediction to Imagination. In A. Abraham (ed.), *The Cambridge Handbook of the Imagination*, Cambridge University Press.

Jones, S.R. 2010. Do we Need Multiple Models of Auditory Verbal Hallucinations? Examining the Phenomenological Fit of Cognitive and Neurological Models. *Schizophrenia Bulletin* 36 (3):566–75.

Jones, S.R. & Fernyhough, C. 2007. Thought as Action: Inner Speech, Self-Monitoring, and Auditory Verbal Hallucinations. *Consciousness and Cognition* 16 (2):391–9. https://doi.org/10.1016/j.concog.2005.12.00

Joyce, R. 2001. *The Myth of Morality*, Cambridge: Cambridge University Press.

Jurcik, T., Chentsova-Dutton, Y.E., Solopieieva-Jurcikova, I. & Ryder, A.G. 2013. Russians in Treatment: The Evidence Base Supporting Cultural Adaptations. *Journal of Clinical Psychology* 69 (7):774–91.

Kalderon, M. 2005. *Moral Fictionalism*, Oxford: Clarendon Press.

Kapur, S. 2003. Psychosis as a State of Aberrant Salience: A Framework Linking Biology, Phenomenology, and Pharmacology in Schizophrenia. *American Journal of Psychiatry* 160 (1):13–23.

Keane, T.M., Zimering, R.T. & Caddell, J.M. 1985. A Behavioral Formulation of Posttraumatic Stress Disorder in Vietnam Veterans. *The Behavior Therapist* 8 (1):9–12.

Kelemen, D. 2004. Are Children "Intuitive Theists"? Reasoning about Purpose and Design in Nature. *Psychological Science* 15 (5):295–301. https://doi.org/10.1111/j.0956-7976.2004.00672.x

Keller, A.S., Leikauf, J.E., Holt-Gosselin, B., Staveland, B.R. & Williams, L.M. 2019. Paying Attention to Attention in Depression. *Translational Psychiatry* 9 (1):1–12.

Kendell, R.E. 1986. What Are Mental Disorders? In A.M. Freedman, R. Brotman, I. Silverman & D. Hutson (eds.), *Issues in Psychiatric Classification: Science, Practice and Social Policy*, Human Sciences Press, pp. 23–45.

Kendler, K.S. & Jablensky, A. 2011. Kraepelin's Concept of Psychiatric Illness. *Psychological Medicine*, 41 (6):1119–26. https://doi.org/10.1017/S0033291710001509

Kendler, K.S., Masterson, C.C. & Davis, K.L. 1985. Psychiatric Illness in First-Degree Relatives of Patients with Paranoid Psychosis, Schizophrenia and Medical Illness. *The British Journal of Psychiatry: The Journal of Mental Science* 147:524–31. https://doi.org/10.1192/bjp.147.5.524

Kerr, N.S. 1884. Society for the Study and Cure of Inebriety. Inaugural Address Delivered in the Medical Society of London's Rooms, April 25, 1884. The Royal College of Surgeons of England, London: H.K. Lewis.

Kingma, E. 2017. Health, Disease and Naturalism: Hausman on the Public Value of Health. *Public Health Ethics* 10 (2):109–21.

Kiverstein, J., Miller, M. & Rietveld, E. 2020. How Mood Tunes Prediction: A Neurophenomenological Account of Mood and its Disturbance in Major Depression. *Neuroscience of Consciousness* 1:1–12. ISSN 2057-2107.

Kokkinidis, L. & Anisman, H. 1981. Amphetamine Psychosis and Schizophrenia: Dual Model. *Neuroscience & Biobehavioral Reviews* 5 (4):449–61.

Korsgaard, C.M. 1983. Two Distinctions in Goodness. *Philosophical Review* 92 (2):169–95.

Laing, R.D. & Esterson, A. 2016. *Sanity, Madness and the Family*, Routledge.

Landis, T., Cummings, J.L., Christen, L., Bogen, J.E. & Imhof, H.G. 1986. Are Unilateral Right Posterior Cerebral Lesions Sufficient to Cause Prosopagnosia? Clinical and Radiological Findings in Six Additional Patients. *Cortex* 22 (2):243–52.

Langdon, R., Jones, S.R., Connaughton, E. & Fernyhough, C. 2009. The Phenomenology of Inner speech: Comparison of Schizophrenia Patients with Auditory Verbal Hallucinations and Healthy Controls. *Psychological Medicine* 39 (4):6553.

Larøi, F., Sommer, I.E., Blom, J.D., Fernyhough, C., Ffytche, D.H., Hugdahl, K., Johns, L.C., McCarthy-Jones, S., Preti, A., Raballo, A., Slotema, C.W., Stephane, M. & Waters, F. 2012. The Characteristic Features of Auditory Verbal Hallucinations in Clinical and Nonclinical Groups: State-of-the-Art Overview and Future Directions. *Schizophrenia Bulletin* 38 (4):724–33. https://doi.org/10.1093/schbul/sbs061

Laruelle, M. & Abi-Dargham, A. 1999. Dopamine as the Wind of the Psychotic Fire: New Evidence from Brain Imaging Studies. *Journal of Psychopharmacology* 13 (4):358–71.

Launay, G. & Slade, P.D. 1981. The Measurement of Hallucinatory Predisposition in Male and Female Prisoners. *Personality and Individual Differences* 2 (3):221–34. https://doi.org/10.1016/0191-8869(81)90027-1

Leshner, A.I. 1997. Addiction is a Brain Disease, and it Matters. *Science (New York, N.Y.)* 278 (5335):45–7. https://doi.org/10.1126/science.278.5335.45

Levy, N. 2013. Addiction is not a Brain Disease (and it Matters). *Frontiers in Psychiatry* 4 (24):1–7.

Levy, N. 2014. Addiction as a Disorder of Belief. *Biology and Philosophy* 29 (3):337–55.

Levy, N. 2019. Addiction: The Belief Oscillation Hypothesis. In H. Pickard et al. (eds.), *The Routledge Handbook of Philosophy and Science of Addiction*, London: Routledge, pp. 54–62.

Lewis, M. 2018. Brain Change in Addiction as Learning, Not Disease. *The New England Journal of Medicine* 379 (16):1551–60. https://doi.org/10.1056/NEJMra1602872

Lucchelli, F. & Spinnler, H. 2007. The Case of Lost Wilma: A Clinical Report of Capgras Delusion. *Neurological Sciences: Official Journal of the Italian Neurological Society and of the Italian Society of Clinical Neurophysiology* 28 (4):188–95. https://doi.org/10.1007/s10072-007-0819-8

Maher, B.A. 1974. Delusional Thinking and Perceptual Disorder. *Journal of Individual Psychology* 30:98–113.

Malacrida, C. 2003. *Cold Comfort: Mothers, Professionals, and Attention Deficit (Hyperactivity) Disorder*, University of Toronto Press. https://doi.org/10.3138/9781442673038

Malatesti, L., Jurjako, M. & Meynen, G. 2020. The Insanity Defence without Mental Illness? Some Considerations. *International Journal of Law and Psychiatry* 71.

Marchetti, G. & Marchetti, S. (eds.). 2018. *Facts and Values: The Ethics and Metaphysics of Normativity*, London and New York: Routledge.

Marr, D. 1982. *Vision*, San Francisco: Freeman Press.

Marshall, J.C. & Halligan, P.W. 1996. Towards a Cognitive Neuropsychiatry. In P.W. Halligan & J.C. Marshall (eds.), *Method in Madness: Case Studies in Cognitive Neuropsychiatry*, Mahwah, NJ: Lawrence Erlbaum Associates.

Mason, O.J., Stevenson, C. & Freedman, F. 2014. Ever-Present Threats from Information Technology: The Cyber-Paranoia and Fear Scale. *Frontiers in Psychology* 5:1298. https://doi.org/10.3389/fpsyg.2014.01298

Matthews, S. 2004. Failed Agency and the Insanity Defence. *International Journal of Law and Psychiatry* 27:413–24.

McCarthy-Jones, S. 2012. *Hearing Voices: The Histories, Causes and Meanings of Auditory Verbal Hallucinations*, Cambridge: Cambridge University Press. https://doi.org/10.1017/CBO9781139017534

McCarthy-Jones, S. & Fernyhough, C. 2011. The Varieties of Inner Speech: Links Between Quality of Inner Speech and Psychopathological Variables in a Sample of Young Adults. *Consciousness and Cognition* 20 (4):1586–93.

McCarthy-Jones, S., Krueger, J., Larøi, F., Broome, M. & Fernyhough, C. 2013. Stop, Look, Listen: The Need for Philosophical Phenomenological Perspectives on Auditory Verbal Hallucinations. *Frontiers in Human Neuroscience* 7:127.

McCarthy-Jones, S., Trauer, T., Mackinnon, A., Sims, E., Thomas, N. & Copolov, D.L. 2014. A New Phenomenological Survey of Auditory Hallucinations: Evidence for Subtypes and Implications for Theory and Practice. *Schizophrenia Bulletin* 40 (1):231–5. https://doi.org/10.1093/schbul/sbs156

McDowell, J. 1985. Functionalism and Anomalous Monism. In B.P. McLaughlin & E. LePore (eds.), *Actions and Events: Perspectives on the Philosophy of Donald Davidson*, Oxford: Blackwell.

McGuigan, F. 1966. Covert Oral Behaviour and Auditory Hallucinations. *Psychophysiology* 3:73–80.

McGuire, P.K., Robertson, D., Thacker, A., David, A.S., Kitson, N., Frackowiak, R.S. & Frith, C.D. 1997. Neural Correlates of Thinking in Sign Language. *Neuroreport* 8 (3):695–8. https://doi.org/10.1097/00001756-199702100-00023

McHugh, P.R. & Slavney, P.R. 1998. *The Perspectives of Psychiatry*, 2nd edition, Johns Hopkins University Press.

Meehl, P.E. 1962. Schizotaxia, Schizotypy, Schizophrenia. *American Psychologist* 17 (12):827.

Mele, A. 2002. Autonomy, Self-Control, and Weakness of Will. In R. Kane (ed.), *The Oxford Handbook of Free Will*, New York, NY: Oxford University Press, pp. 529–48.

Mele, A. 2010. Weakness of Will and Akrasia. *Philosophical Studies* 150 (3):391–404.

Miller, M., Kiverstein, J. & Rietveld, E. 2020. Embodying Addiction: A Predictive Processing Account. *Brain and Cognition* 138:105495. https://doi.org/10.1016/j.bandc.2019.105495

Miller, W.R., Westerberg, V.S., Harris, R.J. & Tonigan, J.S. 1996. What Predicts Relapse? Prospective Testing of Antecedent Models. *Addiction* 91 (12s1):155–72.

Millikan, R.G. 1989. In Defense of Proper Functions. *Philosophy of Science* 56 (June):288–302.

Miyazono, K. 2015. Delusions as Harmful Malfunctioning Beliefs. *Consciousness and Cognition* 33:561–73.

Montague, P.R., Dayan, P. & Sejnowski, T.J. 1996. A Framework for Mesencephalic Dopamine Systems Based on Predictive Hebbian Learning. *The Journal of Neuroscience: The Official Journal of the Society for Neuroscience* 16 (5):1936–47. https://doi.org/10.1523/JNEUROSCI.16-05-01936.1996

Montague, P.R., Dolan, R.J., Friston, K.J. & Dayan, P. 2012. Computational Psychiatry. *Trends in Cognitive Sciences* 16 (1):72–80.

Moreira-Almeida, A., Neto, F.L. & Koenig, H.G. 2006. Religiousness and Mental Health: A Review. *Revista brasileira de psiquiatria (Sao Paulo, Brazil: 1999)* 28 (3):242–50. https://doi.org/10.1590/s1516-44462006000300018

Morgan, C. & Gayer-Anderson, C. 2016. Childhood Adversities and Psychosis: Evidence, Challenges, Implications. *World Psychiatry: Official Journal of the World Psychiatric Association (WPA)* 15 (2):93–102. https://doi.org/10.1002/wps.20330

Morris, J.N. 1978. *The Uses of Epidemiology*, 3rd edition, Edinburgh: Churchill Livingstone.

Moseley, P. & Wilkinson, S. 2014. Inner Speech is not so Simple: A Commentary on Cho and Wu (2013). *Frontiers in Psychiatry* 5:42. https://doi.org/10.3389/fpsyt.2014.00042

Müller, C.P. & Schumann, G. 2011. Drugs as Instruments: A New Framework for Non-Addictive Psychoactive Drug Use. *Behavioral and Brain Sciences* 34 (6):293.

Murphy, D. 2006. *Psychiatry in the Scientific Image*, Cambridge, MA: MIT Press.

Murphy, D. 2010. Explanation in Psychiatry. *Philosophy Compass* 5 (7):602–10.

Murphy, D. 2012. The Folk Epistemology of Delusions. *Neuroethics* 5 (1):19–22.

Murphy, D. 2013. Delusions, Modernist Epistemology and Irrational Belief. *Mind and Language* 28 (1):113–24.

Murray, R.M. 2006. Schizophrenia term use 'invalid'. *BBC News*, 9 October. Available at http://news.bbc.co.uk/2/hi/health/6033013.stm

Nasar, S. 1998. *A Beautiful Mind*, New York: Simon and Schuster.

Nayani, T.H. & David, A.S. 1996. The Auditory Hallucination: A Phenomenological Survey. *Psychological Medicine* 26 (1):177–89. https://doi.org/10.1017/s003329170003381x

Neander, K. 1991. Functions as Selected Effects. *Philosophy of Science* 58:168–84.

Newen, A., De Bruin, L. & Gallagher, S. 2018. *The Oxford Handbook of 4E Cognition*, Oxford: Oxford University Press.

Nielsen, K. 2021. Comparing Two Enactive Perspectives on Mental Disorder. *Philosophy, Psychiatry, and Psychology* 28 (3):175–85.

Norenzayan, A., Shariff, A., Gervais, W., Willard, A., McNamara, R., Slingerland, E. & Henrich, J. 2016. The Cultural Evolution of Prosocial Religions. *Behavioral and Brain Sciences* 39:E1. doi:10.1017/S0140525X14001356

Nozick, R. 1993. *The Nature of Rationality*, Princeton: Princeton University Press.

Osler, L. 2020. Feeling Togetherness Online: A Phenomenological Sketch of Online Communal Experiences. *Phenomenology and the Cognitive Sciences* 19 (3):569–88.

Otgaar, H., Howe, M.L., Patihis, L., Merckelbach, H., Lynn, S.J., Lilienfeld, S.O. & Loftus, E.F. 2019. The Return of the Repressed: The Persistent and Problematic Claims of Long-Forgotten Trauma. *Perspectives on Psychological Science*, 14 (6):1072–95.

Pacherie, E. 2008. Perception, Emotions and Delusions: The Case of the Capgras Delusion. In Tim Bayne & Jordi Fernàndez (eds.), *Delusion and Self-Deception: Affective and Motivational Influences on Belief Formation*, Psychology Press, pp. 107–25.

Pandis, C., Agrawal, N. & Poole, N. 2019. Capgras' Delusion: A Systematic Review of 255 Published Cases. *Psychopathology* 52 (3):161–73. https://doi.org/10.1159/000500474

Parsons, T. 1951. Illness and the Role of the Physician: A Sociological perspective. *American Journal of Orthopsychiatry, 21*(3):452–460. https://doi.org/10.1111/j.1939-0025.1951.tb00003.x

Penfield, W. & Perot, P. 1963. The Brain's Record of Auditory and Visual Experience: A Final Summary and Discussion. *Brain: A Journal of Neurology* 86:595–696. https://doi.org/10.1093/brain/86.4.595

Pezzulo, G. 2014. Why do you Fear the Bogeyman? An Embodied Predictive Coding Model of Perceptual Inference. *Cognitive, Affective & Behavioral Neuroscience* 14 (3):902–11. https://doi.org/10.3758/s13415-013-0227-x

Pickard, H. 2009. Mental Illness is Indeed a Myth. In M. Broome & L. Bortolotti (eds.), *Psychiatry as Cognitive Neuroscience*, Oxford: Oxford University Press.

Pickard, H. 2011. Responsibility without Blame: Empathy and the Effective Treatment of Personality Disorder. *Philosophy, Psychiatry, and Psychology* 18 (3):209–24.

Pickard, H. 2017. Responsibility without Blame for Addiction. *Neuroethics* 10 (1):169–80.

Pickering, M.J. & Clark, A. 2014. Getting Ahead: Forward Models and Their Place in Cognitive Architecture. *Trends in Cognitive Sciences* 18 (9):451–6. https://doi.org/10.1016/j.tics.2014.05.006

Pickering, N. (2005). *The Metaphor of Mental Illness*, Oxford: Oxford University Press.

Pitcher, G. 1971. *A Theory of Perception*, Princeton: Princeton University Press.

Price, A. 2008. *Contextuality in Practical Reason*, Oxford: Oxford University Press.

Pritchard, D. 2004. The Epistemology of Testimony. *Philosophical Issues* 14 (1):326–48.

Proust, J. 2013. *The Philosophy of Metacognition: Mental Agency and Self-Awareness*, Oxford: Oxford University Press.

Psillos, S. 2002. *Causation and Explanation*, London: Routledge.

Putnam, H. 1975. The Meaning of 'Meaning'. In K. Gunderson (ed.), *Minnesota Studies in the Philosophy of Science*, Vol. 7, pp. 131–193.

Putnam, H. 2002. *The Collapse of the Fact/Value Dichotomy and Other Essays*, Cambridge: Harvard University Press.

Radden, J. 2003. Is this Dame Melancholy?: Equating Today's Depression and Past Melancholia. *Philosophy, Psychiatry, and Psychology* 10 (1):37–52.

Radden, J. 2009. Classifying Dysthymia. *APA Newsletter on Philosophy and Medicine* 8 (2):8–12.

Radden, J. 2010. *On Delusion*, London: Routledge.

Radoilska, L. 2013. *Addiction and Weakness of Will,* Oxford: Oxford University Press.

Ramsey, F.P. 1929. *The Foundations of Mathematics and Other Logical Essays*. In R.B. Braithwaite (ed.), London: Routledge.

Rapin, L., Dohen, M., Polosan, M., Perrier, P. & Lœvenbruck, H. 2013. An EMG Study of the Lip Muscles during Covert Auditory Verbal Hallucinations in Schizophrenia. *Journal of Speech, Language, and Hearing Research: JSLHR* 56 (6):S1882–93. https://doi.org/10.1044/1092-4388(2013/12-0210)

Rapoport, A. 1972. Explanatory Power and Explanatory Appeal of Theories. *Synthese* 24 (3–4):321–42.

Ratcliffe, M. 2008. *Feelings of Being: Phenomenology, Psychiatry and the Sense of Reality*, Oxford: Oxford University Press.

Ratcliffe, M. 2014. *Experiences of Depression: A Study in Phenomenology*, Oxford: Oxford University Press.

Ratcliffe, M. 2017a. *Real Hallucinations: Psychiatric Illness, Intentionality, and the Interpersonal World*, Cambridge, MA: MIT Press.

Ratcliffe, M. 2017b. Selfhood, Schizophrenia, and the Interpersonal Regulation of Experience. In *Embodiment, Enaction, and Culture: Investigating the Constitution of the Shared World*, pp.149–71.

Ratcliffe, M., Ruddell, M., & Smith, B. 2014. What is a "sense of foreshortened future?" A phenomenological study of trauma, trust, and time. *Frontiers in psychology, 5,* 1026. https://doi.org/10.3389/fpsyg.2014.01026

Ratcliffe, M. & Wilkinson, S. 2015. Thought Insertion Clarified. *Journal of Consciousness Studies* 22 (11–12):246–69.

Rawls, J. 1980. Kantian Constructivism in Moral Theory. *Journal of Philosophy* 77 (9):515–72.

Read, J., Mosher, L.R. and Bentall, R.P., 2004. Schizophrenia is not an illness. In Read, J., Mosher, L.R. and Bentall, R.P. (eds.) *Models of madness: Psychological, social and biological approaches to schizophrenia.* Psychology Press, pp.3–8

Recanati, F. 2012. *Mental Files*, Oxford: Oxford University Press.

Regier, D.A., Kuhl, E.A. & Kupfer, D.J. 2013. The DSM-5: Classification and Criteria Changes. *World Psychiatry: Official Journal of the World Psychiatric Association (WPA)* 12 (2):92–98. https://doi.org/10.1002/wps.20050

Reznek, L. 1987. *The Nature of Disease*, New York: Routledge.

Roberts, J.R. 2001. Mental Illness, Motivation and Moral Commitment. *Philosophical Quarterly* 51 (202):41–59.

Roberts, T., Krueger, J. & Glackin, S. 2019. Psychiatry Beyond the Brain: Externalism, Mental Health, and Autistic Spectrum Disorder. *Philosophy Psychiatry and Psychology* 26 (3):E51–68.

Robinson, T.E. & Berridge, K.C. 1993. The Neural Basis of Drug Craving: An Incentive-Sensitization Theory of Addiction. *Brain Research: Brain Research Reviews* 18 (3):247–91. https://doi.org/10.1016/0165-0173(93)90013-p

Rogeberg, O. 2004. Taking Absurd Theories Seriously: Economics and the Case of Rational Addiction Theories. *Philosophy of Science* 71 (3):263–85.

Roiser, J.P., Elliott, R. & Sahakian, B.J. 2012. Cognitive Mechanisms of Treatment in Depression. *Neuropsychopharmacology: Official Publication of the American College of Neuropsychopharmacology* 37 (1):117–36. https://doi.org/10.1038/npp.2011.183

Roiser, J.P. & Sahakian, B.J. 2013. Hot and Cold Cognition in Depression. *CNS Spectrums* 18 (3):139–49. https://doi.org/10.1017/S1092852913000072

Ross, D. 2010. Economic Models of Pathological Gambling. In D. Ross, D. Kincaid, D. Spurrett & P. Collins (eds.), *What is Addiction?* Cambridge: MIT Press, pp. 131–58.

Ross, D.C. 1995. Using Severity Rather Than Randomization for Assignment in Clinical Trials: Monte Carlo Tests and Practical Illustrations of the Robbins-Zhang Method. *Journal of Psychiatric Research* 29 (4):315–32. https://doi.org/10.1016/0022-3956(95)00022-w

Ross, R. & McKay, R. 2017. Why Is Belief in God Not a Delusion? *Religion, Brain & Behavior* 7 (4):316–9. doi:10.1080/2153599X.2016.1249917

Ruscio, A.M. & Ruscio, J. 2002. The Latent Structure of Analogue Depression: Should the Beck Depression Inventory be Used to Classify Groups? *Psychological Assessment* 14 (2):135–45. https://doi.org/10.1037/1040-3590.14.2.135

Ruscio, J. & Ruscio, A.M. 2000. Informing the Continuity Controversy: A Taxometric Analysis of Depression. *Journal of Abnormal Psychology* 109 (3):473–87. https://doi.org/10.1037/0021-843X.109.3.473

Ruse, M. 1988. Taking Darwin Seriously. *Ethics* 98 (2):400–2.

Ryle, G. 1932. Systematically Misleading Expressions. *Proceedings of the Aristotelian Society* 32:139–70.

Ryle, G. 1949. *The Concept of Mind*, New York: Barnes & Noble.

Saha, S., Chant, D., Welham, J. & McGrath, J. 2005. A Systematic Review of the Prevalence of Schizophrenia. *PLoS Medicine* 2 (5):e141. https://doi.org/10.1371/journal.pmed.0020141

Samuelson, P.A. 1948. Consumption Theory in Terms of Revealed Preference. *Economica* 15 (60):243–53.

Sass, L.A. & Parnas, J. 2003. Schizophrenia, Consciousness, and the Self. *Schizophrenia Bulletin* 29 (3):427–44.

Saunders. J. 2001. The Practice of Clinical Medicine as an Art and as a Science. *Western Journal of Medicine* 174 (2):137–41.

Schaler, J.A. 2004. *Szasz Under Fire the Psychiatric Abolitionist Faces His Critics*, Open Court Publishing.
Schotte, C.K.W., Maes, M., Cluydts, R., De Doncker, D. & Cosyns, P. 1997. Construct Validity of the Beck Depression Inventory in a Depressive Population. *Journal of Affective Disorders* 46 (2):115–25.
Schultz, W. 1992. Activity of Dopamine Neurons in the Behaving Primate. *Seminars in Neuroscience* 4 (2):129–38, New York: Academic Press.
Schwartz, P. 2007. Decision and Discovery in Defining 'Disease'. In H. Kincaid & J. McKitrick (eds.), *Establishing Medical Reality*, Amsterdam: Springer, pp. 47–63.
Searle, J.R. 2001. *Rationality in Action*, Cambridge: MIT Press.
Sharma, D., Albery, I.P. & Cook, C. 2001. Selective Attentional Bias to Alcohol Related Stimuli in Problem Drinkers and Non-Problem Drinkers. *Addiction (Abingdon, England)* 96 (2):285–95. https://doi.org/10.1046/j.1360-0443.2001.96228512.x
Shergill, S.S., Bullmore, E., Simmons, A., Murray, R. & McGuire, P. 2000. Functional Anatomy of Auditory Verbal Imagery in Schizophrenic Patients with Auditory Hallucinations. *The American Journal of Psychiatry* 157 (10):1691–3. https://doi.org/10.1176/appi.ajp.157.10.1691
Shoemaker, D. 2007. Moral Address, Moral Responsibility, and the Boundaries of the Moral Community. *Ethics* 118 (1):70–108.
Simon, J. 2007. Beyond Naturalism and Normativism: Reconceiving the 'Disease' Debate. *Philosophical Papers* 36 (3):343–70.
Sit, D., Rothschild, A.J. & Wisner, K.L. 2006. A Review of Postpartum Psychosis. *Journal of Women's Health* 15 (4):352–68.
Smith, B. 2013. Depression and Motivation. *Phenomenology and the Cognitive Sciences* 12 (4):615–35.
Sperber, D. 1985. Anthropology and Psychology: Towards an Epidemiology of Representations. *Man* 20 (1):73–89. https://doi.org/10.2307/2802222
Sperber, D., Clément, F., Heintz, C., Mascaro, O., Mercier, H., Origgi, G. & Wilson, D. 2010. Epistemic Vigilance. *Mind and Language* 25 (4):359–93.
Sperry, R.W. 1950. Neural Basis of the Spontaneous Optokinetic Response Produced by Visual Inversion. *Journal of Comparative and Physiological Psychology* 43 (6):482–9. https://doi.org/10.1037/h0055479
Sprevak, M. 2011. Neural Sufficiency, Reductionism, and Cognitive Neuropsychiatry. *Philosophy, Psychiatry, and Psychology* 18 (4):339–44.
Stanghellini, G., Broome, M., Fernandez, A.V., Fusar-Poli, P., Raballo, A. & Rosfort, R. (eds.). 2019. *The Oxford Handbook of Phenomenological Psychopathology*, Oxford: Oxford University Press.
Stefanatos, G.A. 2008. Regression in Autistic Spectrum Disorders. *Neuropsychology Review*, 18 (4):305–19. https://doi.org/10.1007/s11065-008-9073-y
Stephane, M., Polis, I. & Barton, S.N. 2001. A Subtype of Auditory Verbal Hallucinations Responds to Fluvoxamine. *The Journal of Neuropsychiatry and Clinical Neurosciences* 13 (3):425–7. https://doi.org/10.1176/jnp.13.3.425
Stephens, G.L. & Graham, G. 2000. *When Self-Consciousness Breaks: Alien Voices and Inserted Thoughts*, Cambridge, MA: MIT Press.
Stoljar, D. 2016. The Semantics of 'What It's Like' and the Nature of Consciousness. *Mind* 125 (500):1161–98.
Stone, T. & Young, A.W. 1997. Delusions and Brain Injury: The Philosophy and Psychology of Belief. *Mind & Language* 12:327–64.

Strawson, P. 1962. Freedom and Resentment. *Proceedings of the British Academy* 48: 187–211.

Stroop, J.R. 1935. Studies of Interference in Serial Verbal Reactions. *Journal of Experimental Psychology* 18 (6):643.

Styron, W. 2001. *Darkness Visible*, London: Vintage.

Sullivan-Bissett, E. 2020. Unimpaired Abduction to Alien Abduction: Lessons on Delusion Formation. *Philosophical Psychology* 33 (5):679–704.

Sveinsdóttir, Á.K. 2013. The Social Construction of Human Kinds. *Hypatia* 28 (4):716–32.

Svenaeus, F. 2013. Anorexia Nervosa and the Body Uncanny: A Phenomenological Approach. *Philosophy, Psychiatry, and Psychology* 20 (1):81–91.

Szasz, T.S. 1962. *The Myth of Mental Illness: Foundations of a Theory of Personal Conduct*, New York: Harper & Row.

Szasz, T.S. 1976. Schizophrenia: The Sacred Symbol of Psychiatry. *The British Journal of Psychiatry* 129:308–16. https://doi.org/10.1192/bjp.129.4.308

Taylor, C. 1976. Responsibility for Self. In A.O. Rorty (ed.), *The Identities of Persons*, Berkeley: University of California Press, pp. 281–99.

Taylor, D. 2016. Prescribing According to Diagnosis: How Psychiatry Is Different. *World Psychiatry: Official Journal of the World Psychiatric Association (WPA)* 15 (3):224–5. https://doi.org/10.1002/wps.20343

Taylor, S.E. 1989. *Positive Illusions: Creative Self-Deception and the Healthy Mind*, New York: Basic Books/Hachette Book Group.

Tekin, Ş. 2011. Self-Concept Through the Diagnostic Looking Glass: Narratives and Mental Disorder. *Philosophical Psychology* 24 (3):357–80.

Terr, L.C. 1991. Childhood Traumas: An Outline and Overview. *American Journal of Psychiatry* 148 (1):10–20.

Toon, A. (2016). Fictionalism and the Folk. *The Monist* 99 (3):280–95.

Tumulty, M. 2012. Delusions and Not-Quite-Beliefs. *Neuroethics* 5 (1):29–37.

Twaddle, A. 1994. Disease, Illness and Sickness Revisited. In A. Twaddle & L. Nordenfelt (eds.), *Disease, Illness and Sickness: Three Central Concepts in the Theory of Health*, Linkoping: Studies on Health and Society, pp. 1–18.

Van Fraassen, B.C. 1980. *The Scientific Image*, Oxford: Oxford University Press.

Van Os, J. 2016. "Schizophrenia" Does Not Exist. *BMJ: British Medical Journal* 352.

Velleman, D. 2000. *The Possibility of Practical Reason*, New York: Oxford University Press.

Vincent, N.A. 2008. Responsibility, Dysfunction and Capacity. *Neuroethics* 1 (3):199–204.

Vinogradov, S., Willis-Shore, J., Poole, J.H., Marten, E., Ober, B.A. & Shenaut, G.K. 1997. Clinical and Neurocognitive Aspects of Source Monitoring Errors in Schizophrenia. *American Journal of Psychiatry* 154 (11):1530–7.

Volkow, N.D., Hitzemann, R., Wolf, A.P., Logan, J., Fowler, J.S., Christman, D., Dewey, S.L., Schlyer, D., Burr, G. & Vitkun, S. 1990. Acute Effects of Ethanol on Regional Brain Glucose Metabolism and Transport. *Psychiatry Research* 35 (1):39–48. https://doi.org/10.1016/0925-4927(90)90007-s

Vos, T., Abajobir, A.A., Abate, K.H. et al. 2017. Global, Regional, and National Incidence, Prevalence, and Years Lived with Disability for 328 Diseases and Injuries for 195 Countries, 1990–2016: A Systematic Analysis for the Global Burden of Disease Study 2016. *Lancet* 390:1211–59. https://doi.org/10.1016/S0140-6736(17)32154-2

Vygotsky, L.S. (1934/1987). In R.W. Rieber & A.S. Carton (eds.), *The Collected Works of L. S. Vygotsky, Vol. 1. Problems of General Psychology*, Plenum Press.

Wakefield, J.C. 1992. Disorder as Harmful Dysfunction: A Conceptual Critique of DSM-III-R's Definition of Mental Disorder. *Psychological Review* 99 (2):232–47.

Wakefield, J.C. 2007. The Concept of Mental Disorder: Diagnostic Implications of the Harmful Dysfunction Analysis. *World Psychiatry: Official Journal of the World Psychiatric Association (WPA)* 6 (3):149–56.

Wakefield, J.C. 2009. Mental Disorder and Moral Responsibility: Disorders of Personhood as Harmful Dysfunctions, with Special Reference to Alcoholism. *Philosophy, Psychiatry, and Psychology* 16 (1):91–99.

Wallace, M. 2022. Mental Fictionalism: A Foothold Amid Deflationary Collapse. In Tamas Demeter, Adam Toon & T. Parent (eds.), *Mental Fictionalism: Philosophical Explorations*, pp. 27–51

Walther, G. 1923. Zur Ontologie der sozialen Gemeinschaften. *Jahrbuch für Philosophie und phänomenologische Forschung* 6:1–158.

Warner, L.A., Kessler, R.C., Hughes, M., Anthony, J.C. & Nelson, C.B. 1995. Prevalence and Correlates of Drug Use and Dependence in the United States. Results from the National Comorbidity Survey. *Archives of General Psychiatry* 52 (3):219–29. https://doi.org/10.1001/archpsyc.1995.03950150051010

Waters, F.A., Badcock, J.C., Maybery, M.T. & Michie, P.T. 2003. Inhibition in Schizophrenia: Association With Auditory Hallucinations. *Schizophrenia Research* 62 (3):275–80. https://doi.org/10.1016/s0920-9964(02)00358-4

Waters, F.A., Badcock, J.C., Michie, P.T. & Maybery, M.T. 2006. Auditory Hallucinations in Schizophrenia: Intrusive Thoughts and Forgotten Memories. *Cognitive Neuropsychiatry* 11:65–83.

Waters, F.A., Maybery, M.T., Badcock, J.C. & Michie, P.T. 2004. Context Memory and Binding in Schizophrenia. *Schizophrenia Research* 68 (2–3):119–25.

Watkins, K.E., Strafella, A.P. & Paus, T. 2003. Seeing and Hearing Speech Excites the Motor System Involved in Speech Production. *Neuropsychologia* 41:989–94.

Watson, G. 1975. Free Agency. *Journal of Philosophy* 72 (April):205–20.

Watts, C.A. 1956. The Incidence and Prognosis of Endogenous Depression. *British Medical Journal* 1 (4980):1392–7. https://doi.org/10.1136/bmj.1.4980.1392

Wedgwood, R. 2002. The Aim of Belief. *Philosophical Perspectives* 16:267–97.

West, L.J. 1962. *Hallucinations*, New York, NY: Grune & Stratton.

Westerhausen, R., Kompus, K. & Hugdahl, K. 2011. Impaired Cognitive Inhibition in Schizophrenia: A Meta-Analysis of the Stroop Interference Effect. *Schizophrenia Research* 133 (1–3):172–81. https://doi.org/10.1016/j.schres.2011.08.025

Wilkinson, S. 2013a. Beyond Believing Badly. *Teorema: International Journal of Philosophy* 32 (3):105–19.

Wilkinson, S. 2013b. Egocentric and Encyclopedic Doxastic States in Delusions of Misidentification. *Review of Philosophy and Psychology* 4 (2):219–34.

Wilkinson, S. 2014. Accounting for the Phenomenology and Varieties of Auditory Verbal Hallucination Within a Predictive Processing Framework. *Consciousness and Cognition* 30:142–55.

Wilkinson, S. 2015. Forward Models and Passive Psychotic Symptoms. *Frontiers in Human Neuroscience* 9:22. https://doi.org/10.3389/fnhum.2015.00022

Wilkinson, S. 2016. A Mental Files Approach to Delusional Misidentification. *Review of Philosophy and Psychology* 7 (2):389–404.

Wilkinson, S. 2019. Hearing Soundless Voices. *Philosophy, Psychiatry, and Psychology* 26 (3):27–34.

Wilkinson, S. 2020a. Expressivism About Delusion Attribution. *European Journal of Analytic Philosophy* 16 (2):59–77.

Wilkinson, S. 2020b. The Agentive Role of Inner Speech in Self-Knowledge. *Teorema: International Journal of Philosophy* (2):7–26.

Wilkinson, S. 2022. Psychiatric Fictionalism. In Tamas Demeter, Adam Toon & T. Parent (eds.), *Mental Fictionalism: Philosophical Explorations*, pp. 235–52.

Wilkinson, S. & Bell, V. 2016. The Representation of Agents in Auditory Verbal Hallucinations. *Mind and Language* 31 (1):104–26.

Wilkinson, S., Dodgson, G. & Meares, K. 2017. Predictive Processing and the Varieties of Psychological Trauma. *Frontiers in Psychology* 8.

Wilkinson, S. & Fernyhough, C. 2017. Auditory Verbal Hallucinations and Inner Speech: A Predictive Processing Perspective. In Z. Radman (ed.), *Before Consciousness: In Search of the Fundamentals of Mind*, Exeter: Imprint Academic, Ltd.

Wilkinson, S. & Fernyhough, C. 2018. When Inner Speech Misleads. In Langland-Hassan and Vicente (eds.), *Inner Speech*, New York: Oxford University Press.

Wilkinson, S. & Krueger, J. (forthcoming). The Phenomenology of Voice-Hearing and Two Concepts of Voice. In A. Woods, B. Alderson-Day & C. Fernyhough (eds.), *Voices in Psychosis: Interdisciplinary Perspective*.

Williams, B. 1973. Deciding to Believe. In *Problems of the Self*, Cambridge University Press, pp. 136–51.

Williams, D. 2020. Socially adaptive belief. *Mind and Language* 36 (3):333–354.

Wittgenstein, L. 1953. *Philosophical Investigations*, New York, NY: Wiley-Blackwell.

Wittgenstein, L. 1969. *On Certainty (ed. Anscombe and von Wright)*, New York and London: Harper Torchbooks.

Wolf, S. 1987. Sanity and the Metaphysics of Responsibility. In F.D. Schoeman (ed.), *Responsibility, Character, and the Emotions: New Essays in Moral Psychology*, Cambridge: Cambridge University Press, pp. 46–62.

Wolpert, D. 1997. Computational Approaches to Motor Control. *Trends in Cognitive Sciences* 1 (6):209–16.

Woods, A., Jones, N., Alderson-Day, B., Callard, F. & Fernyhough, C. 2015. Experiences of Hearing Voices: Analysis of a Novel Phenomenological Survey. *The Lancet Psychiatry* 2 (4):323–31. https://doi.org/10.1016/S2215-0366(15)00006-1

Woodward, J. 2003. *Making Things Happen: A Theory of Causal Explanation*, Oxford: Oxford University Press.

Wu, W. 2012. Explaining Schizophrenia: Auditory Verbal Hallucination and Self-Monitoring. *Mind and Language* 27 (1):86–107.

Yablo, S. 2000. A Paradox of Existence. In A. Everett & T. Hofweber (eds.), *Empty Names, Fiction and the Puzzles of Non-Existence*, Stanford: CSLI Publications, pp. 275–312.

Yablo, S. 2002. Abstract Objects: A Case Study. *Philosophical Issues* 12 (1):220–40.

Ylikoski, P. & Kuorikoski, J. 2010. Dissecting Explanatory Power. *Philosophical Studies* 148 (2):201–19.

Zachar, P. & Kendler, K.S. 2007. Psychiatric Disorders: A Conceptual Taxonomy. *The American Journal of Psychiatry* 164 (4):557–65. https://doi.org/10.1176/ajp.2007.164.4.557

Index

Lightning Source UK Ltd.
Milton Keynes UK
UKHW022251241122
412746UK00011B/114

9 781138 212688